Methods in Molecular Biology™

Series Editor
John M. Walker
School of Life Sciences
University of Hertfordshire
Hatfield, Hertfordshire, AL10 9AB, UK

For further volumes:
http://www.springer.com/series/7651

Genomic Imprinting

Methods and Protocols

Edited by

Nora Engel

Fels Institute/Biochemistry, School of Medicine, Temple University, Philadelphia, PA, USA

Editor
Nora Engel
Fels Institute/Biochemistry
School of Medicine
Temple University
Philadelphia, PA, USA

ISSN 1064-3745 ISSN 1940-6029 (electronic)
ISBN 978-1-62703-010-6 ISBN 978-1-62703-011-3 (eBook)
DOI 10.1007/978-1-62703-011-3
Springer New York Heidelberg Dordrecht London

Library of Congress Control Number: 2012943828

Printed on acid-free paper

Humana Press is a brand of Springer
Springer is part of Springer Science+Business Media (www.springer.com)

Preface

> While timorous knowledge stands considering, audacious ignorance hath done the deed.
>
> -Samuel Daniel

Genomic imprinting has been fascinating us for over three decades and has provided many emerging scientists with the chance to hit their stride in a frontier posing many unexpected questions and even more surprising answers. Imprinting is the process by which the non-equivalence of the paternal and maternal genomes is established, leading to parent-of-origin-specific effects. The most ostensible effects in mammals of parental-specific marks—and to date, the most accessible to study—are the differential outcomes in gene expression between the paternal and maternal alleles. During the first two decades, the field grew hand in hand with technological innovations in embryology and gene targeting, mainly in the mouse. In fact, advances in imprinting and other unique regulatory mechanisms were instrumental in establishing Epigenetics as the "umbrella organization," as Davor Solter so wittily calls it (1). Many of the broader principles of epigenetic regulation were unearthed by studying imprinted domains (2) and their alterations in cancer and developmental diseases. As technology has moved forward into the "genome-wide" and "high-throughput" arenas, many imprinted regions have been even more fully characterized—with an abundance of information on the epigenetic modifications occurring at specific domains and throughout development. The availability of genome sequences and their variations have moved the field forward enormously. We now know that imprinted genes tend to occur in clusters, that the mechanisms by which the inactive genes are silenced vary from one region to another, that establishment and erasure of the imprints occur at different developmental stages for male and female germ cells, and that DNA methylation is the most consistent candidate for the imprint, at least in the embryo. Clusters of imprinted genes are regulated in *cis* by long-range control elements, designated as imprinting control regions, and these are the sequences bearing the memory of parental origin. Moreover, noncoding RNAs with regulatory roles are present in all imprinted domains.

It is interesting to note, however, that we have yet to answer some of the fundamental questions that the discovery of imprinting posed when it was first described—i.e., how widespread is imprinting across the animal and plant kingdoms, how does the imprinting process vary across genotypes and species, how is the imprint targeted to specific DNA sequences, how is the marking erased, what is the mechanism of tissue-specific and stage-specific imprinting (3), and what is the functional role and origin of imprinting (4). The huge amounts of genome-wide epigenetic data are correlative and have not provided an answer to the question of whether the marks are the cause or consequence of gene expression state, nor have we gained insight into how chromatin-modifying enzymes are targeted to specific sequences. Still to be achieved is the feat of conferring imprinting on a normal gene by transferring a specific sequence into its vicinity. A host of candidate imprinted genes await validation by site-specific molecular studies. Taking advantage of the combined genomic and epigenomic data, we now need more detailed mechanistic models to be tested. In addition, new questions have emerged on the variability of imprinting marks in the population, the effects of culture and in vitro fertilization on imprints, the nature of imprinting in extraembryonic tissues, and the role of noncoding RNAs, among others.

Genomic Imprinting: Methods and Protocols is a survey of the technologies that are being applied to advance the study of imprinting. It includes new technologies that are accelerating the pace of discovery of imprinted genes and characterization of their epigenetic profile, bioinformatic procedures for prediction and comparative analyses of imprinted genes, as well as methods in embryology and basic molecular biology that have been employed for many years, some appearing in new versions for small cell numbers. Undoubtedly, focusing on individual imprinting clusters has uncovered many novel mechanisms in gene regulation, and doing so with traditional but ever more sensitive molecular biology tools will continue to be essential in elucidating the molecular logic of imprint establishment and erasure.

Since many of the compelling questions of the field will require querying very small numbers of cells, we anticipate that the newer technologies will eventually be scaled down to meet this requirement. Also, bioinformatics will continue to expand its influence in the field to bring new insights into the evolutionary history of imprinting. Hopefully, we will also begin to see more of an impact of our imprinting research on other parent-of-origin effects (5). Although attempts are continuously being made to synthesize and generalize our knowledge of imprinted genes, the fact remains that each imprinted domain is unique in some respects, and there is still much to be explored at the molecular level. There is no doubt the next few years will unveil both much-awaited answers and new questions to keep us busy for many exciting years to come.

I thank all the authors for their outstanding contributions to this volume.

Philadelphia, PA, USA *Nora Engel*

References

1. Solter D (1998) Imprinting. Int J Dev Biol 42:951–4
2. Barlow DP (2011) Genomic imprinting: a mammalian epigenetic discovery model. Annu Rev Genet 45:379–403
3. Latham KE (1995) Stage-specific and cell type-specific aspects of genomic imprinting effects in mammals. Differentiation 59:269–82
4. Hurst LD (1997) Evolutionary theories of genomic imprinting. In: Reik W, Surani A (ed) Genomic imprinting. Frontiers in molecular biology, 18. IRL Press
5. Pardo-Manuel de Villena F, de la Casa-Esperon E and Sapienza C (2000) Natural selection and the function of genome imprinting: beyond the silenced minority. Trends Genet 16:573–579

Contents

Contributors

WADO AKAMATSU • *Department of Physiology, School of Medicine, Keio University, Tokyo, Japan*
MD ALMAMUN • *Division of Animal Sciences, University of Missouri, Columbia, MO, USA*
DASARI AMARNATH • *Department of Biochemistry, The Fels Institute for Cancer Research and Molecular Biology, Temple University School of Medicine, Philadelphia, PA, USA*
TOMAS BABAK • *Department of Biology, Stanford University, Stanford, CA, USA*
JULIE BORGEL • *Institute of Molecular Genetics, UMR 5535, Université Montpellier 2, Université Montpellier 1, CNRS, Montpellier, France*
CHELSEA BRIDEAU • *Nuffield Department of Surgical Sciences, Gray Institute for Radiation Oncology and Biology, The University of Oxford, Oxford, UK*
ELENA DE LA CASA-ESPERÓN • *Albacete Science and Technology Park, Regional Center for Biomedical Research (C.R.I.B.), University of Castilla-La Mancha, Albacete, Spain*
YONG CHENG • *Department of Biochemistry, The Fels Institute for Cancer Research and Molecular Biology, Temple University School of Medicine, Philadelphia, PA, USA*
MICHAEL C. GOLDING • *Veterinary Physiology and Pharmacology, Texas A&M University, College Station, TX, USA*
SYLVAIN GUIBERT • *Institute of Molecular Genetics, UMR 5535, Université Montpellier 2, Université Montpellier 1, CNRS, Montpellier, France; UMR 7242 Biotechnology and Cell Signalling, Université de Strasbourg, CNRS, ESBS, Illkirch, France*
SARAH ROSE HUFFMAN • *Division of Animal Sciences, University of Missouri, Columbia, MO, USA*
MASANORI IMAMURA • *Department of Physiology, School of Medicine, Keio University, Tokyo, Japan*
YOKO ITO • *Department of Oncology, CRUK Cambridge Research Institute, University of Cambridge, Cambridge, UK*
YOSHIMI KAWAMURA • *Department of Physiology, School of Medicine, Keio University, Tokyo, Japan*
GAVIN KELSEY • *Epigenetics Programme, The Babraham Institute, Cambridge, UK; Centre for Trophoblast Research, University of Cambridge, Cambridge, UK*
AHMAD M. KHALIL • *Department of Genetics, Center for RNA Molecular Biology, Case Western Reserve University School of Medicine, Cleveland, OH, USA*
KEITH E. LATHAM • *Department of Biochemistry, The Fels Institute for Cancer Research & Molecular Biology, Temple University School of Medicine, Philadelphia, PA, USA*

Louis Lefebvre • *Department of Medical Genetics, Molecular Epigenetics Group, Life Sciences Institute, University of British Columbia, Vancouver, BC, Canada*
Zachary Yu-Ching Lin • *Department of Physiology, School of Medicine, Keio University, Tokyo, Japan*
Yumi Matsuzaki • *Department of Physiology, School of Medicine, Keio University, Tokyo, Japan*
Victoria A. Moran • *Department of Genetics, Center for RNA Molecular Biology, Case Western Reserve University School of Medicine, Cleveland, OH, USA*
Adele Murrell • *Department of Oncology, CRUK Cambridge Research Institute, University of Cambridge, Cambridge, UK*
Ryusuke Nakajima • *Department of Physiology, School of Medicine, Keio University, Tokyo, Japan*
Raffaella Nativio • *Department of Oncology, CRUK Cambridge Research Institute, University of Cambridge, Cambridge, UK; Laboratory of Receptor Biology and Gene Expression, National Cancer Institute, Bethesda, MD, USA*
Courtney N. Niland • *Department of Genetics, Center for RNA Molecular Biology, Case Western Reserve University School of Medicine, Cleveland, OH, USA*
Hideyuki Okano • *Department of Physiology, School of Medicine, Keio University, Tokyo, Japan*
Hirotaka James Okano • *Department of Physiology, School of Medicine, Keio University, Tokyo, Japan*
Hironobu Okuno • *Department of Physiology, School of Medicine, Keio University, Tokyo, Japan*
Andrew Pask • *Department of Molecular and Cell Biology, University of Connecticut, Storrs, CT, USA*
Martina Paulsen • *Life Sciences, Saarland University, Saarbrücken, Germany*
Mathieu Rederstorff • *Université de Lorraine, Biopôle, CNRS UMR 7214 AREMS, Vandoeuvre-lès-Nancy*
James L. Resnick • *Department of Molecular Genetics and Microbiology, College of Medicine, University of Florida, Gainesville, FL, USA*
Rocío Melissa Rivera • *Division of Animal Sciences, University of Missouri, Columbia, MO, USA*
Erika Sasaki • *Department of Physiology, School of Medicine, Keio University, Tokyo, Japan; Laboratory of Applied Developmental Biology, Marmoset Research Department, Central Institute for Experimental Animals, Kawasaki, Japan; PRESTO Japan Science and Technology Agency, Tokyo, Japan*
Andrew J. Sharp • *Department of Genetics and Genomic Sciences, Mount Sinai School of Medicine, New York, NY, USA*
Purnima Singh • *Department of Molecular and Cellular Biology, City of Hope National Medical Center and Beckman Research Institute, Duarte, CA, USA*
Sébastien A. Smallwood • *Epigenetics Programme, The Babraham Institute, Cambridge, UK; Centre for Trophoblast Research, University of Cambridge, Cambridge, UK*
Emily Y. Smith • *Department of Molecular Genetics and Microbiology, College of Medicine, University of Florida, Gainesville, FL, USA*

PAUL SOLOWAY • *Division of Nutritional Sciences, Cornell University, Ithaca, NY, USA*
PIROSKA E. SZABÓ • *Department of Molecular and Cellular Biology, City of Hope National Medical Center and Beckman Research Institute, Duarte, CA, USA*
IKUO TOMIOKA • *Department of Physiology, School of Medicine, Keio University, Tokyo, Japan; Laboratory of Applied Developmental Biology, Marmoset Research Department, Central Institute for Experimental Animals, Kawasaki, Japan*
MICHAEL WEBER • *Institute of Molecular Genetics, UMR 5535, Université Montpellier 2, Université Montpellier 1, CNRS, Montpellier, France; UMR 7242 Biotechnology and Cell Signalling, Université de Strasbourg, CNRS, ESBS, Illkirch, France*
WENYAN XIAO • *Department of Biology, Saint Louis University, St. Louis, MO, USA*

Part I

Parent-of-Origin Effects

Chapter 1

Uniparental Embryos in the Study of Genomic Imprinting

Yong Cheng, Dasari Amarnath, and Keith E. Latham

Abstract

Nuclear transplantation has been used to study genomic imprinting. Available nuclear transfer methods include pronuclear transfer (PNT), intracytoplasmic sperm injection, and round spermatid injection. By generating uniparental embryos that have exclusively paternal or maternal genomes, it is possible to study the functions of the parental genomes separately. It is possible to compare functions in haploid and diploid states. In addition, nuclear transfer allows the effects of the ooplasm, including mitochondria, to be distinguished from effects of the maternally inherited chromosomes. PNTs can also be used to study epigenetic modifications of the parental genomes by the ooplasm. This chapter reviews the methods employed to generate uniparental embryonic constructs for these purposes.

Key words: Pronuclear transfer, Androgenone, Gynogenone, Parthenogenone, Uniparental embryo, Imprinting, Ooplasm

1. Introduction

Green algae, carrots, salamanders, frogs, sea urchins, and mammals—all of these organisms have been used in a remarkable series of experiments dating back for over a century to study nuclear potency via nuclear transplantation, embryo splitting, and cellular reprogramming (1), leading ultimately to the demonstration of nuclear totipotency of somatic cells by the end of the last century (2, 3). These studies demonstrated that in most organisms the hereditary material is contained within the nucleus and remains intact during development, establishing the basic foundation for our current concept of cellular differentiation via epigenetic regulation of the genome.

Epigenetic regulation of the embryonic genome begins at conception and proceeds throughout the life of the organism. The earliest steps in this long legacy are mediated by the ooplasm acting upon the maternal and paternal genomes within the newly formed zygote.

Nora Engel (ed.), *Genomic Imprinting: Methods and Protocols*, Methods in Molecular Biology, vol. 925, DOI 10.1007/978-1-62703-011-3_1, © Springer Science+Business Media, LLC 2012

These early epigenetic processes are responsible for creating and activating the embryonic genome, maintaining parent-of-origin-specific information (imprints), modifying imprints, establishing or modifying epigenetic information, providing adaptive responses to environmental stressors, and initiating the developmental program.

While we accept that these processes occur in the early embryo, the nature and extent of the resulting modifications remains an active and interesting area of study. The techniques of somatic cell hybridization, heterokaryons, and induction of pluripotency using combinations of transcription factors (4, 5) have provided insight into mechanisms that can be employed in vitro to modify epigenetic information. The degree to which such in vitro methods recapitulate normal embryonic processes is uncertain. Nuclear transplantation remains a key approach to understanding epigenetic mechanisms and processes during early development.

With nuclear transplantation, it is possible to dissect the various roles played by the ooplasm, mitochondria, maternal genome, and paternal genome in early development: (1) nuclear transplantation allows the epigenetic aspects of the parental genomes to be studied within the context of a normally fertilized cell; more specifically, it permits the ontogeny of imprinting information to be studied; (2) it allows genetic and evolutionary questions to be addressed by creating interstrain or interspecies nuclear-cytoplasmic hybrids; (3) matrilineal effects can be characterized as either ooplasmic or genomic; (4) interactions between the nucleus and other cellular organelles and structures can be examined; (5) the reversibility and timing of epigenetic changes can be determined; and (6) it potentially allows the effects of reduced ooplasm quality to be overcome.

Thus, nuclear transplantation remains an important tool for understanding developmental changes in the epigenetic control of genome function. This chapter provides examples of different applications of nuclear transplantation methods to study epigenetic genome regulation.

1.1. Uniparental Embryos

The existence of genomic imprinting in mammals was initially discovered through a combination of pronuclear transfer (PNT) studies and genetic studies. Genetic studies involving mice bearing Robertsonian translocations revealed that certain chromosome segments affected phenotype differently when of maternal or paternal origin (6). PNT studies revealed lethal phenotypes in embryos possessing exclusively maternal or paternal chromosomes (7–9). Studies of parthenogenetic embryos likewise indicated defects related to the absence of a paternal set of chromosomes.

Through application of single PNT methods developed initially by McGrath and Solter (10, 11), it was found that diploid androgenones (exclusively paternal chromosomes) and gynogenones (exclusively maternal chromosomes) are unable to develop to term (7–9). Abnormal phenotypes are somewhat complementary, with the extraembryonic tissues being most highly affected in gynogenones, as compared to greater deficiencies in blastocyst development and development of the embryo proper with androgenesis.

Pronuclear transplantation and uniparental diploid embryos also provided new insight into the biology of imprinting when individual genes were examined at early stages of development. For example, the *Igf2r* gene, which is expressed from the maternal allele in most somatic tissues of the mouse, is nevertheless expressed in androgenetic preimplantation stage embryos (12). The *Ascl2* gene (a.k.a. *Mash2*) was likewise biallelically expressed in preimplantation stage uniparental embryos (13). Subsequent studies confirmed stage-dependent and/or tissue-dependent gene silencing of many imprinted genes. More recent studies revealed acquisition of imprinted gene methylation and histone acetylation patterns during preimplantation development (14, 15) as well as changes in non-imprinted genes (14).

More recently, nuclear transfer has been used to study the ontogeny of imprints during gametogenesis. Nuclear transfer in early stage oocytes yielded oocytes that could be activated parthenogenetically and display enhanced developmental potential. Combined with targeted gene modifications, term development of parthenogenetic mice was achieved (16, 17). These studies, combined with developmental studies of acquisition of DNA methylation patterns (18), provided novel insight into the timing of establishment of imprints during oogenesis, indicating that individual gene imprints are acquired at different times. In contrast, nuclear transfer using round spermatids and both primary and secondary spermatocyte nuclei into oocytes supported term development, indicating that paternal imprinting is established at least by the primary spermatocyte stage (19, 20), although other developmental factors affect faithful chromosome segregation and can limit early development (19).

The ontogeny of imprinting information is also relevant to understanding post-fertilization modification of epigenetic information. It is well established that epigenetic inheritance is modified after fertilization, including, for example, strain-specific oocyte modifiers of gene function (21), active global DNA demethylation state and changes in other chromatin properties of the paternal genome (22–28), and differential modification of microinjected transgenes regulated by the initial methylation state (29). The question invariably arises whether such post-fertilization modifications are related to imprinting, as many choose to define imprinting as a strictly gametogenic process. However, differential modification of parental genomes (which remain physically separated) after fertilization will yield the equivalent result of gametogenic imprints, namely, parental chromosome-specific modifications.

PNT yielded early evidence for differential modification of the paternal genome by the ooplasm. Androgenones display differences in developmental potential determined by the ooplasm strain-of-origin (C57Bl/6 and DBA/2) (21). This effect is not observed with gynogenetic embryos, and is due to a stable modification imposed by the time pronuclei are formed, indicating that the ooplasm specifically modifies the paternal genome during the period immediately following fertilization. This difference was mapped to two separate genetic loci (30, 31). A separate set of genetic loci controlling effects of Balb/c oocyte modifiers on transgenes was mapped (32). These observations collectively support a model wherein genomic imprinting information from the father may be subject to an editing function of the ooplasm, possibly to compensate for genetic variation in maternal imprints. More recently ooplasm transfer combined with intracytoplasmic sperm injection (ICSI) to generate diploid androgenones revealed that the developmental potential of the paternal genome could be affected by transferring ooplasm from the low developing strain, but the reciprocal enhancement or rescue with ooplasm from the high developing stain could not be achieved (33).

Uniparental embryos also have been valuable in studying X chromosome regulation. Examining mouse 2-cell stage embryos revealed early expression of the paternal allele of the *Xist* gene (34). Subsequent studies in androgenetic and gynogenetic embryos revealed early repression of genes lying near the paternal X chromosome inactivation center, with spreading to more distal regions as development proceeded (34), manifested as differences in gene expression between the androgenetic, gynogenetic, and fertilized control embryos. Another area in which nuclear transfer has been useful has been to investigate the interaction between parent of origin and haploidy (35).

One very interesting, emerging area of study relates to understanding transgenerational inheritance. Mouse 2-cell stage embryos display a genetic variation in predisposition to blastomere fragmentation. PNT to vary the combination of maternal and paternal origin of ooplasm, maternal genome, and paternal genome revealed that the maternal pronucleus was the main determinant of fragmentation (36). Interestingly, different effects were seen for reciprocal F1 hybrid maternal pronuclei, indicating an effect of the maternal grandpaternal allele. Such insights can be realized when the effects of maternal genome and ooplasm are separated microsurgically.

1.2. Production of Uniparental Embryos

The above discussion illustrates the value of uniparental embryos in studying epigenetic processes during early development, particularly genomic imprinting. There are many different ways to produce uniparental embryos. Parthenogenesis has been employed to study maternal imprinting and to search for novel imprinted genes (37). Though production of diploid parthenogenones is generally simple to perform in comparison to microsurgery, consideration needs to be given to the possible effects of the chosen method of oocyte activation on gene regulation (38). Androgenones and gynogenones prepared by PNT are an attractive, classical source of uniparental embryos for study. Androgenones can also be produced by removing the oocyte spindle–chromosome complex (SCC) and then injecting either two sperm or two spermatids. The following sections describe routine methodologies for PNT, sperm injection, and spermatocyte/spermatid nuclear transfer.

2. Materials

The methods described here are all well established and widely applied, and have been the subject of many recent laboratory protocol publications. However, considerable variability exists in the specific equipment, solutions, media, and procedural details that can be incorporated. Our goal here is to describe procedures that will be effective, offer choices in some of these details, and provide information about the potential impact of some of these procedural variations.

2.1. Equipment

1. Stereo microscope (see Note 1).
2. Inverted microscope with micromanipulators (see Note 2).
3. Microforge (see Note 3).
4. Pipet Puller (see Note 4).
5. Pipet beveler (see Note 5).
6. Microinjectors (see Note 6).
7. Piezo pipet driver (see Notes 7 and 8).
8. Electrofusion device (e.g., ECM 2001, BTX Inc., San Diego, CA, USA).
9. Temperature-controlled, humidified CO_2 cell culture incubator.
10. Billups-Rothenberg (Del Mar, CA) modular incubators or equivalent.

2.2. Culture and Culture Media (See Notes 9 and 10)

1. HEPES-buffered CZB (HCZB) or modified M2 medium as described (39).
2. KSOM medium or sequential media (e.g., CZB followed by Whitten's medium or M16; see refs. 40–44).
3. Activation medium: Ca^{2+}-free CZB or KSOM supplemented with 10 mM $SrCl_2$.
4. PVP supplemented media: HCZB with10% PVP or 7% PVP.
5. Electrofusion medium: 275 mM mannitol, 0.05 mM $CaCl_2$, 0.1 mM $MgSO_4$, and 0.3% BSA.

6. Dulbecco's PBS containing 5.6 mM glucose and 5.4 mM sodium lactate (GL-PBS).
7. Erythrocyte lysis buffer (ELB): 155 mM NH_4Cl, 10 mM $NaHCO_3$, 2 mM EDTA, pH 7.2.

2.3. Solutions and Chemicals (See Note 11)

1. Cytochalasin B (Sigma, 5 mg/ml 1000×stock in ethanol).
2. Demecolcine (Sigma, 0.2 mg/ml 1000×stock).
3. Equine (pregnant mare) chorionic gonadotropin (eCG, a.k.a. PMSG) (Calbiochem, EMD Chemicals, Gibbstown, NJ).
4. Human Chorionic gonadotropin (hCG) (Sigma, St. Louis, MO).

2.4. Pipets (See Note 12)

1. Embryo transfer pipet connected to aspiration device.
2. Holding pipets (see Note 13).
3. Spindle removal pipets (see Note 14).
4. PNT pipets (see Note 15).
5. Sperm ICSI pipets (see Note 16).
6. Spermatocyte and round spermatid nuclear transfer pipets.

3. Methods

3.1. Oocyte Isolation and Culture

1. Isolate MII stage oocytes from females after either spontaneous ovulation or, more commonly, induced superovulation (5 IU eCG followed 46–48 h later with 5 IU hCG). Oocytes are best isolated near the time of ovulation at approximately 14 h post-hCG injection, and then manipulated promptly, followed by embryo culture or activation procedure if needed.
2. Release oocytes from the ampullae into either HCZB or M2 medium containing 4.16 mM bicarbonate.
3. For microsurgical manipulations, remove cumulus cells by brief, gentle treatment with hyaluronidase (Sigma, H3506, stock concentration 600 U/ml diluted to 100 U/ml when applied) at room temperature as rapidly as possible. Oocytes are then cultured in the medium of choice (e.g., CZB medium). Once manipulated and activated, the constructs are washed and cultured in the appropriate medium, depending on embryo type.

3.2. Embryo Isolation and Culture

1. Using similar procedures to those described for oocytes, fertilized zygotes are isolated from mated females, typically at 19–20 h post hCG injection.
2. Culture embryos in medium of choice (e.g., KSOM). Select high-quality fertilized embryos (most easily recognized by the presence

of pronuclei) of appropriate morphology and granularity for manipulation. Microsurgical manipulations can be performed in HCZB, M2, or HEPES-buffered KSOM.

3.3. Pronuclear Transfer

1. The basic PNT technique (10) involves removing a plasma membrane-bound "karyoplast" containing one pronucleus and placing it under the zona pellucida of the recipient zygote, followed by fusion to complete the PNT. The following setup and procedure are presented as appropriate for constructing androgenones and gynogenones using an inverted microscope system and electrofusion to accomplish karyoplast fusion (see Note 17). Variations in setup can be made as needed for other purposes. Major steps in the procedure are shown in Fig. 1.

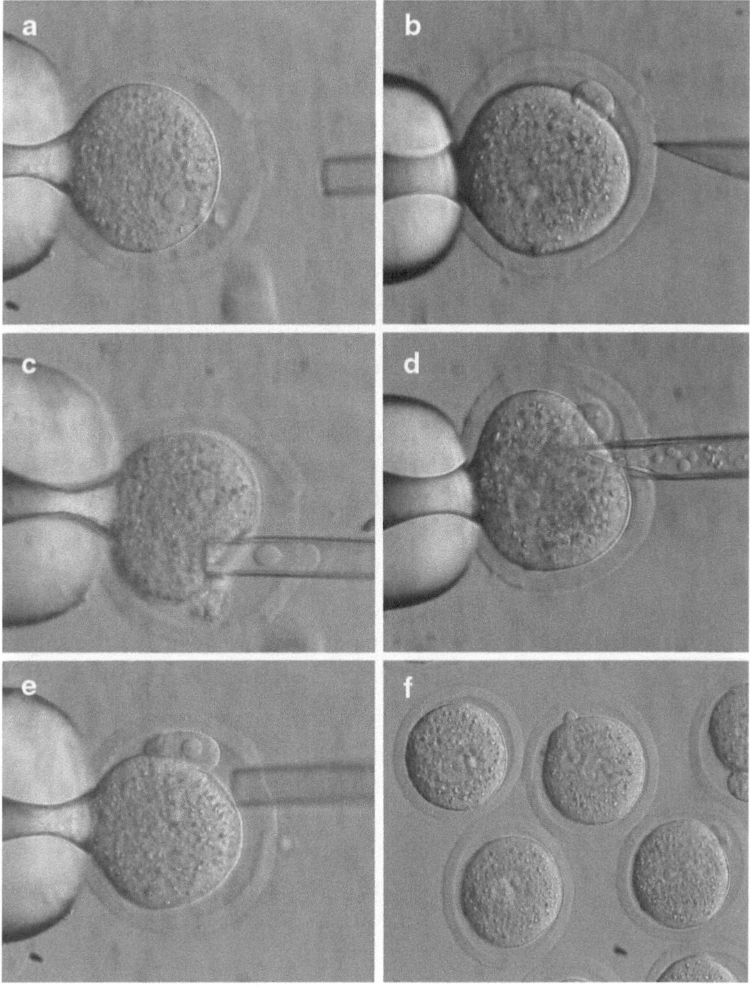

Fig. 1. Pronuclear transfer procedure. Panels show embryos before manipulation (**a**) using blunt pipet or (**b**) beveled pipet, followed by pronucleus aspiration using (**c**) a blunt pipet or (**d**) a beveled pipet, (**e**) karyoplast inserted into the perivitelline space, and (**f**) embryos after karyoplast fusion.

2. Place two rows of drops of HEPES-buffered manipulation medium (e.g., HCZB containing 5 μg/ml CB and 0.2 μg/ml demecolcine) on the plastic dish under mineral oil (see Note 18), most conveniently in a staggered arrangement to allow easy access to each drop from both sides.
3. Before manipulation, treat zygotes for at least 30 min with 5 μg/ml CB and 0.2 μg/ml demecolcine at 37 °C in the incubator. One zygote is added to each drop. It is preferable that all embryos loaded be manipulated in no longer than 30 min time on the microscope. This often equates to 10–12 drops of embryos per round.
4. Hold the first zygote on the holding pipet with a slight negative pressure. Rotate the zygote until it is oriented so that both pronuclei are visible and in the same plane of focus, and the polar body between 10 and 2 o'clock positions. The maternal pronucleus is typically smaller and located closer to the polar body than the paternal pronucleus, although the size difference may vary with strain. The pronucleus to be removed should be oriented for easy access to the tip of the PNT pipet (e.g., between 3 and 6 o'clock with PNT pipet on right hand side).
5. While maintaining slight to moderate negative pressure on the holding pipet, insert the PNT pipet (bevel oriented toward the 6 o'clock position) through the zona pellucida without penetrating the plasma membrane. This can be accomplished with a flat-tip pipet inserted through a slit cut in the zona pellucida with a sharp glass needle, with a beveled pipet sharpened using the "broken spike" method, or with a piezo pipet driver and unsharpened, beveled pipet. In the latter case, care must be taken to keep the intensity of pulses as low as possible and to avoid transmitting the pulse to the oolemma, which would lyse the cell.
6. Once through the zona, press the tip of the PNT pipet inward and position it adjacent to the target pronucleus. The pronucleus can be nudged to ensure that it is in position.
7. Apply negative pressure to draw the intervening plasma membrane, minimal cytoplasm, and pronucleus gradually into the pipet. Withdraw the pipet from the perivitelline space. The karyoplast and plasma membrane will seal themselves. The first pronucleus removed in the experiment is discarded, thus forming the first recipient. Simultaneously with removal of each pronucleus, the operator can remove the polar body if desired.
8. Using the same approach, obtain a donor karyoplast from the next zygote. Once the karyoplast is drawn into the pipet, a small volume of medium is drawn in to prevent the karyoplast from contacting the mineral oil between drops. The operator returns to the previously manipulated zygote (now the recipient). The recipient zygote is reacquired onto the holding pipet so

that the original opening in the zona pellucida is oriented at the 3 o'clock position. The opening is often easily visible by a small strand of cytoplasm, or as a slit seen in profile. The tip of the PNT pipet is reinserted into the perivitelline space, and the karyoplast is gently expelled. If desired, the PNT pipet tip can be withdrawn from the perivitelline space as soon as the pronucleus portion of the karyoplast is expelled in order to reduce transfer of any excess cytoplasm, which can then be expelled separately.

9. Alternately transfer maternal and paternal pronuclei to successive recipients to yield both gynogenones and androgenones. If no loss occurs, this leaves androgenones in one row and gynogenones in the other, with one haploid zygote to serve as recipient for the next round of embryos. It is recommended that the operator maintain an ongoing record to confirm the constructs in each drop during each round of manipulations.
10. Wash PNT constructs and return to embryo culture without cytoskeletal inhibitors for at least 30 min at 37 °C. Electrofusion is then performed using a suitable apparatus and a dish containing electrodes about 1 mm apart. Because of density differences between the electrofusion and embryo culture media, embryos should be washed through electrofusion medium or equilibrated with this medium inside the pipet before loading between electrodes. With the BTX system, a brief AC pulse (50 V/cm) can be given to orient the constructs between the electrodes (membranes at point of contact between karyoplast and recipient cell parallel to the electrodes). Immediately after orientation, a single DC pulse of 900 V/cm is delivered (see Note 19). Embryos are then washed through several drops of culture medium and allowed to recover in the incubator. Fusion should be completed within about 1 h.

3.4. Intracytoplasmic Sperm Injection

ICSI involves injecting a spermatozoon into the ooplasm of matured eggs to achieve fertilization. ICSI bypasses the need for sperm motility, zona penetration, binding, and fusion to the oocyte. ICSI in the mouse requires piezo-actuated micromanipulation to avoid lysing the oocyte (45). ICSI can be used to produce uniparental androgenone embryos by removing the maternal chromosomes and injecting one (haploid) or two (diploid) spermatozoa. It is also useful where a high efficiency of fertilization is needed, or where IVF may be problematic.

1. Prepare injection pipet. The diameter of injection pipet for sperm varies with different strains. For example, DBA/2 strain sperm heads are a little larger than C57BL/6 sperm heads. Correspondingly, the diameter of the injection pipet for injecting DBA/2 sperm is bigger than that for injecting C57BL/6 sperm (see Note 16). A small bead of mercury is introduced into the injection pipet in order to increase the mass loaded

and decrease the lateral oscillations that may damage the oocyte when using the piezo driver.

2. Enucleate MII stage oocytes (for preparing androgenones): Place eggs in the drop of M2 medium with 5 μg/ml CB for 3 min. The SCC in matured eggs is visible as a nongranular "clear" region within the ooplasm under Hoffman modulation contrast optics. Gently aspirate the egg onto the holding pipet, rotate it (can use fluid flow in and out of holding pipet and contact with the enucleation pipet to turn the oocyte) to the position with the spindle at 3 o'clock, and then stabilize the position using negative pressure in the holding pipet. Move the enucleation pipet to the outer surface of zona pellucida at 3 o'clock. A couple of piezo pulses are applied to allow the enucleation pipet to penetrate through the zona pellucida into the perivitelline space. Piezo pulses should be terminated as the inner surface of the zona pellucida is approached to avoid lysing the oocyte. Position the enucleation pipet adjacent to the spindle, and observe the spindle move as this is achieved. Gently increase the negative pressure on the spindle removal pipet to aspirate the spindle into the pipet. Withdraw the pipet from the perivitelline space to remove the SCC as a membrane-bound "karyoplast." Push the spindle out of the pipet and release the enucleated oocyte (cytoplast) to complete one enucleation procedure. Continuously remove the spindles of other oocytes as rapidly as possible, within about 10 min for experienced operator to remove 20–30 spindles in one round. Completely rinse the cytoplasts in fresh CZB medium and allow them to recover in the incubator at least 15 min.

3. Prepare capacitated sperm for injection. We suggest using capacitated sperm from adult males for ICSI in order to obtain highest fertilization rates. A 200 μl CZB medium drop in 65 mm diameter culture dish is covered with the mineral oil and equilibrated in the incubator for at least 30 min. The cauda epididymes are dissected from one adult male and immediately placed in the CZB drop to allow the sperm to swim out freely. It is helpful to squeeze the cauda epididymes with a pair of sterile fine forceps to increase sperm quantity. After the sperm become active, sperm at the edge of the medium drop are collected and transferred into 7% PVP drop in the manipulation plate for injection.

4. Prepare the ICSI micromanipulation dish. Manipulation solutions in the dish for ICSI consist of three kinds of drops. One drop of 10% PVP is used to lubricate the inner wall of the injection pipet by repeated aspiration (see Note 20). A couple of 7% PVP drops are added, in which to place capacitated sperm. Several HCZB medium drops are applied to the dish, in which to place the enucleated cytoplasts. In 7% PVP solution, sperm swim gently and slowly and can be captured easily.

5. Perform sperm head injection. Aspirate the sperm tail first using the injection pipet and apply several pulses immediately at the junction between the sperm head and principal piece of tail to separate sperm head. Blow out the tail and aspirate sperm heads individually into the injection pipet. It is important not to accumulate sperm heads touching each other. Holding the cytoplast with the holding pipet, insert the injection pipet through the zona pellucida by applying a couple of pulses at intensity of 3–6 and frequency at 2. Once the injection pipet passes into the perivitelline space, put the injection tip gently touching the ooplasm membrane and at the same time, pushing sperm heads forward to the pipet tip. Press the injection pipet against the ooplasm membrane and then toward the opposite side near the holding pipette. Promptly give the pulse (intensity and frequency settings of "1") to penetrate the ooplasm membrane. Operator should observe backward rebound of the oocyte membrane to confirm successful penetration. Gently push sperm head(s) into the ooplasm and immediately withdraw the pipet to complete the ICSI procedure. Once all oocytes in the group are injected, leave them in the injection dish for about 5 min to recover.
6. Collect the manipulated eggs in and rinse with HEPES-free CZB medium completely. Transfer the embryos to embryo culture medium, such as CZB or KSOM, to observe pronucleus formation and embryo development.

3.5. Round Spermatid Nuclear Transfer

Viable offspring have been produced from round spermatid injection (ROSI) in mouse, rat, rabbit, and humans (46, 47). Round spermatids are immature haploid cells characterized by the presence of a decondensed nucleus. The difference in nuclear status between spermatid and spermatozoa, which are decondensed and condensed, respectively, affects ICSI and ROSI protocols. In a standard ICSI protocol, a spermatozoon is simply injected into an MII oocyte. In ROSI, however, injected oocytes must be artificially activated before or after the injection of the round spermatid. Similar to ICSI, ROSI can be used to produce uniparental androgenic embryos by injecting two round spermatid nuclei.

1. Oocyte preparation for ROSI. While ROSI can be used to make androgenones using SCC-depleted MII oocytes followed by chemical activation, the proportion of oocytes surviving injection of two round spermatids is better (48) if spermatids are injected into preactivated intact MII oocytes (progressing to telophase) followed by removal of maternal pronucleus within 4 h after activation. The injections should be completed within 70–80 min of activation (48, 49). Activation of mouse oocytes can be achieved efficiently by exposing oocytes to 5 mM of $SrCl_2$ in Ca^{2+}-free CZB/KSOM medium for 20 min.

The procedure for ROSI is almost the same as described above for ICSI except for the following changes.

2. Prepare ROSI pipets with an inner diameter of 3.5 μm (adjust for sperm donor strain as needed) and place a small bead of mercury inside as described under ICSI.
3. Collect the testes from mature males into GL-PBS and remove the tunica albuginea using a pair of fine forceps. Allow the seminiferous tubules to spread into the buffer and cut into minute pieces using a pair of sharp scissors. Avoid contaminating testis and seminiferous tubules with blood. If necessary testes and seminiferous tubules may be washed in ELB briefly (155 mM NH_4Cl, 10 mm $KHCO_3$, 2 mM EDTA, pH 7.2) briefly before placing them in GL-PBS. Gently pipet the mixture repeatedly to disperse spermatozoa and spermatogenic cells into the collection medium. Filter the cell suspension through a 50 μm nylon mesh and wash three times by centrifugation at 200 g for 5 min. Resuspend the cells in GL-PBS and keep them at 4 °C. The isolated cell suspension stored at 4 °C should be viable for several hours.
4. Prepare the ROSI micromanipulation dish as in ICSI described above. Mix a small aliquot of 1–3 μl of cell suspension (from step 3) with 10 μl of PVP as described above in ICSI.
5. Round spermatids can be recognized easily by their small size (~10 μm) and a distinct centrally located chromatin mass. For injection, draw a single spermatid into the injection pipet. Move the spermatid in and out the pipet until the plasma membrane is ruptured and the spermatid nucleus is separated from the cytoplasm. While only one or two spermatids are injected into a single oocyte, loading several (ten or more) spermatids into the pipet improves speed of production. Inject the spermatid nucleus into a telophase or metaphase II stage oocyte using a piezo-driven pipet as described under ICSI. Release the injected oocyte and repeat with another oocyte until all the spermatids drawn into the injection pipet are finished. Usually 10–15 oocytes can be injected within 15–20 min with practice. Allow the injected oocytes to recover for 5–10 min before returning them to the culture dish containing CZB droplets. Briefly wash the injected oocytes to remove HEPES before transferring them to the HEPES-free CZB drops in the incubator.
6. Following ROSI the maternal of Telophase II oocytes the maternal pronucleus is removed along with second polar body in the presence of cytochalasin B and demecolcine as described for PNT. Allow 5–10 min for the recovery of oocytes after pronucleus removal, and then wash in cytochalasin B-free medium

and return to the incubator. Transfer the embryos to embryo culture medium, such as CZB or KSOM, to observe pronucleus formation and embryo development.

4. Notes

1. Stereomicroscope providing from 10.5× to 105× magnification, with mirror (e.g., Gimbal mount) suitable for providing oblique illumination so that intracellular detail (e.g., pronuclei) can be seen.
2. The different procedures described here may impose different optical requirements on the microscope to be used for microsurgery or microinjection. Fixed stage, upright microscopes can be used in conjunction with a hanging drop system for specimen mounting. More commonly, inverted microscopes are used in conjunction with plastic dishes with specimens in medium droplets under oil. An excellent system that will be easy to use for all of the procedures described here is the Olympus Inverted model IX71 with modulation contrast optics and available magnifications of 4×, 20×, and 40× objectives and 10× eyepieces.
3. There are two main kinds of microforge that can be used. One is a DeFonbrune style microforge that provides fine control of filament temperature using a combination of rheostat, fan speed, and flow restriction. Alternatively, Narishige provides a widely used model (MF900) with excellent control for crafting fine gauge microneedles.
4. We find the Sutter Instrument line of pipet pullers easily programmable and capable of yielding pulled pipets with consistent geometry, which can be varied easily through choice of filament size and type, with guidance from the merchant-supporting literature.
5. We use the Sutter Instruments pipet beveler, which can come with a choice of grinding surface and is easily maintained.
6. One microinjector is required for aspiration of pronuclei and spindles, microinjection transfer of karyoplasts or cells beneath the zona, or injection of nuclei or sperm heads into the ooplasm. A second microinjector can be used to control pressure on the holding pipet, or an air-filled 50 cm^3 syringe also works well for this. The Narishige model IM9B/5B injector works well for microinjection and the IM9A/5A model works well for the holding pipet.
7. A common choice for micromanipulator has been the Narishige hydraulic system (MN4). Eppendorf also provides a widely

used system. Sutter Instruments offers an attractive new system of manipulators that provides excellent motion control and that avoids problems with leaking of hydraulic fluid.

8. We prefer the PMM piezo-drill micromanipulation controller PMM-150 (Prime Tech Ltd, Ibaraki, Japan).
9. A widely preferred medium for mouse embryo culture is KSOM (40). This medium is an excellent choice for embryos because of documented gains in retaining embryo phenotype very similar to in vivo developing embryos.
10. Two-step systems can also provide excellent results, such as using CZB (42) through the 8-cell stage followed by culture in Whitten's medium. Other media (e.g., HCZB and M2 medium) are preferred for culturing MII stage oocytes during in vitro manipulation. CZB and KSOM medium are used to culture MII stage oocytes inside incubator before and after manipulation.
11. Cytochalasin B is diluted in 100% ethanol to prepare 1,000× stock (5 mg/ml). Demecolcine is diluted in Millipore water to prepare 1,000× stock (0.2 mg/ml). Both of them are stored as aliquots at −70 °C.
12. Pipets to be used on an inverted microscope are typically bent on the microforge by about 20° at a distance of about 1–2 mm from the tip.
13. Holding pipets are constructed by pulling pipets with an elongated geometry, approximately 90–100 μm outer diameter (O.D.). The pipet needs to be cut on the microforge with an end that is as close to flat and 90° angle to the sides as possible, and then heat polished on the microforge to yield an opening of about 15–25 μm.
14. Spindle removal pipets are prepared with a flat tip and an inner diameter of 10 μm. Adhesion to the cell membrane can be reduced by washing with 10% PVP.
15. PNT pipets can be prepared with either flat or beveled tips, 20–30 μm inner diameter (I.D.) depending on operator preference and skill. Smaller pipets are easier to penetrate the zona pellucida but larger pipets will offer less adhesion to the membrane of the karyoplast. The tip is beveled at an angle of 45°. The flat-tip pipet needs to be washed three times in 10% PVP to lubricate the inner wall prior to use. Beveled pipets are washed with 20% hydrofluoric acid quickly for three times (excessive time dissolves too much glass), washed with MilliQ water for five times, and then washed with 95% ethanol for three times before use. If no piezo driver is to be used in the PNT, a very fine spike is drawn at the tip using the micoforge, and the spike is then broken at its base at the time of use. Prior to use, the PNT pipet should be treated by aspirating Igepal

CA-630 several times aspirating Igepal CA-630 (Sigma, Cat. I-3021), then washed at least ten times with vigorous agitation in a large beaker of MilliQ water, and then air dried. This treatment minimizes adhesion of the karyoplast membrane.

16. Sperm injection pipets are blunt, and cut at a tip with outer diameter of 6 to 9 μm (varies with strain of sperm) and an inner diameter slightly larger than diameter of sperm head to avoid adhesion. The PVP in the sperm suspension medium reduces adhesion.
17. For PNT, karyoplast fusion was originally accomplished by McGrath and Solter using inactivated Sendai virus introduced simultaneously with insertion of the karyoplast under the zona pellucida. This method works very well with a suitable preparation of virus. However, in the absence of such a virus preparation, the electrofusion method also works very well with no apparent detriment to the construct, provided that repeated fusogenic pulses are avoided.
18. As with all culture system components, mineral oil should be quality tested to ensure that it supports maximum embryo viability. Once suitable lots of oil and other culture system components are identified, it is recommended that a supply of these lots sufficient to last for a prolonged period is acquired.
19. For electrofusion only a single pulse should be given. Additional constructs may fuse with additional pulses. However, the additional pulses may compromise embryo quality.
20. PVP is added to HCZB media at 10% as a lubricant to minimize adherence of cell membranes inside pipets and to facilitate nuclear aspiration prior to injection. We use Sigma PVP -360 (molecular weight 360,000). PVP containing medium for suspending nuclei or sperm should contain a maximum of 10%w/v (diluted to about 7% after sperm loading). Due to possible PVP toxicity, the PVP concentration should be kept as low as possible.

References

1. Di Berardino MA (1997) Genomimc potential of differentiated cells. Columbia University Press, New York
2. Campbell KH, McWhir J, Ritchie WA et al (1996) Sheep cloned by nuclear transfer from a cultured cell line. Nature 380:64–66
3. Wakayama T, Perry AC, Zuccotti M et al (1998) Full-term development of mice from enucleated oocytes injected with cumulus cell nuclei. Nature 394:369–374
4. Davidson RI (1974) Gene expression in somatic cell hybrids. Annu Rev Genet 8: 195–218
5. Takahashi K, Yamanaka S (2006) Induction of pluripotent stem cells from mouse embryonic and adult fibroblast cultures by defined factors. Cell 126:663–676
6. Cattanach BM, Kirk M (1985) Differential activity of maternally and paternally derived chromosome regions in mice. Nature 315:496–498
7. McGrath J, Solter D (1984) Completion of mouse embryogenesis requires both the maternal and paternal genomes. Cell 37:179–183
8. Barton SC, Surani MA, Norris ML (1984) Role of paternal and maternal genomes in mouse development. Nature 311:374–376

9. Surani MA, Barton SC, Norris ML (1984) Development of reconstituted mouse eggs suggests imprinting of the genome during gametogenesis. Nature 308:548–550
10. McGrath J, Solter D (1983) Nuclear transplantation in mouse embryos. J Exp Zool 228: 355–362
11. McGrath J, Solter D (1983) Nuclear transplantation in the mouse embryo by microsurgery and cell fusion. Science 220:1300–1302
12. Latham KE, Doherty AS, Scott CD et al (1994) Igf2r and Igf2 gene expression in androgenetic, gynogenetic, and parthenogenetic preimplantation mouse embryos: absence of regulation by genomic imprinting. Genes Dev 8:290–299
13. Rossant J, Guillemot F, Tanaka M et al (1998) Mash2 is expressed in oogenesis and preimplantation development but is not required for blastocyst formation. Mech Dev 73:183–191
14. Borgel J, Guibert S, Li Y et al (2010) Targets and dynamics of promoter DNA methylation during early mouse development. Nat Genet 42:1093–1100
15. Kim JM, Ogura A (2009) Changes in allele-specific association of histone modifications at the imprinting control regions during mouse preimplantation development. Genesis 47: 611–616
16. Kono T (2006) Genomic imprinting is a barrier to parthenogenesis in mammals. Cytogenet Genome Res 113:31–35
17. Wu Q, Kumagai T, Kawahara M et al (2006) Regulated expression of two sets of paternally imprinted genes is necessary for mouse parthenogenetic development to term. Reproduction 131:481–488
18. Hiura H, Obata Y, Komiyama J et al (2006) Oocyte growth-dependent progression of maternal imprinting in mice. Genes Cells 11:353–361
19. Kimura Y, Tateno H, Handel MA et al (1998) Factors affecting meiotic and developmental competence of primary spermatocyte nuclei injected into mouse oocytes. Biol Reprod 59:871–877
20. Ogura A, Yanagimachi R (1995) Spermatids as male gametes. Reprod Fertil Dev 7:155–158, discussion 158–159
21. Latham KE, Solter D (1991) Effect of egg composition on the developmental capacity of androgenetic mouse embryos. Development 113:561–568
22. Yeo S, Lee KK, Han YM et al (2005) Methylation changes of lysine 9 of histone H3 during preimplantation mouse development. Mol Cells 20:423–428
23. Santos F, Peters AH, Otte AP et al (2005) Dynamic chromatin modifications characterise the first cell cycle in mouse embryos. Dev Biol 280:225–236
24. Santos F, Hendrich B, Reik W et al (2002) Dynamic reprogramming of DNA methylation in the early mouse embryo. Dev Biol 241: 172–182
25. Park JS, Jeong YS, Shin ST et al (2007) Dynamic DNA methylation reprogramming: active demethylation and immediate remethylation in the male pronucleus of bovine zygotes. Dev Dyn 236:2523–2533
26. van der Heijden GW, Dieker JW, Derijck AA et al (2005) Asymmetry in histone H3 variants and lysine methylation between paternal and maternal chromatin of the early mouse zygote. Mech Dev 122:1008–1022
27. McLay DW, Clarke HJ (2003) Remodelling the paternal chromatin at fertilization in mammals. Reproduction 125:625–633
28. Lepikhov K, Walter J (2004) Differential dynamics of histone H3 methylation at positions K4 and K9 in the mouse zygote. BMC Dev Biol 4:12
29. Howell CY, Steptoe AL, Miller MW et al (1998) cis-Acting signal for inheritance of imprinted DNA methylation patterns in the preimplantation mouse embryo. Mol Cell Biol 18:4149–4156
30. Latham KE (1994) Strain-specific differences in mouse oocytes and their contributions to epigenetic inheritance. Development 120: 3419–3426
31. Latham KE, Sapienza C (1998) Localization of genes encoding egg modifiers of paternal genome function to mouse chromosomes one and two. Development 125:929–935
32. Pickard B, Dean W, Engemann S et al (2001) Epigenetic targeting in the mouse zygote marks DNA for later methylation: a mechanism for maternal effects in development. Mech Dev 103:35–47
33. Liang CG, Han Z, Cheng Y et al (2009) Effects of ooplasm transfer on paternal genome function in mice. Hum Reprod 24:2718–2728
34. Latham KE, Rambhatla L (1995) Expression of X-linked genes in androgenetic, gynogenetic, and normal mouse preimplantation embryos. Dev Genet 17:212–222
35. Latham KE, Akutsu H, Patel B et al (2002) Comparison of gene expression during preimplantation development between diploid and haploid mouse embryos. Biol Reprod 67: 386–392
36. Han Z, Chung YG, Gao S et al (2005) Maternal factors controlling blastomere fragmentation

in early mouse embryos. Biol Reprod 72: 612–618

37. Kaneko-Ishino T, Kuroiwa Y, Miyoshi N et al (1995) Pegl/Mest imprinted gene on chromosome 6 identified by cDNA subtraction hybridization. Nat Genet 11:52–59
38. Ozil JP, Banrezes B, Toth S et al (2006) Ca2+ oscillatory pattern in fertilized mouse eggs affects gene expression and development to term. Dev Biol 300:534–544
39. Kuretake S, Kimura Y, Hoshi K et al (1996) Fertilization and development of mouse oocytes injected with isolated sperm heads. Biol Reprod 55:789–795
40. Lawitts JA, Biggers JD (1991) Optimization of mouse embryo culture media using simplex methods. J Reprod Fertil 91:543–556
41. Summers MC, Biggers JD (2003) Chemically defined media and the culture of mammalian preimplantation embryos: historical perspective and current issues. Hum Reprod Update 9:557–582
42. Chatot CL, Ziomek CA, Bavister BD et al (1989) An improved culture medium supports development of random-bred 1-cell mouse embryos in vitro. J Reprod Fertil 86:679–688
43. Chung YG, Gao S, Latham KE (2006) Optimization of procedures for cloning by somatic cell nuclear transfer in mice. Methods Mol Biol 348:111–124
44. Latham KE, Westhusin ME (2000) Nuclear transplantation and cloning in mammals. Methods Mol Biol 136:405–425
45. Kimura Y, Yanagimachi R (1995) Intracytoplasmic sperm injection in the mouse. Biol Reprod 52:709–720
46. Ogura A, Ogonuki N, Miki H et al (2005) Microinsemination and nuclear transfer using male germ cells. Int Rev Cytol 246:189–229
47. Yanagimachi R (2005) Intracytoplasmic injection of spermatozoa and spermatogenic cells: its biology and applications in humans and animals. Reprod Biomed Online 10:247–288
48. Miki H, Hirose M, Ogonuki N et al (2009) Efficient production of androgenetic embryos by round spermatid injection. Genesis 47:155–160
49. Kishigami S, Wakayama S, Nguyen VT et al (2004) Similar time restriction for intracytoplasmic sperm injection and round spermatid injection into activated oocytes for efficient offspring production. Biol Reprod 70:1863–1869

Chapter 2

Derivation of Induced Pluripotent Stem Cells by Retroviral Gene Transduction in Mammalian Species

Masanori Imamura, Hironobu Okuno, Ikuo Tomioka, Yoshimi Kawamura, Zachary Yu-Ching Lin, Ryusuke Nakajima, Wado Akamatsu, Hirotaka James Okano, Yumi Matsuzaki, Erika Sasaki, and Hideyuki Okano

Abstract

Pluripotent stem cells can provide us with an enormous cell source for in vitro model systems for development. In 2006, new methodology was designed to generate pluripotent stem cells directly from somatic cells, and these cells were named induced pluripotent stem cells (iPSCs). This method consists of technically simple procedures: donor cell preparation, gene transduction, and isolation of embryonic stem cell-like colonies. The iPSC technology enables cell biologists not only to obtain pluripotent stem cells easily but also to study the reprogramming events themselves. Here, we describe the protocols to generate iPSCs from somatic origins by using conventional viral vectors. Specifically, we state the usage of three mammalian species: mouse, common marmoset, and human. As mouse iPSC donors, fibroblasts are easily prepared, while mesenchymal stem cells are expected to give rise to highly reprogrammed iPSCs efficiently. Common marmoset (*Callithrix jacchus*), a nonhuman primate, represents an alternative model to the usual laboratory animals. Finally, patient-specific human iPSCs give us an opportunity to examine the pathology and mechanisms of dysregulated genomic imprinting. The iPSC technology will serve as a valuable method for studying genomic imprinting, and conversely, the insights from these studies will offer valuable criteria to assess the potential of iPSCs.

Key words: Genomic imprinting, Induced pluripotent stem cells, Embryonic stem cells, Reprogramming, Pluripotency, Epigenetics, Germ cells, Cell culture, Common marmoset, Disease model

1. Introduction

Induced pluripotent stem cells (iPSCs) can be generated by transduction of various sets of defined factors into somatic cells (1, 2). Molecular and cellular properties of iPSCs are quite similar to those of embryonic stem cells (ESCs), and they have pluripotency in vivo and in vitro. Among pluripotent stem cells, there is a great advantage in utilizing iPSCs: facile derivation from individuals.

Nora Engel (ed.), *Genomic Imprinting: Methods and Protocols*, Methods in Molecular Biology, vol. 925,
DOI 10.1007/978-1-62703-011-3_2,

Because of this, iPSC technology holds an enormous utility as a cell source for studying many genetic mutants, including those involved in human disease. In humans, dysregulation of imprinted genes is correlated with tumorigenesis or various disorders such as Beckwith–Wiedemann syndrome, Prader–Willi syndrome, and Angelman syndrome. Therefore, generation of patient-specific iPSCs could bring us better understanding of the pathology of genomic imprinting-related disorders (3). Moreover, iPSCs have become valuable cell sources for parthenogenesis, which is a unique model for studying genomic imprinting. It has been shown that parthenogenetic blastocysts and ESCs exhibit partial loss of imprinting (4). Recently, bimaternal parthenogenetic iPSCs were established from mouse neural stem cells (5). The iPSCs have lost the parthenogenetic imprinting pattern despite their origin, suggesting an attenuation of parthenogenetic imprinting through reprogramming process. Given that parthenogenetic cells generally exhibit growth defects, the iPSC technology would also provide ideal materials and methods to dissect this phenomenon as well.

Epigenetic regulation is intimately tied to the input and output of artificial reprogramming by somatic cell nuclear transfer (SCNT), cell fusion, and iPSC technology (6). For example, SCNT sometimes results in abnormal embryogenesis correlated with dysregulated genomic imprinting (7). Since iPSCs with somatic origin do not receive any germline-derived factors during reprogramming, it is important to examine the genomic imprinting pattern for quality assessment of iPSCs. Indeed, recent studies have revealed an important role of genomic imprinting in developmental potential of iPSCs. Gene expression profiling found aberrant silencing of the *Dlk1-Dio3* imprinted locus in mouse iPSCs, although overall gene expression was indistinguishable with that of ESCs (8, 9). The activation of the *Dlk1-Dio3* imprinted locus is positively correlated with a fully reprogrammed status, and notably, the developmental potency in partially reprogrammed iPSCs can be rescued by reactivation of this locus (8). Similarly, human iPSCs also exhibit aberrations in imprinted genes such as *H19* and *PEG3* in their allele-specific expression pattern, expression intensity, and DNA methylation status (10). Thus, the proper genomic imprint could serve as a vital marker to identify fully reprogrammed and clinically applicable iPSCs lines.

Obviously, the advantage of pluripotent stem cells in life sciences is their pluripotency and as an in vitro system to elucidate the mechanisms of development and differentiation. Since the pioneering work on germ cell production from ESCs in culture (11–13), it has been revealed that this potential is commonly observed among pluripotent stem cell lines (14–16). Furthermore, this propensity is valid for iPSCs as well; presumptive germ cells can be induced by in vitro differentiation of mouse and human iPSCs

(17–19). In these studies, genomic imprinting is a standard subject of analyses because germ cells undergo a dynamic alteration of genomic imprinting status in a developmental phase-specific manner. Hence, the imprinting status is one of the landmarks used to define iPSC-derived cells as "germ cells."

Here, we describe the protocols to obtain iPSCs from somatic cells derived from three mammalian species: mouse, common marmoset (*Callithrix jacchus*), and human. Because, to date, iPSCs have been successfully established in various animals using basically similar protocols (1, 20–25), the current methods could be applicable for other mammals of interest. We believe that this information will help accelerate elucidation of genomic imprinting with molecular and cellular biological approaches.

2. Materials

2.1. Mouse iPSCs from Fibroblasts

2.1.1. General Equipment

1. Tissue culture plates and dishes: 100-mm, 6-, 24-, and 96-well (BD Falcon).
2. Conical tubes: 15- and 50-ml (BD Falcon).
3. Plastic disposable pipettes: 1-, 5-, 10-, and 25-ml (BD Falcon).
4. 0.22-µm Bottle-top filter (Techno Plastic Products, Trasadingen, Switzerland).
5. 0.22-µm Pore size filter (Millipore, Billerica, MA, USA).
6. 10-ml Disposable syringe (Terumo, Tokyo, Japan).
7. Cell-freezing container (Nalgene, Rochester, NY, USA).
8. Cryovial (Nunc, Waltham, MA, USA).

2.1.2. Cell Culture

1. PBS.
2. 0.25% (w/v) Trypsin/EDTA solution (Invitrogen).
3. Recovery Cell Culture Freezing Medium (Invitrogen).
4. mDMEM/10% FBS: DMEM containing 4.5 g/l glucose (Nacalai Tesque, Kyoto, Japan) supplemented with 10% (v/v) Fetal bovine serum (FBS; Invitrogen, Carlsbad, CA, USA), 50 U/ml penicillin and 50 mg/ml streptomycin (Invitrogen). Filter with a 0.22-µm bottle-top filter and store at 4°C up to a week.
5. mESC medium: DMEM supplemented with 15% (v/v) FBS, 2 mM L-Glutamine (Invitrogen), 0.1 mM non-essential amino acids (Invitrogen), 0.1 mM 2-mercaptoethanol (Invitrogen) (see Note 1), 50 U/ml penicillin and 50 mg/ml streptomycin, and 1,000 U/ml ESGRO (Millipore). Filter with a bottle-top filter and store at 4°C up to a week.

6. Gelatin-coated culture dishes: To prepare 10× stock solution, dissolve 1 g of gelatin powder (Sigma, St. Louis, MO, USA) in 100 ml of distilled water, autoclave, and store at 4°C for 2 months. To prepare 1× gelatin solution, warm the 10× gelatin stock to 37°C, add 50 ml of the stock to 450 ml of distilled water. Filter the solution with a bottle-top filter and store at 4°C up to 2 weeks. Add 0.1% (w/v) gelatin solution to cover the entire area of culture dishes. Incubate for at least 30 min at 37°C. Aspirate the solution immediately before plating cells.

2.1.3. Preparation of Fibroblasts from Mouse Embryos and Adult Mouse Tail

1. Sterilized forceps and scissors.

2.1.4. Retrovirus Production

1. pMXs vectors containing the cDNAs of *Oct4* (Plasmid 13366), *Sox2* (Plasmid 13367), *Klf4* (Plasmid 13370), *c-Myc* (Plasmid 13375), and *DsRed* (Plasmid 22724) (Addgene, Cambridge, MA, USA).
2. Plat-E packaging cells (Available from Dr. Toshio Kitamura at the University of Tokyo; kitamura@ims.u-tokyo.ac.jp).
3. Puromycin: Dissolve puromycin powder (Sigma) in distilled water at 10 mg/ml concentration, and filter it through a 0.22-μm filter (Millipore). Aliquot and store at −20°C.
4. Blastocidin S: Dissolve blastocidin S hydrochloride (Funakoshi, Tokyo, Japan) in distilled water at 10 mg/ml concentration, and filter it through a 0.22-μm filter. Aliquot and store at −20°C.
5. 0.05% Trypsin/EDTA: Mix 10 ml of 0.25% (w/v) Trypsin/EDTA solution (Invitrogen) and 40 ml of PBS. Store at −20°C.
6. Opti-MEM I Reduced-Serum Medium (Invitrogen).
7. FuGENE 6 transfection reagent (Promega, Madison, WI).
8. 0.45-μm cellulose acetate filter (Schleicher & Schuell, Keene, NH, USA).
9. Polybrene solution: To prepare the stock solution at 8 mg/ml concentration, dissolve 80 mg of polybrene (Nacalai Tesque) in 10 ml of distilled water and filter it through a 0.22-μm filter. Store at 4°C.

2.1.5. iPSCs' Derivation from Mouse Fibroblasts

1. SNL medium: DMEM supplemented with 7% (v/v) FBS, 2 mM L-Glutamine, 50 U/ml penicillin and 50 mg/ml streptomycin. Filter with a bottle-top filter and store at 4°C up to a week.
2. SNL feeder cells: SNL cells (SNL76/7; DS Pharma Biomedical, Osaka, Japan) are a derivative of STO cells, which express

neomycin-resistant gene and leukemia inhibitory factor (LIF). Cultivate the cells with SNL medium on gelatin-coated culture dishes. At 80–90% confluency in 100-mm culture dishes, add 0.3 ml of 0.4 mg/ml mitomycin C (Kyowa Hakko Kirin, Tokyo, Japan) solution to the culture medium and incubate at 37°C for 2.5 h (see Note 2). After washing with 10 ml of PBS twice, trypsinize, and count the cell number. Resuspend with SNL medium and seed the cells on gelatin-coated culture dishes at 1×10^6 cells per 100-mm culture dish. Use within 2 weeks.

2.2. Mouse iPSCs from Mesenchymal Stem Cells

2.2.1. General Equipment

1. Tissue culture plates and dishes: 100- and 60-mm dish (BD Falcon).
2. Conical tubes: 15- and 50-ml (BD Falcon).
3. Plastic disposable pipettes: 1-, 5-, 10-, and 25-ml.
4. 0.22-μm bottle-top filter (Techno Plastic Products).

2.2.2. Animals

1. *Nanog*$^{\text{GFP-IRES-Puro}}$ mice (available from RIKEN BioResource Center, Tsukuba, Japan).

2.2.3. Cell Culture

1. PBS.
2. 2× PBS containing 4% FBS: 10× PBS is diluted five times with sterile water and supplemented with 4% (v/v) FBS.
3. HBSS$^+$: HBSS (Nacalai Tesque) supplemented with 2% (v/v) FBS, 10 mM HEPES, and 50 U/ml penicillin and 50 mg/ml streptomycin (Invitrogen).
4. Mesenchymal stem cell (MSC) medium: MEM Alpha + GlutaMAX-I (GIBCO) supplemented with 10% (v/v) FBS, 10 mM HEPES, and 50 U/ml penicillin and 50 mg/ml streptomycin.
5. mESC medium: see Subheading 2.1.2, item 5.

2.2.4. Preparation of Bone Marrow Cell Suspension

1. Enzymatic dissociation solution: 0.2% (w/v) collagenase (Wako, Osaka, Japan) in DMEM containing 1.0 g/l glucose supplemented with 10 mM HEPES and 50 U/ml penicillin and 50 mg/ml streptomycin.
2. Cell strainer: 70-μm pore size (BD Falcon).
3. Sterile water.

2.2.5. Purification of MSCs

1. Fluorescently conjugated antibodies (eBioscience, San Diego, CA, USA): PE-conjugated CD45 (Clone: 30-F11, 12-0451), TER119 (Clone: TER-119, 12-5921), APC-conjugated PDGFRα (Clone: APA5, 17-1401), and FITC-conjugated Sca-1 (Ly6A/E, Clone: D7, 11-5981).

2. Fluorescently conjugated isotype controls (eBioscience): Rat IgG2b K Isotype Control PE (12-4031), Rat IgG2a K Isotype Control APC (17-4321), and Rat IgG2a K Isotype Control FITC (11-4321).
3. Propidium iodide solution (Sigma).
4. Triplelaser cell sorter such as MoFlo (Dako) and JSAN (Bay Bioscience).

2.2.6. Retrovirus Production

1. pMXs vectors: see Subheading 2.1.4, item 1.
2. Plat-E packaging cells (available from Dr. Toshio Kitamura at the University of Tokyo; kitamura@ims.u-tokyo.ac.jp).
3. mDMEM/10% FBS: see Subheading 2.1.2, item 4.
4. 0.05% Trypsin/EDTA: see Subheading 2.1.4, item 5.
5. FuGENE 6 transfection reagent (Promega).
6. 0.45-μm Cellulose acetate filter (Schleicher & Schuell).
7. Polybrene solution: see Subheading 2.1.4, item 9.

2.2.7. Induction of iPSCs from PαS Cells

1. SNL medium: see Subheading 2.1.5, item 1.
2. SNL feeder cells: see Subheading 2.1.5, item 2.
3. Puromycin: see Subheading 2.1.4, item 3.

2.3. Marmsoet iPSCs from Fetal Liver Cells

2.3.1. General Equipment Required Through Experiments

1. Tissue culture plates and dishes: 100-mm (Greiner bio-one, Frickenhausen, Germany) and 96-well (Iwaki, Tokyo, Japan).
2. Gelatin-coated culture dishes: 100-mm and 12-well (Iwaki).
3. Conical tubes: 15- and 50-ml (BD Falcon).
4. Plastic disposable pipettes: 1-, 5-, 10- (BD Falcon, 357551), and 25-ml (BD Falcon).

2.3.2. Cell Culture

1. cjDMEM/10% FBS: Dulbecco's modified Eagle's medium (Wako) supplemented with 10% (v/v) FBS (Biowest) and 1% (v/v) antibiotic–antimycotic solution (Invitrogen).
2. cjESC medium: Knockout DMEM (Gibco) supplemented with 10% (v/v) Knockout Serum Replacement (Invitrogen), 1 mM L-glutamine (Invitrogen), 0.1 mM MEM nonessential amino acids (Invitrogen), 0.1 mM 2-mercaptoethanol (Sigma), and 1% (v/v) antibiotic–antimycotic solution (Gibco).
3. Trypsin solution for ESCs: 0.25% (v/v) Difco trypsin 250 (BD, Baltimore, MD, USA), 1 mM $CaCl_2$, and 20% (v/v) KSR.
4. Hank's buffered salt solution without calcium or magnesium (Gibco).

2.3.3. Virus Production

1. pMXs retroviral vectors carrying human *OCT4* (Addgene, Plasmid 17217), *SOX2* (Addgene, Plasmid 17218), *KLF4* (Addgene, Plasmid 17219), *C-MYC* (Addgene, Plasmid 17220), *NANOG* (kindly provided by Dr. Yamanaka), *LIN28* (kindly provided by Dr. Yamanaka), and *GFP* (kindly provided by Dr. Yamanaka) (see Note 3).
2. pVSV-G vector and GP-2 cells (Retroviral Gene Transfer and Expression; TaKaRa, Shiga, Japan).
3. Opti-MEM I Reduced-Serum Medium (Invitrogen).
4. FuGENE 6 transfection reagent.
5. 0.45-μm pore-size cellulose acetate filter (Sartorius, Goettingen, Germany).
6. Poly-L-lysine (Sigma).

2.3.4. Preparation of Fetal Liver Cells

1. Sterilized forceps and scissors.
2. Collagenase solution: Dissolve Collagenase type I in DMEM at 0.5% (w/v) concentration.

2.3.5. Retroviral Infection of Marmoset Cells

1. Polybrene (Nacalai Tesque).
2. Mitomycin C-treated or irradiated MEF feeder cell plates.

2.3.6. Passage of iPSCs

1. Cell strainer, 100-μm nylon (BD Falcon).
2. Mitomycin C-treated or irradiated MEF feeder cell plates.

2.3.7. Storage of Established iPSCs

1. Cell Banker 2 (ZENOAQ, Koriyama, Fukushima, Japan).
2. 2-ml plastic cryogenic vial (Iwaki).

2.4. Human iPSCs from Fibroblasts

2.4.1. General Equipment

1. Tissue culture plates and dishes: 100-mm (FPI, Kobe, Japan), 6-, 24-, and 96-well (Nunc).
2. Conical tubes: 15- and 50-ml (Greiner).
3. Plastic disposable pipettes: 2-, 5-, 10-, 25-, and 50-ml (Greiner).
4. 0.22-μm bottle-top filter (Techno Plastic Products).
5. 0.22-μm pore size filter (Millipore).
6. 10-ml disposable syringe (Terumo).
7. Cryovial (Nunc).

2.4.2. Cell Culture

1. PBS.
2. 2.5% Trypsin.
3. 0.25% Trypsin/EDTA solution and 0.05% Trypsin/EDTA solution.
4. Water (Sigma).
5. Gelatin-coated culture dishes: see Subheading 2.1.2, item 6.
6. mDMEM/10% FBS: see Subheading 2.1.2, item 4.

2.4.3. Preparation and Culture of Human Dermal Fibroblasts

1. Dermapunch (Maruho, Osaka, Japan).
2. Sterilized forceps and scissors.
3. Cell Banker 2 (ZENOAQ).

2.4.4. Lentivirus Production

1. pLenti6/UbC vector containing mouse *Slc7a1* gene (Plasmid 17224, Addgene).
2. 293FT cells (Invitrogen).
3. CalPhos Mammalian Transfection kit (TaKaRa).
4. Virapower Lentiviral expression system (Invitrogen).
5. Solution A: Dilute 3 μg of Virepower packaging mix (pLP1, pLP2, and pLP/VSVG mixture) and 1 μg of pLenti6/UbC/mSlc7a1 in 12.4 μl of 2 M Calcium Solution, and add up to 100 μl with sterile water.
6. Solution B: Transfer 100 μl of 2× HBS into a 60-mm dish.
7. 0.45-μm pore size cellulose acetate filter.
8. Blastocidin S hydrochloride: see Subheading 2.1.4, item 4.

2.4.5. Retrovirus Production

1. pMXs retrovial vectors containing the cDNAs of human *OCT4* (Plasmid 17217), human *SOX2* (Plasmid 17218), human *KLF4* (Plasmid 17219), and human *C-MYC* (Plasmid 17220) (Cell biolabs, Inc., San Diego, CA, USA; http://www.cellbiolabs.com/).
2. Plat-E packaging cells (available from Dr. Toshio Kitamura at the University of Tokyo; kitamura@ims.u-tokyo.ac.jp).
3. OPTI-MEM I.
4. FuGENE 6 transfection reagent.
5. Polybrene solution: see Subheading 2.1.4, item 9.
6. Puromycin: see Subheading 2.1.4, item 3.
7. Blastocidin S hydrochloride: see Subheading 2.1.4, item 4.

2.4.6. Induction of iPSCs from Human Fibroblasts

1. SNL medium: see Subheading 2.1.5, item 1.
2. SNL cells: see Subheading 2.1.5, item 2.
3. 0.4 mg/ml mitomycin C: Dissolve 10 mg of mitomycin C in 25 ml of water. Filter through a 0.22-μm pore size filter, aliquot, and store at −20°C.
4. SNL feeder cells: Incubate SNL cells at 80–90% confluency with 12 μg/ml mitomycin C for 2 h and 15 min in 37°C, 5% CO_2 incubator. Wash the cells with 4.5 ml of PBS twice, trypsinize, and count the cell number. Plate the cells at 2.6×10^4 per cm^2 onto gelatin-coated culture dishes; 1.5×10^6 cells/dish (100-mm culture dish), 2.5×10^5 cells/well (6-well culture plate), and 5.2×10^4 cells/well (24-well plate) (see Note 4).

5. hESC medium: DMEM/F12 containing 20% KSR, 2 mM L-glutamine, 0.1 mM non-essential amino acids (Sigma), 0.1 mM 2-mercaptoethanol (Sigma), and 50 U and 50 mg/ml penicillin and streptomycin. Filter through a 0.22-μm bottle-top filter and store at 4°C up to 2 weeks.
6. 50 μg/ml FGF-2: Dissolve 1 mg of FGF-2 (PeproTech, Rocky Hill, NJ, USA) in 20 ml of hESC medium. Aliquot and store at −20°C.

2.4.7. Picking, Expanding, Freezing, and Thawing Human iPSCs

1. 10 mg/ml Collagenase IV: Dissolve 1 g of collagenase IV (Invitrogen) in 100 ml of water, and filter through a 0.22-μm pore size filter. Aliquot and store at −20°C.
2. 0.1 M $CaCl_2$: Dissolve 555 mg of $CaCl_2$ in 50 ml of water, and filter through a 0.22-μm pore size filter. Store at 4°C.
3. CTK solution: Mix 10 ml of 2.5% Trypsin, 10 ml of 10 mg/ml collagenase IV, 1 ml of 100 mM $CaCl_2$, and 20 ml of KSR with 59 ml of PBS (26). Aliquot and store at −20°C. Avoid repeated freezing and thawing. Store at 4°C up to 1 week.
4. Cell scraper (Iwaki).
5. 10 mM Y-27632: Dissolve 5 mg of Y-27632 (Wako) in 1.48 ml of sterile water. Aliquot and store at −20°C.
6. DAP213 solution: Mix 1.42 ml of DMSO (Sigma), 0.59 g of acetamide (Sigma), and 2.2 ml of propylene glycol (Sigma) with 6 ml of hESC medium. Filter through a 0.22-μm pore size filter and store at −80°C.

3. Methods

3.1. Mouse iPSCs from Fibroblasts

In most experiments of iPSC generation, reprogrammed cells have been selected based on the expression of fluorescence protein or drug-resistance genes driven by the promoter of pluripotency-related genes such as *Nanog* and *Oct4*. Although this helps to select highly reprogrammed cells, it is not always necessary to take advantage of the system. For a wider usage of the iPSC technique, in this part, we described the mouse iPSCs' generation from embryonic and adult fibroblasts without reporter-dependent selection.

3.1.1. Preparation of Fibroblasts from Mouse Embryos

1. Euthanize female mice on the day 13.5 of pregnancy by cervical dislocation (see Note 5). Wipe with 70% ethanol, and isolate uteri using sterilized forceps and scissors into 100-mm culture dishes containing PBS. Separate the embryos from their placenta and wash them with PBS twice. Remove the embryo's head, visceral tissues, and gonads.

2. Transfer the remaining bodies to a new 100-mm culture dish containing PBS and mince them into small pieces. Transfer into a 50-ml conical tube containing 0.25% Trypsin/EDTA solution (3 ml per embryo) and incubate at 37°C for 20 min. Then, add an additional 0.25% Trypsin/EDTA solution (3 ml per embryo) and incubate at 37°C for 20 min. Invert the tube gently several times and add an equal amount of mDMEM/10% FBS (6 ml per embryo). Pipette up and down to dissociate the tissues.
3. Centrifuge at 200 g for 5 min, discard the supernatant, and resuspend the pellet in mDMEM/10% FBS. Count the cell number and plate 1×10^7 cells per 100-mm gelatin-coated culture dish containing 10 ml of mDMEM/10% FBS. Incubate at 37°C with 5% CO_2 overnight (passage 1), and the next day replace the medium to remove floating cells.
4. When the cells grow to confluency, split or freeze them at 1:4 dilution. Remove the medium, wash once with PBS, and trypsinize with 1 ml of 0.25% Trypsin/EDTA at 37°C for 5 min. Then, add 9 ml of mDMEM/10% FBS and resuspend by pipetting. For passage, split the cells to new 100-mm gelatin-coated culture dishes at 1:4 dilution (passage 2) (see Note 6).
5. To prepare the freeze stocks, transfer the cell suspension to 15-ml conical tubes and centrifuge at 200 g for 5 min. Discard the supernatant and resuspend the cells with Recovery Cell Culture Freezing Medium. Aliquot 1 ml of the cell suspension per freezing vial. Keep the vials in a cell-freezing container at −80°C overnight and then transfer them into a liquid nitrogen tank.

3.1.2. Preparation of Fibroblasts from Adult Mouse Tail

1. Cut the tail from an adult mouse and wash with PBS (see Note 5). Incise using sterilized scissors, peel superficial dermis by hand, and mince the remaining tail into 1-cm pieces with scissors. Place two pieces per well of 6-well gelatin-coated plates, add 2 ml of mDMEM/10% FBS, and incubate at 37°C with 5% CO_2 for 5 days.
2. Remove the tissues of tails and replace the medium with 2 ml of fresh mDMEM/10% FBS. When they reach confluency, aspirate the medium, wash twice with 2 ml of PBS, add 0.3 ml of 0.25% Trypsin/EDTA, and incubate at 37°C for 10 min. Add 2 ml of mDMEM/10% FBS, suspend the cells, and transfer to a 15-ml conical tube. Centrifuge the cells at 200 g for 5 min.
3. Discard the supernatant, resuspend the cells with 10 ml of mDMEM/10% FBS, and plate to a 100-mm gelatin-coated culture dish (passage 2). When the cells become confluent, trypsinize with 1 ml of 0.25% Trypsin/EDTA at 37°C for 5 min, and resuspend with 9 ml of mDMEM/10% FBS.

Passage to new 100-mm gelatin-coated culture dishes at 1:4 dilution (passage 3). These cells usually become confluent within 3–4 days (see Note 6).

3.1.3. Retrovirus Production

1. Thaw a vial of Plat-E cells in 37°C water bath. Resuspend the cells with 10 ml of mDMEM/10% FBS and transfer to a 100-mm gelatin-coated culture dish. Incubate the cells in 37°C, 5% CO_2 incubator. From the next day onwards, cultivate the cells in 10 ml of mDMEM/10% FBS supplemented with 1 μg/ml puromycin and 10 μg/ml blastocidin S. Split the cells at 1:5 dilution when they reach confluency.
2. Twenty-four hours before transfection, aspirate the medium, gently wash with PBS once, and add 1 ml of 0.05% Trypsin/EDTA. After incubation at room temperature for 5 min, suspend with 10 ml of mDMEM/10% FBS, and transfer to a 50-ml conical tube. Count the cell number and plate the cells in mDMEM/10% FBS at 3.6×10^6 cells per 100-mm culture dish, 1.5×10^6 cells per 60-mm culture dish, or 6×10^5 cells per well of a 6-well culture plate. For the four iPSC factors to be transduced, prepare five culture dishes to transfect the five plasmids pMXs-Oct4, Sox2, Klf4, c-Myc, and DsRed separately.
3. Transfer 0.3 ml of Opti-MEM I Reduced-Serum Medium to 1.5-ml plastic tubes. Add 27 μl of FuGENE 6 transfection reagent, mix gently by tapping, and incubate at room temperature for 5 min. Then, add 9 μg of pMXs plasmid DNA, mix gently by finger tapping, and incubate at room temperature for 15 min (see Note 7).
4. Add the DNA/FuGENE 6 mixture to the Plat-E cell culture dishes dropwise and incubate at 37°C, 5% CO_2 overnight. Replace the medium with 10 ml of fresh mDMEM/10% FBS and further incubate overnight.
5. Collect the supernatants from the Plat-E cell culture dishes and filter them through a 0.45-μm cellulose acetate filter (Fig. 1). Combine an equal volume of the virus supernatants containing each factor. For the transduction of four iPSC factors, mix the supernatants of *Oct4*, *Sox2*, *Klf4*, *c-Myc*, and *DsRed* at 1:1:1:1:4 ratio. For three iPSC factors without *c-Myc*, mix the supernatants of *Oct4*, *Sox2*, *Klf4*, and *DsRed* at 1:1:1:3. Add polybrene solution to the virus supernatant mixture at the final concentration of 4 μg/ml and mix gently. Use immediately for transduction (see Note 8).

3.1.4. iPSCs' Derivation from Mouse Fibroblasts

1. Twenty-four hours before retroviral gene transduction, trypsinize MEF or TTF within passage 3, and plate 8×10^5 cells per 100-mm gelatin-coated culture dishes. Prepare one extra culture dish for transduction of *DsRed* in addition to those for the iPSC factors. The next day, aspirate the medium and add

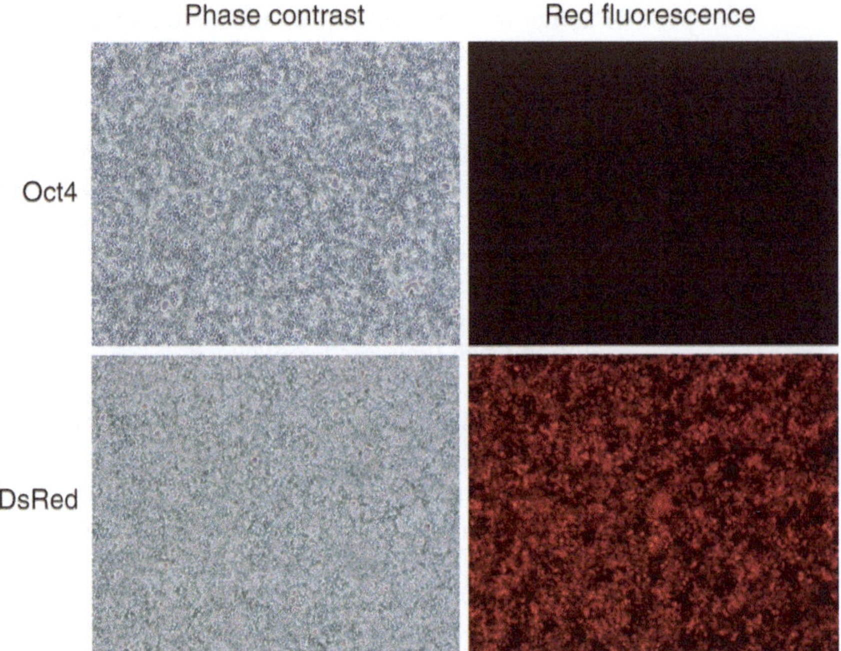

Fig. 1. Plat-E packaging cells after transfection of pMXs retrovirus plasmids. Phase and fluorescence images of Plat-E cells just before collection of virus supernatants (*Oct4* and *DsRed*). The Plat-E cells with pMXs-DsRed transfection show high Red fluorescence when the virus is properly produced.

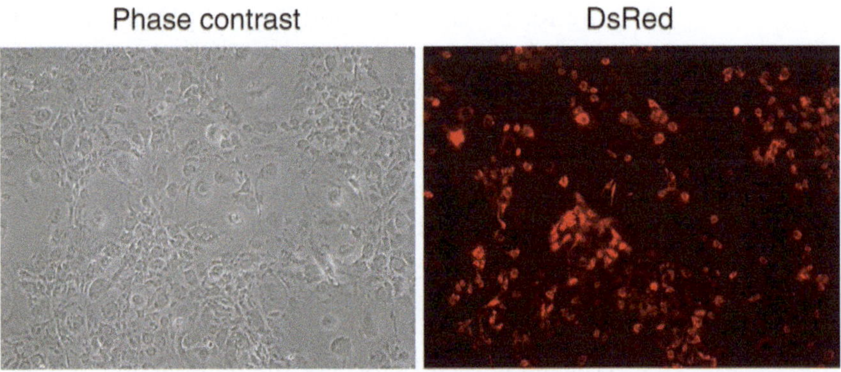

Fig. 2. Fibroblasts with successful gene transduction. Retroviral gene transduction can be monitored by red fluorescence in mouse fibroblasts infected with the pMXs-DsRed retrovirus. The image was photographed after replating onto SNL feeder cells.

the retrovirus supernatant mixture prepared at step 5 of Subheading 3.1.3. Incubate the cells at 37°C, 5% CO_2 overnight and replace the medium with 10 ml of fresh mDMEM/10% FBS. Two days later, exchange the medium with 10 ml of fresh mDMEM/10% FBS again (see Note 9) (Fig. 2).

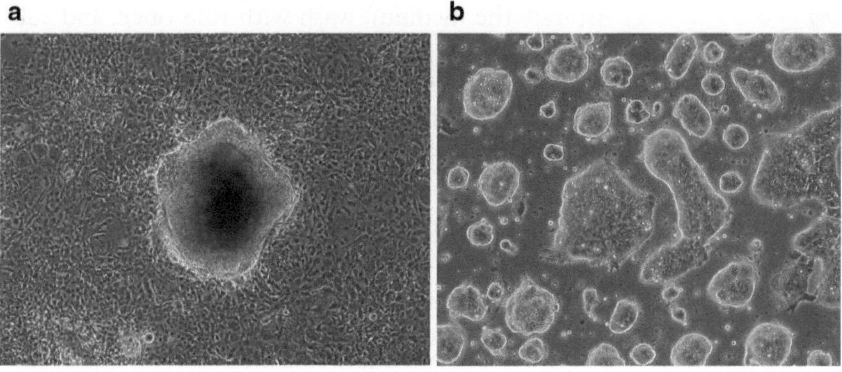

Fig. 3. Derivation of mouse fibroblast-derived iPSCs. (**a**) Morphology of mouse iPSC colony derived from fibroblasts (with three iPSC factors: *Oct4, Sox2, Klf4*) just before picking. (**b**) Expansion culture of fibroblast-derived mouse iPSCs. The image was photographed at passage 8 on gelatin-coated culture dish.

2. On the day 8 after transduction, trypsinize the cells with 0.25% Trypsin/EDTA. Resuspend with 10 ml of mDMEM/10% FBS and count the cell number. Replate them at a density of $0.5–5 \times 10^4$ cells (with four iPSC factors) or 3.5×10^5 cells (with three iPSC factors) per 100-mm culture dishes with SNL feeder cells (see Note 10). Incubate the cells at 37°C, 5% CO_2 overnight and replace the medium with 10 ml of mESC medium the next day.
3. Exchange the medium with 10 ml of fresh mESC medium every other day until the emerging iPSC colonies grow large enough to be picked (Fig. 3a).

3.1.5. Picking and Expanding iPSC Colonies

1. Aliquot 20 µl of 0.25% Trypsin/EDTA per well of a 96-well culture plate. Then, aspirate the mESC medium and wash the cells with 10 ml of PBS once. Add 5 ml of PBS, pick colonies with ESC-like morphology using a 10-µl pipette, and transfer each colony to one well of the 96-well culture plate. When picking iPSC colonies, choose the colonies without DsRed fluorescence to select highly reprogrammed cells (see Note 11). Incubate at 37°C for 15 min.
2. Add 180 µl of mESM medium to each well and pipette up and down to dissociate the colony into single cells. Transfer each cell suspension into one well of a 24-well plate with SNL feeder cells and add 300 µl of mESC medium. Incubate at 37°C, 5% CO_2 until the cells grow to 50–60% confluency.
3. To expand the iPSCs, aspirate the medium, wash with PBS once, and add 0.1 ml of 0.25% Trypsin/EDTA per well of a 24-well culture plate. Incubate at 37°C for 10 min and add 0.4 ml of mESC medium. Carefully pipette up and down to obtain a single-cell suspension and transfer to a well of 6-well culture plates. Add 1.5 ml of mESC medium and cultivate at 37°C, 5% CO_2 until the cells reach 80–90% confluency (Fig. 3b).

3.1.6. Freezing and Thawing iPSCs

1. Aspirate the medium, wash with PBS once, and add 0.3 ml of 0.25% Trypsin/EDTA per well of 6-well culture plates. Incubate at 37°C for 10 min. Add 2 ml of mESC medium and carefully pipette up and down to obtain single-cell suspension. Transfer the cell suspension to a 15-ml conical tube.
2. Centrifuge the tube at 200 g for 5 min. Discard the supernatant and resuspend the cells with Recovery Cell Culture Freezing Medium at $1–2 \times 10^6$ cells/ml. Aliquot 1 ml of the cell suspension per freezing vial. Keep the vials in a cell-freezing container at −80°C overnight and then transfer them into a liquid nitrogen tank the next day.
3. To thaw the iPSC freeze stocks, warm the vials in 37°C water bath until half of the ice crystals disappear. Transfer the cell suspension into a 15-ml conical tube containing 9 ml of mDMEM/10% FBS. Centrifuge at 200 g for 5 min and gently resuspend the cells with 2 ml of mESC medium. Plate the cells to a well of a 6-well culture plate with SNL feeder cells.

3.2. Mouse iPSCs from Mesenchymal Stem Cells

MSCs are defined as plastic-adherent, fibroblast-like cells which undergo sustained in vitro growth and can give rise to multiple mesenchymal lineages (bone, adipose and cartilage tissue, etc.). We previously established a method for isolating highly enriched MSCs from adult murine bone marrow based on their expression of PDGFRα and Sca-1 (27). The iPSCs generated from purified MSCs (PαS) by *Oct4*, *Sox2*, and *Klf4* seem to be the closest equivalent to ESCs by global gene profile and germline transmission, compared with those from $PDGFR\alpha^-$/$Sca\text{-}1^-$ osteoprogenitors and tail-tip fibroblasts (28). These results suggest that tissue stem cells could be a promising cell source for producing high-quality iPSCs.

3.2.1. Preparation of Bone Marrow Cell Suspension

1. Dissect femurs and tibias from adult mice (3–20 mice) and remove residual tissues from the bones. Wash with PBS three times.
2. Put the bones on a mortar and crush them with a pestle (see Note 12). Wash the crushed bones several times with $HBSS^+$ to remove the hematopoietic cells.
3. Incubate the bone fragments in 20 ml of enzymatic dissociation solution in 50-ml conical tube for 1 h at 37°C with shaking (110 rpm/min). Filter the suspension through a cell strainer (70-μm pore size), and collect the cells by centrifugation at 280 g for 7 min at 4°C.
4. Discard the supernatant. Resuspend the pellet with 1 ml of sterile water for 5–10 s to burst red blood cells, and add 1 ml of 2× PBS containing 4% FBS. At this step, cell debris can be seen. Then, resuspend the cells in 10 ml of $HBSS^+$. To remove

the debris, filter the cell suspension through a cell strainer (70-μm pore size).

5. Collect the cells by centrifugation at 280 g for 7 min at 4°C. Discard the supernatant, and resuspend the cells in 1 ml of $HBSS^+$.

3.2.2. Purification of MSCs from Bone Marrow Cell Suspension

1. To compensate the fluorescence interference, transfer approximately 1×10^5 cells into five 1.5-ml plastic tubes as "control tubes." Add a single fluorescence-conjugated antibody (PE, APC, or FITC) to the "control tubes" one by one. The final concentration of antibodies is 0.8 (anti-CD45), 1 (anti-TER119), 1.4 (anti-PDGFRα), and 2.5 μg/ml (anti-Sca-1). Prepare one extra tube for an unstained (negative) control. Avoid light exposure and incubate the tubes at 4°C for 30 min.
2. Add all fluorescence-conjugated antibodies to the remaining cell suspension in 50-ml conical tube as "sample tube." Avoid light exposure and incubate the tube at 4°C for 30 min.
3. Add 500 μl and 9 ml of $HBSS^+$ to the "control tubes" and "sample tube," respectively. Centrifuge the "control tubes" at 800 g for 3 min at 4°C. For the "sample tube," centrifuge at 280 g for 7 min at 4°C. Discard the supernatant, and add 300–500 μl and 5–9 ml of $HBSS^+$ containing 2 μg/ml PI to the "control tubes" and the "sample tube," respectively.
4. For sorting cells, use a triplelaser cell sorter such as MoFlo (Dako) or JSAN (Bay Bioscience), following to the instrument calibration and standardization by the protocols established in your laboratory. Compensate each laser using the "control tubes," measure PI fluorescence, and define a live cell gate excluding PI-positive cells. Define additional gates to isolate the cells positive for PDGFRα and Sca-1 but negative for CD45 and TER119, according to the isotype control fluorescence intensity (see Note 13). Routinely, 0.1–2% of bone marrow cells are $CD45^-/TER119^-$, and 10–20% of $CD45^-/TER119^-$ cells are $PDGFR\alpha^+/Sca\text{-}1^+$ (Fig. 4).
5. Collect the sorted $PDGFR\alpha^+/Sca\text{-}1^+/CD45^-/TER119^-$ cells by centrifugation at 280 g for 7 min at 4°C. Discard the supernatant and resuspend the cells with 1 ml of MSC medium. Repeat this washing step again.
6. Plate the sorted cells onto a 100-mm tissue culture dish containing 10 ml of MSC medium. Exchange the medium every 3 days until the cells reach confluency (Fig. 5a).

3.2.3. Retrovirus Production

1. Seed Plat-E cells at 8×10^6 cells per 100-mm dish.
2. On the next day, introduce 9 μg of pMX-based retroviral vectors for *DsRed*, *Oct4*, *Sox2*, *Klf4*, and *c-Myc* individually into

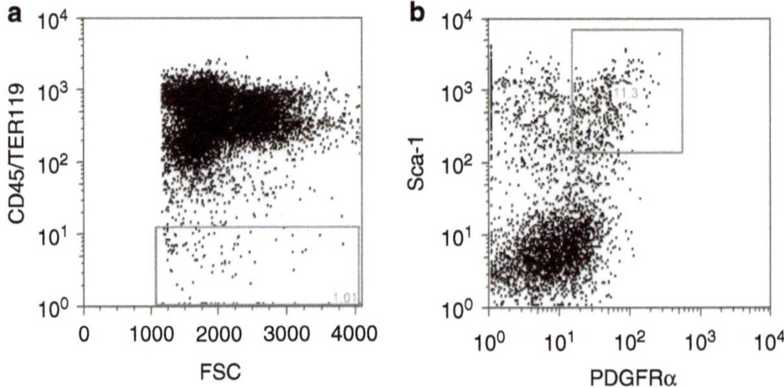

Fig. 4. Isolation of mouse PαS MSCs from adult bone marrow. (**a**) Cell sorter profile of CD45⁻/TER119⁻ non-blood cells in whole bone marrow cells. In this experiment, 1.01% of cells are CD45⁻/TER119⁻. (**b**) Cell sorter profile of PDGFRα⁺/Sca-1⁺ MSCs in CD45⁻/TER119⁻ cells. PDGFRα⁺/Sca-1⁺ cells were separated after gating on CD45⁻ and TER119⁻. In this experiment, 11.3% of cells are PDGFRα⁺/Sca-1⁺. In toto, 0.01–0.4% of cells are usually isolated as PαS MSCs from bone marrow.

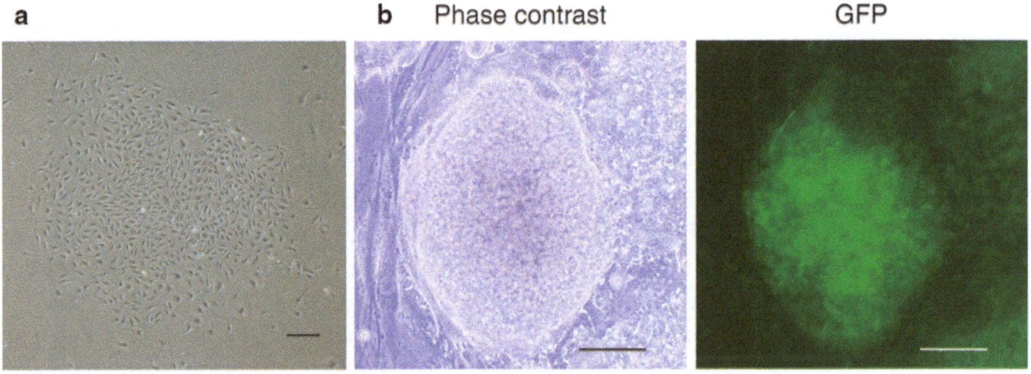

Fig. 5. Derivation of mouse PαS-derived iPSCs. (**a**) Morphology of PαS MSCs purified from adult bone marrow. (**b**) Phase and fluorescence images of PαS-derived iPSC colonies. When using transgenic mice with *Nanog*$^{GFP-IRES-Puro}$, fully reprogrammed iPSCs are visualized by GFP fluorescence driven by the promoter of pluripotency marker gene *Nanog*.

separate dishes of Plat-E cells using 27 μl of FuGENE 6 transfection reagent.

3. After 24 h, replace the medium with 10 ml of mDMEM/10% FBS and incubate overnight. Collect the virus-containing supernatants from the Plat-E cultures and filter them through 0.45-μm cellulose acetate filters.
4. Make a mixture of equal volumes of supernatants containing four or three (without c-Myc) iPSC factors and DsRed retroviruses. DsRed is used as a marker for infection efficiency and transgene silencing. Add polybrene solution into the filtered virus-containing supernatants at the final concentration of 4 μg/ml. Use immediately for transduction.

3.2.4. Induction of iPSCs from PαS Cells

1. Twenty-four hours before retroviral gene transduction, trypsinize PαS cells to seed at 1×10^4 cells per 60-mm culture dish covered with SNL feeder cells.
2. Remove the medium from PαS cell culture dishes and add the virus/polybrene-containing medium. Incubate the cells for 24 h and replace the medium with 5 ml of MSC medium.
3. Two days after infection, exchange the medium with 5 ml of mESC medium. The next day, plate 1×10^4 DsRed$^+$-infected cells per 100-mm culture dish covered with SNL feeder cells.
4. To select highly reprogrammed cells by *Nanog*$^{\text{GFP-IRES-Puro}}$, add puromycin to the culture medium at the final concentration of 1.5 μg/ml after 3 weeks (four iPSC factors) or 4 weeks (three iPSC factors: *OCT4*, *SOX2*, and *KLF4*) since the infection. Change the medium every day until iPSC colonies grow large enough to be picked (Fig. 5b).

3.3. Marmoset iPSCs from Fetal Liver Cells

In this chapter, we describe a protocol to establish common marmoset iPSCs from fetal liver cells via retrovirus-mediated introduction of six human transcription factors, i.e., *OCT4*, *SOX2*, *KLF4*, *C-MYC*, *NANOG*, and *LIN28* (24). We found that *LIN28*, in addition to Yamanaka's four transcription factors, improved the efficiency of iPSCs' establishment in marmosets. The availability of marmosets, and their ease of breeding, may provide an alternative to the use of traditional Old World nonhuman primates. In the future, common marmosets and their iPSCs could provide a powerful preclinical model for the study of regenerative medicine and possibly increase interest in the field.

3.3.1. Virus Production

Retroviruses carrying the transcription factors were produced using the Retroviral Gene Transfer and Expression System according to the manufacturer's instructions.

1. Seed GP-2 cells at 3×10^6 cells per 100-mm poly-L-lysine-coated dish 1 day prior to transfection (see Note 14).
2. Mix 27 μl of FuGENE 6 transfection reagent with 0.3 ml of OPTI-MEM I in a 1.5-ml tube, and incubate at room temperature for 5 min.
3. Combine 6 μg of each pMX vector (carrying human *OCT4*, *SOX2*, *KLF4*, *C-MYC*, *NANOG*, *LIN28*, and *GFP*) and 6 μg of pVSV-G vector with the FuGENE 6 and OPTI-MEM I mixture. Mix gently, and incubate at room temperature for 15 min.
4. Add the DNA/FuGENE 6 complex to the GP-2 cell dish culture in 10 ml of OPTI-MEM I, and incubate at 37°C, 5% CO_2 overnight. The next day, replace the medium containing the DNA/FuGENE 6 complex with 10 ml of cjDMEM/10% FBS.

5. At 48 and 72 h post transfection, collect the medium as a virus-containing supernatant, and filter with a 0.45-μm pore-size cellulose acetate filter. Aliquot and store the virus-containing supernatant at −80°C (see Note 15).

3.3.2. Preparation of Fetal Liver Cells

Common marmoset fetal liver cells were isolated from a miscarried female fetus.

1. Remove the fetus liver and mince (with sterilized scissors) on a 100-mm cell culture dish after washing twice with HBSS. Add 5 ml of collagenase solution, and transfer the cell suspension to a 50-ml centrifuge tube. Incubate at 37°C for 30 min with shaking.
2. Add 30 ml of cjDMEM/10% FBS, centrifuge at 190 g for 5 min, and discard the supernatant. Resuspend the cells with 10 ml of cjDMEM/10% FBS, plate onto 100-mm cell culture dishes, and culture at 37°C with 5% CO_2. Change the medium every other day.

3.3.3. Retroviral Infection of Marmoset Cells

1. Seed the liver cells at 1×10^6 cells per 100-mm cell culture dish 1 day prior to infection. The cells will reach 70–80% confluency the following day.
2. Mix equal volumes of each virus-containing supernatant with *OCT4*, *SOX2*, *KLF4*, *C-MYC*, *NANOG*, *LIN28*, and *GFP*. The final volume of the mixture is 8–12 ml. Add polybrene into the virus-containing mixture to the final concentration of 4 μg/ml.
3. Replace the culture medium with the virus-containing mixture, and incubate the cells for a minimum of 4 h (maximum overnight) at 37°C, 5% CO_2.
4. After infection, replace the virus-containing mixture with cjDMEM/10% FBS (Fig. 6a) (see Note 16). Replace the medium every other day.

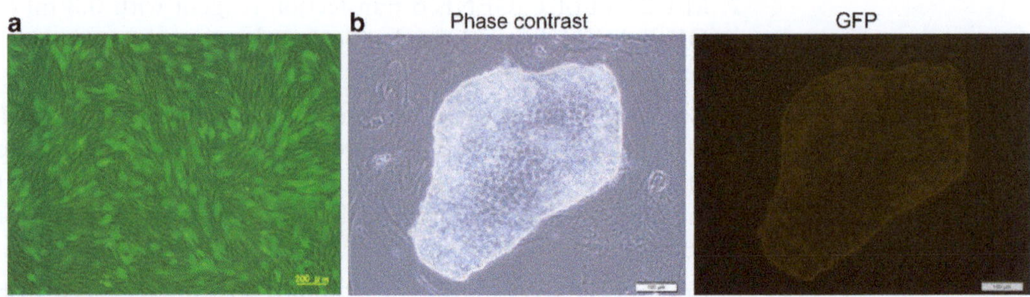

Fig. 6. Derivation of marmoset iPSCs. (**a**) GFP fluorescence 2 days after viral infection. Approximately 33% of the visible cells fluoresced. (**b**) Phase and fluorescence images of marmoset iPSCs emerged 3–5 weeks post infection with the six iPSC factors. All iPSCs exhibited flat, packed, tight colony morphology, and a high nucleus-to-cytoplasm ratio. Fully reprogrammed iPSCs are GFP negative under UV light because of the transgene silencing.

5. Seven days after transfection, harvest the cells by trypsinization and plate onto MEF feeder cells at $1-2 \times 10^5$ cells per 100-mm gelatin-coated dish. Replace the cjDMEM/10% FBS with cjESC medium. Change the medium every other day.

3.3.4. Picking Colonies

Three to 5 weeks after introducing the six transcription factors, several colonies resembling ESCs will emerge (Fig. 6b) (see Note 17).

1. Aliquot 20 μl of cjESC medium (per well) into a 96-well plate. Pick each colony in the 96-well plate using a 20-μl pipette, and dissociate the colony to small clumps by repeated pipetting.
2. Transfer the cell suspension onto MEF feeder cells in gelatin-coated 12-well plates and culture the cells in cjESC medium. Change the medium every other day.

3.3.5. Passage of iPSCs

Seven to 10 days after picking the colonies, the iPSC colonies develop to approximately 100–200 μm in a diameter.

1. Aspirate the culture medium and wash the cells twice with HBSS. Add 0.2 ml of Trypsin solution for the ESCs per well of the 12-well plate, and incubate at 37°C for 5 min.
2. Add 1 ml of cjESC medium and remove colonies from the feeder cells by repeated pipetting. Transfer the cell suspension to a 15-ml centrifuge tube, centrifuge at 190 g for 5 min, and discard the supernatant (see Note 18).
3. Dissociate the colonies by repeated pipetting to small clumps of 20–30 cells. Replate on new MEF feeder cells in a 100-mm gelatin-coated dish (see Note 19).

3.3.6. Storage of Established iPSCs

1. Aspirate culture medium and wash the cells twice with HBSS. Add 2 ml of Trypsin solution for the ESCs to a 100-mm cell culture dish, and incubate at 37°C for 5 min.
2. Add 10 ml of cjESC medium and remove colonies from the feeder cells by repeated pipetting. Transfer the cell suspension to a 15-ml centrifuge tube, centrifuge at 190 g for 5 min, and discard the supernatant.
3. Add 3 ml of Cell Banker 2 and aliquot into 2-ml plastic cryogenic vials (see Note 20). Store the vials at −80°C.

3.4. Human iPSCs from Fibroblasts

In our laboratory, we have generated human iPSCs by retroviral transduction of four reprogramming factors (*Oct4*, *Sox2*, *Klf4*, and *c-Myc*), which was initially introduced by Shinya Yamanaka in 2007 (29). A unique step of Yamanaka method is to introduce mouse solute carrier family 7 member 1 (Slc7a1) gene, which encodes an ecotropic retrovirus receptor, into human cells. Although there are currently several strategies to deliver reprogramming factors (e.g., various virus vectors, nonviral DNAs, and miRNA), we usually utilize retrovirus vectors because of convenience and high efficiency. Here, we described our protocol, especially focusing on the problems that we encountered.

3.4.1. Preparation of Culture Human Dermal Fibroblasts

1. Obtain primary human fibroblasts from skin biopsy using a 5-mm dermapunch (see Note 21). Place the biopsy specimen immediately in mDMEM/10% FBS on ice, and transport it to the laboratory.
2. Transfer the biopsy sample to a 60-mm culture dish, and eliminate the outer layer of the skin. Cut the inner skin into 1 mm pieces using sterilized forceps and scissors. Place the four pieces per 60-mm culture dish. Routinely, three to four dishes with skin pieces can be prepared from a biopsy specimen.
3. When the pieces adhere to the culture dish, add 5 ml of mDMEM/10% FBS into the dish. If some pieces do not adhere, aspirate the medium and try this procedure again.
4. Incubate the cells in 37°C, 5% CO_2 incubator and leave them still for a week. When outgrowth of fibroblasts appears, exchange the medium twice a week.
5. When the cells grow to 30–50% confluency, split them at 1:3 dilution. Aspirate the medium, wash twice with PBS, and trypsinize with 0.5 ml of 0.05% Trypsin/EDTA at 37°C for 7 min. Add 3 ml of mDMEM/10% FBS and resuspend by pipetting. Split the cells to new 60-mm culture dishes at 1:3 dilution (see Note 22).
6. Prepare the freeze stocks when the cells grow to 80% confluency. Trypsinize with 1 ml of 0.05% Trypsin/EDTA at 37°C for 7 min. Add 6 ml of mDMEM/10% FBS and resuspend by pipetting. Transfer the cell suspension to a 15-ml conical tube, centrifuge at 160 g for 5 min, and discard the supernatant.
7. Resuspend the cells with Cell Banker 2 at 1×10^6 cells/ml approximately. Aliquot 1 ml of the cell suspension per freezing vial. Keep the vials in a freezing container at –80°C overnight, and transfer them to the gas phase in a liquid nitrogen tank.
8. To thaw the cell stocks, warm the vials in 37°C water bath until most (but not all) cells are thawed. Transfer the cells into a 15-ml conical tube containing 9 ml of mDMEM/10% FBS. Centrifuge at 160 g for 5 min, discard the supernatant, and resuspend the cells with 10 ml of mDMEM/10% FBS. Plate the cells into a 100-mm culture dish. Change the medium every other day.

3.4.2. Lentiviral Production

1. Thaw a vial of 293FT cells in 37°C water bath. Transfer the cells to a 15-ml conical tube containing 9 ml of mDMEM/10% FBS. Centrifuge at 160 g for 5 min and discard the supernatant.

Resuspend the cells with 10 ml of mDMEM/10% FBS, and transfer to a 100-mm culture dish. Incubate the cells in 37°C, 5% CO_2 incubator until the cells grow to 80–90% confluency. Exchange the medium every other day.

2. When the cells grow to 70–80% confluency, trypsinize with 1 ml of 0.25% Trypsin/EDTA at 37°C for 3 min and resuspend with mDMEM/10% FBS. Split the cells at 1:3 to 1:5 dilution.
3. The day before transfection, plate the cells at 1×10^6 cells per 60-mm culture dish. The next day, prepare Solution A and B, and drop Solution A onto Solution B dropwise. Incubate the solution mixture at room temperature for 20 min.
4. Add the solution mixture to 293FT cell culture dish, and incubate at 37°C, 5% CO_2 overnight (see Note 23). Then, replace the medium with 5 ml of mDMEM/10% FBS.
5. Twenty-four hours after the medium change, collect the supernatant from the 293FT cell culture and filter it through a 0.45-μm cellulose acetate filter. Store at −80°C.

3.4.3. Lentiviral Transduction of Fibroblasts

1. Twenty-four hours before transduction, plate the fibroblasts at 3.2×10^5 cells per 60-mm culture dish. Incubate at 37°C, 5% CO_2 overnight.
2. Replace the medium with 5 ml of the lentivirus supernatant supplemented with 4 μg/ml polybrene. Incubate at 37°C, 5% CO_2 overnight (see Note 24).
3. Twenty-four hours after transduction, aspirate the virus-containing medium and add 5 ml of fresh mDMEM/10% FBS.
4. When the cells grow to 70–80% confluency, passage the cells to two new 100-mm culture dishes (see Note 25).

3.4.4. Retrovirus Production

1. The day before transfection, seed Plat-E cells at 3.6×10^6 cells per 100-mm culture dish, and incubate at 37°C, 5% CO_2 overnight (see Note 26).
2. On the next day, mix 27 μl of FuGENE 6 transfection reagent with 0.3 ml of OPTI-MEM I in a 1.5 ml tube (see Note 26). Incubate at room temperature for 5 min.
3. Add 9 μg of pMXs vectors (encoding *OCT4*, *SOX2*, *KLF4*, *C-MYC*, and *GFP*) one by one into the FuGENE 6/OPTI-MEM I mixture. Mix gently and incubate at room temperature for 15 min.
4. Add the DNA/FuGENE 6 complex dropwise into the Plat-E cell culture dishes, and incubate at 37°C, 5% CO_2 overnight (see Note 27). After 24 h, replace the medium with 10 ml of mDMEM/10% FBS and incubate further overnight.

5. On the next day, collect the supernatant from each Plat-E cell culture, and filter through a 0.45-μm pore size cellulose acetate filter. Add polybrene solution into the filtered virus-containing medium at the final concentration of 4 μg/ml.
6. Make a mixture of equal volume of the supernatants containing each retrovirus (see Note 28).

3.4.5. Induction of iPSCs from Human Fibroblasts

1. The day before transduction, plate the fibroblasts expressing mouse *Slc7a1* which encodes an ecotropic retrovirus receptor at 3.2×10^5 cells per 60-mm culture dish (Fig. 7a) (see Note 29). Incubate at 37°C, 5% CO_2 overnight.
2. Aspirate the medium and add 5 ml of retrovirus mixture prepared at step 6 of Subheading 3.4.4. Incubate the cells at 37°C, 5% CO_2 overnight, and replace the medium with mDMEM/10% FBS. Exchange the medium every other day (Fig. 7b).
3. On the day 11 after infection, trypsinize the cells with 0.5 ml of 0.05% Trypsin/EDTA at 37°C for 5 min. Resuspend with 4 ml of mDMEM/10% FBS, and count the cell number.
4. Seed 5×10^4 or 5×10^5 cells onto 100-mm culture dishes covered with SNL feeder cells containing 10 ml of mDMEM/10% FBS. Incubate at 37°C, 5% CO_2 overnight.
5. The next day, replace the medium with 10 ml of hESC medium. Culture them in 37°C, 3% CO_2 incubator. Exchange the medium to 10 ml of hESC medium supplemented with 4 ng/ml FGF-2 every other day, until the iPSC colonies become large enough to be picked (see Note 30). Routinely, iPSC colonies are observed 2–3 weeks after the retroviral infection (Fig. 8) (see Note 31).

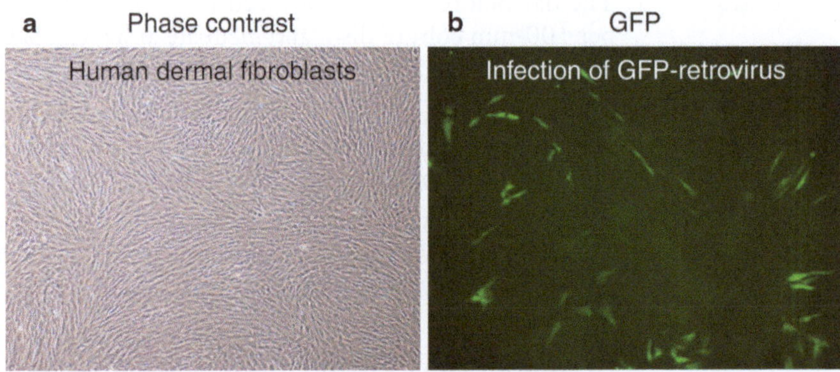

Fig. 7. Retroviral transduction of human dermal fibroblasts. (**a**) Human fibroblasts before the retroviral infection. Replate the cells at 3.2×10^5 cells on 60-mm culture dishes the day before infection. (**b**) Human fibroblasts 7 days after infection. The infection efficiency can be evaluated by transduction of GFP-retrovirus.

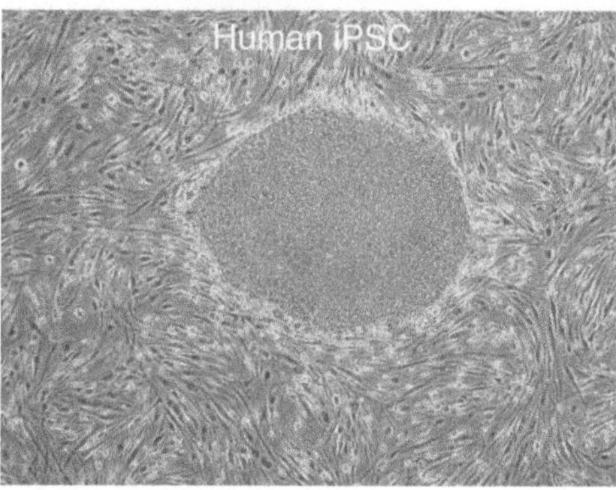

Fig. 8. Human iPSCs derived from fibroblasts. At this experiment, 5×10^4 cells of fibroblasts were replated onto SNL feeder cells in a 100-mm culture dish. ESC-like iPSC colonies emerge by day 30 after retroviral infection.

3.4.6. Picking and Expanding Human iPSCs

1. Aliquot 100 μl of hESC medium with FGF-2 per well of a 96-well culture plate. Pick iPSC colonies from the culture dish under the stereomicroscope using a 20-μl pipette, and transfer each colony to each well of the 96-well culture plate (see Note 32). Pipette up and down to dissociate the colonies to cell clumps composed of 20–30 cells (see Note 33).
2. Add 400 μl of hESC medium with FGF-2 per well, and transfer the cell suspensions into a 24-well plate with SNL feeder cells. Culture them in 37°C, 3% CO_2 incubator until the cells grow to 80–90% confluency.
3. To passage the iPSCs, aspirate the medium, wash with 0.5 ml of PBS, and add 0.1 ml of CTK solution. Aspirate an excess of CTK immediately, and incubate at 37°C for 5 min.
4. Add 0.5 ml of hESC medium with FGF-2 and transfer the cells into a 1.5-ml plastic tube without pipetting. Centrifuge at 160 g for 5 min at room temperature and discard the supernatant. Add 1 ml of hESC medium with FGF-2 and pipette carefully to obtain cell clumps composed of 20–30 cells.
5. Transfer the cell suspension to a well of 6-well culture plates with SNL feeder cells. Add 1 ml of hESC medium with FGF-2, and incubate in 37°C, 3% CO_2 incubator until cells grow to 80–90% confluency. Exchange the medium every day.
6. For further passages, aspirate the medium, wash with 2 ml/well of PBS, and add 0.5 ml of CTK solution. Incubate at 37°C for 2–5 min. Then, aspirate CTK solution and wash with 2 ml of PBS twice.
7. Add 2 ml of hESC medium with FGF-2 and detach iPSCs by using a cell scraper. Dissociate the iPSC colonies to cell clumps composed of 20–30 cells by pipetting. Add 8 ml of hESC

medium with FGF-2, and plate the cells into a 100-mm culture dish with SNL feeder cells. Culture in 37°C, 3% CO_2 incubator until the cells grow to 80–90% confluency again (see Note 34).

3.4.7. Freezing and Thawing Human iPSCs

To make the iPSC freeze stocks, prepare the cells which grow to 80–90% confluency. It is recommended to store the iPSCs at early passages. We usually use Y-27632, a specific inhibitor for p160-Rho-associated coiled-coil kinase (ROCK), to enhance a viability of the frozen cells (30).

1. Aspirate the medium and wash the cells with 6 ml of PBS. Add 1 ml of CTK solution and aspirate the excess immediately. Then, incubate at 37°C for 5 min.
2. Add 6 ml of hESC medium. Detach the iPSC colonies from dish by using a cell scraper, and transfer the cell suspension to two 15-ml conical tubes per 100-mm culture dish.
3. Centrifuge the cells at 160 g for 5 min. Remove the supernatant and resuspend the pellet with 0.2 ml of DAP213 solution by pipetting (see Note 33). Transfer the cell suspension to freezing vials. Put the vials quickly into liquid nitrogen (see Note 35).
4. To thaw the freeze stocks, warm 10 ml of hESC medium in 37°C water bath. Add 0.8 ml of pre-warmed hESC medium into each frozen viral, and thaw quickly by pipetting two to three times.
5. Transfer the cell suspension to the 15-ml conical tube containing hESC medium. Centrifuge at 160 g for 5 min at room temperature.
6. Aspirate the supernatant and add 4 ml of hESC medium supplemented with 4 μl of 10 mM Y-27632 and FGF-2. Plate the cells into 100-mm culture dishes with SNL feeder cells. Culture them in 37°C, 3% CO_2 incubator until the cells grow to 80–90% confluency. Do not move the dish for the initial 48 h, and then exchange the medium every day.

4. Notes

1. 2-Mercaptoethanol is toxic. Avoid inhalation, ingestion, or contact with skin and eye. Use protective gloves and safety glasses when handling.
2. Mitomycin C is toxic. Avoid inhalation, ingestion, or contact with skin and eye. Use protective gloves and safety glasses when handling.

3. Retroviral pMX vectors for human *OCT4*, *SOX2*, *KLF4*, *C-MYC*, *NANOG*, and *LIN28* were kindly provided by Dr. Yamanaka (29). These vectors except *NANOG* and *LIN28* are available from Addgene (http://www.addgene.org/Shinya Yamanaka).
4. Use SNL feeder cells within 3 days. Otherwise, the feeder cells might detach from the culture dish during iPSCs' induction culture. It is possible to utilize frozen stocks of SNL feeder cells kept at −80°C.
5. Experiments involving use of animals must be approved by the international and institutional regulations. Technically, all the procedures should be conducted aseptically.
6. We recommend using early passage fibroblasts (passage 3) as donors for iPSC derivation because the prolonged culture causes replicative senescence, which results in low efficiency of the iPSC derivation.
7. Prepare one tube of DNA/FuGENE 6 mixture for each pMXs plasmid. At this time, it is essential to prepare the proper control to confirm successful gene transduction. We normally utilize pMXs retroviral vector of red fluorescence protein DsRed to evaluate the transduction efficiency. Thus, in the case of four iPSC factors' transduction, a total five tubes of DNA/FuGENE 6 mixture are necessary for the production of *Oct4*, *Sox2*, *Klf4*, *c-Myc*, and *DsRed* viruses.
8. The viral medium should be used soon after collection. Storing of viral supernatant causes a significant reduction of transfection efficiency and consequently a lower number of iPSC colonies. Because the retrovirus infects Plat-E packaging cells themselves, the successful viral production can be conventionally evaluated by DsRed fluorescence in Plat-E cells transfected with pMXs-DsRed (Fig. 1).
9. Check the transduction efficiency by monitoring DsRed fluorescence. High efficiency of gene transduction is essential for successful iPSC derivation. We routinely observe >80% transfection efficiency.
10. The number of replated cells dramatically affects the frequency of the iPSC derivation. For the transduction of four iPSC factors, replating too many cells results in overgrowth of transformed cells and difficulty in isolation of highly reprogrammed iPSCs. On the other hand, without *c-Myc*, the iPSCs' frequency becomes much lower so that all the cells can be replated. Considering the number of iPSC colonies fluctuates among experiments, it is better to validate the optimal cell number by replating with a dilution series.

11. Highly reprogrammed iPSCs can be distinguished by a silencing of retroviral transgenes. By retroviral transduction of the iPSC factors in combination with *DsRed*, we can visually monitor the transgene silencing based on the fluorescence. The iPSC colonies without DsRed fluorescence are most likely to be highly reprogrammed iPSCs.
12. Crush the bones in pieces but do not grind them.
13. To adjust the isotype controls, arrange the preparatory experiment cells similarly to the situation of the fluorescence compensation using "control tubes." Once the adjustment is done, there is no need to check for the following experiments.
14. The cells reach 80–90% confluency the following day.
15. We can generate marmoset iPSCs using frozen viral stocks stored for up to 1 month. Do not repeatedly freeze/thaw the viral stocks to avoid reducing the viral titer.
16. Infection rates as assessed by GFP fluorescence should be more than 30% (Fig. 6a).
17. This timing of iPSC colony appearance is consistent with the study on rhesus monkey iPSCs (31). All colonies morphologically resembling marmoset ESCs (32) are GFP negative (Fig. 6b). Although various colony types appear approximately 2 weeks post infection, these are not iPSC colonies. ESC-like, clear-edged colonies should appear within 3–5 weeks post infection.
18. When the differentiated cells occupy the majority of the dishes, collect the undifferentiated colonies using a 20-μl pipette, or remove the large clumps of differentiated cells by filtering the cell suspension using a 100-μm nylon cell strainer.
19. Do not dissociate the cell clumps into single cells because too much dissociation could trigger cell death. Cells are passaged approximately every 5–7 days.
20. Do not dissociate the colonies to small clumps prior to freezing.
21. For biopsy of human dermal fibroblasts, you must obtain a proper informed consent from the donors.
22. Be careful not to plate the cells at low cell density to prevent replicative senescence.
23. Transfection of the lentiviral vector should be done when 293FT cells grow to 70% confluency in a 100-mm culture dish.
24. The overnight incubation with lentivirus is sometimes toxic to fibroblasts. In that case, shorten the incubation duration down to 5 h.

25. To evaluate the successful transduction, use a GFP-encoding lentiviral vector as a control. Alternatively, it is possible to select the infected cells by culturing with 10 μg/ml blastocidin S, because pLenti6/UbC/mSlc7a1 vector carries a blastocidin-resistance gene.
26. Prepare one cell culture dish and one 1.5-ml plastic tube per plasmid. Five cell culture dishes and tubes are required for transfection of *OCT4*, *SOX2*, *KLF4*, *C-MYC*, and *GFP* retrovirus vectors.
27. Do not transfect more than two plasmids into a Plat-E cell culture dish. It causes a low-efficient generation of resultant iPSCs.
28. Use immediately for transduction. Do not freeze the virus supernatants.
29. The efficiency of retroviral transduction markedly decreases when using fibroblasts at older passages. It is recommended to use passage 8–10 fibroblasts for iPSC induction.
30. Fuzzy-edged cell colonies appear approximately 2 weeks after retroviral transduction, but they are not iPSC colonies. Keep the culture until clear-edged, hESC-like colonies are observed.
31. Check the emergence of iPSC colonies carefully, because the timing differs in each experiment even when using the same lot of fibroblasts.
32. We usually pick 20–30 iPSC colonies from a culture dish.
33. Do not dissociate iPSC colonies into single cells.
34. To keep undifferentiated iPSC culture, remove the differentiated colonies by aspirating during passaging procedure. Transfer only undifferentiated iPSC colonies to a new culture dish.
35. To ensure high cell viability, these procedures should be done within 15 s.

Acknowledgements

This work was supported by grants from the Ministry of Education, Culture, Sports, Science and Technology of Japan (MEXT); the Ministry of Health, Labor, and Welfare; the Japan Society for the Promotion of Science (JSPS); the National Institute of Biomedical Innovation; the Project for Realization of Regenerative Medicine, MEXT; the Funding Program for World-leading Innovative R&D in Science and Technology (FIRST), JSPS; and Grant-in-Aid for Young Scientists (B).

References

1. Stadtfeld M, Hochedlinger K (2010) Induced pluripotency: history, mechanisms, and applications. Genes Dev 24:2239–2263
2. Takahashi K, Yamanaka S (2006) Induction of pluripotent stem cells from mouse embryonic and adult fibroblast cultures by defined factors. Cell 126:663–676
3. Chamberlain SJ et al (2010) Induced pluripotent stem cell models of the genomic imprinting disorders Angelman and Prader-Willi syndromes. Proc Natl Acad Sci U S A 107:17668–17673
4. Horii T et al (2008) Loss of genomic imprinting in mouse parthenogenetic embryonic stem cells. Stem Cells 26:79–88
5. Do JT et al (2009) Generation of parthenogenetic induced pluripotent stem cells from parthenogenetic neural stem cells. Stem Cells 27:2962–2968
6. Takahashi K (2010) Direct reprogramming 101. Dev Growth Differ 52:319–333
7. Humpherys D et al (2001) Epigenetic instability in ES cells and cloned mice. Science 293:95–97
8. Stadtfeld M et al (2010) Aberrant silencing of imprinted genes on chromosome 12qF1 in mouse induced pluripotent stem cells. Nature 465:175–181
9. Liu L et al (2010) Activation of the imprinted Dlk1-Dio3 region correlates with pluripotency levels of mouse stem cells. J Biol Chem 285:19483–19490
10. Pick M et al (2009) Clone- and gene-specific aberrations of parental imprinting in human induced pluripotent stem cells. Stem Cells 27:2686–2690
11. Toyooka Y et al (2003) Embryonic stem cells can form germ cells in vitro. Proc Natl Acad Sci U S A 100:11457–11462
12. Geijsen N et al (2004) Derivation of embryonic germ cells and male gametes from embryonic stem cells. Nature 427:148–154
13. Hubner K et al (2003) Derivation of oocytes from mouse embryonic stem cells. Science 300:1251–1256
14. Eguizabal C et al (2009) Generation of primordial germ cells from pluripotent stem cells. Differentiation 78:116–123
15. Hayashi K, Surani MA (2009) Self-renewing epiblast stem cells exhibit continual delineation of germ cells with epigenetic reprogramming in vitro. Development 136:3549–3556
16. Lavagnolli TM et al (2009) Presumptive germ cells derived from mouse pluripotent somatic cell hybrids. Differentiation 78:124–130
17. Imamura M et al (2010) Induction of primordial germ cells from mouse induced pluripotent stem cells derived from adult hepatocytes. Mol Reprod Dev 77:802–811
18. Park TS et al (2009) Derivation of primordial germ cells from human embryonic and induced pluripotent stem cells is significantly improved by coculture with human fetal gonadal cells. Stem Cells 27:783–795
19. Kim JB et al (2009) Oct4-induced pluripotency in adult neural stem cells. Cell 136:411–419
20. Honda A et al (2010) Generation of induced pluripotent stem cells in rabbits: potential experimental models for human regenerative medicine. J Biol Chem 285:31362–31369
21. Shimada H et al (2010) Generation of canine induced pluripotent stem cells by retroviral transduction and chemical inhibitors. Mol Reprod Dev 77:2
22. Nagy K et al (2011) Induced pluripotent stem cell lines derived from equine fibroblasts. Stem Cell Rev 7(3):693–702
23. Bao L et al (2011) Reprogramming of ovine adult fibroblasts to pluripotency via drug-inducible expression of defined factors. Cell Res 21(4):600–8
24. Tomioka I et al (2010) Generating induced pluripotent stem cells from common marmoset (Callithrix jacchus) fetal liver cells using defined factors, including Lin28. Genes Cells 15:959–969
25. Sumer H et al (2011) NANOG is a key factor for induction of pluripotency in bovine adult fibroblasts. J Anim Sci 89(9):2708–2716
26. Fujioka T et al (2004) A simple and efficient cryopreservation method for primate embryonic stem cells. Int J Dev Biol 48:1149–1154
27. Morikawa S et al (2009) Prospective identification, isolation, and systemic transplantation of multipotent mesenchymal stem cells in murine bone marrow. J Exp Med 206:2483–2496
28. Niibe K et al (2011) Purified mesenchymal stem cells Are an efficient source for iPS cell induction. PLoS One 6:e17610
29. Takahashi K et al (2007) Induction of pluripotent stem cells from adult human fibroblasts by defined factors. Cell 131:861–872
30. Watanabe K et al (2007) A ROCK inhibitor permits survival of dissociated human embryonic stem cells. Nat Biotechnol 25:681–686
31. Liu H et al (2008) Generation of induced pluripotent stem cells from adult rhesus monkey fibroblasts. Cell Stem Cell 3:587–590
32. Sasaki E et al (2005) Establishment of novel embryonic stem cell lines derived from the common marmoset (Callithrix jacchus). Stem Cells 23:1304–1313

Chapter 3

Generation of Trophoblast Stem Cells

Michael C. Golding

Abstract

The isolation and culture of both embryonic and extraembryonic stem cells provide an enormous opportunity to study the molecular processes that establish and maintain lineage-specific, monoallelic patterns of gene expression. This chapter describes the isolation an culture of trophectoderm stem cells from mouse blastocyst stage embryos. Using this powerful *in vitro* system, scientists can now begin to tease apart the epigenetic processes that result in placental patterns of imprinted gene expression and begin to better understand the role these genes play in development and disease.

Key words: Placental stem cell, Trophectoderm, Extraembryonic lineage, TS cell

1. Introduction

Genomic imprinting is a specialized transcriptional regulatory mechanism that restricts expression to the maternally or paternally inherited allele (1). Misregulation of these lineage-specific patterns of monoalleleic gene expression has been associated with numerous developmental disorders and cancer (2, 3). Recently, we have come to recognize the extreme importance of imprinted gene expression to the proper development and function of the placenta, through observed defects in the development of embryos produced through assisted reproductive technologies (2, 4–10).

To better define these and other molecular events driving mammalian embryogenesis, pluripotent stem cells from each of the three distinct lineages present within the preimplantation blastocyst have been derived (11–14). Embryonic (ES), trophectoderm (TS), and extraembryonic endoderm (XEN) stem cells each possess the developmental potential of their founding lineages and exhibit distinct patterns of imprinted X-inactivation and gene expression (14–17).

Nora Engel (ed.), *Genomic Imprinting: Methods and Protocols*, Methods in Molecular Biology, vol. 925,
DOI 10.1007/978-1-62703-011-3_3, © Springer Science+Business Media, LLC 2012

However, the molecular basis for the establishment and maintenance of these differing monoallelic patterns of gene expression remains poorly defined.

Trophectoderm stem cells represent a powerful system with which to study the developmental origins of the extraembryonic tissues that give rise to the placenta (13). In the analysis of patterns of imprinted gene expression it is essential to be able to identify parent-of-origin expression. To this end, genetic crosses between distinct strains of mice allow the tracking and identification of allele-specific patterns of gene expression through the identification of single-nucleotide polymorphisms either through direct sequencing or using restriction enzyme digestion (18). This chapter describes the derivation of TS cells from embryos potentially derived from these crosses.

2. Materials

2.1. Production of Embryonic Feeders

1. A source of mouse embryonic fibroblasts—either commercial or primary cultures.
2. DMEM with 10% Fetal Bovine Serum and 1% Antibiotic/Antimycotic.
3. Phosphate-buffered saline.
4. 0.1% (w/v) trypsin–1 mM EDTA.
5. Mitomycin C.
6. DMSO.
7. T175 Flasks or 15 cm tissue culture dishes.
8. Freezing vials.
9. –80 °C Freezer.
10. Liquid nitrogen tank or –160 °C freezer.
11. 37 °C Water bath.

2.2. Derivation of Trophectoderm Stem Cells

1. TS Cell Medium:

   500 ml RPMI (Sigma).

   6 ml Pen/strep (50 μg/ml each final concentration, Sigma).

   6 ml 100 mM sodium pyruvate (Invitrogen 11360070, final concentration 1 mM).

   6 ml 10 mM B-mercaptoethanol (Sigma, final concentration 100 μM) (14.3 M BME use 35 μl in 50 ml to make 100× stock).

   1,000× FGF4 (R&D Systems).

   1,000× FGF Basic (R&D Systems).

1,000× Heparin (1mg/ml).

6 ml 200 mM L-glutamine (Sigma, final concentration 2 mM).

FCS to final volume of 15% (Hyclone ES Serum, Fisher Scientific).

2. Low-wall tissue culture dishes.
3. Mitomycin C-treated MEF feeders.
4. Source of Mouse blastocysts.

2.3. Culture of Trophectoderm Stem Cells in the Absence of Feeders

1. Mitomycin C-treated MEFs.
2. DMEM/10% FBS.
3. 15 cm Dishes.
4. TS medium.
5. 0.45 μm Filter.
6. 10–15 ml Syringe.
7. 37 °C Water bath.

2.4. Freezing TS Cell Cultures

1. TS Cell Medium:

   500 ml RPMI (Sigma).

   6 ml Pen/strep (50 μg/ml each final conc Sigma).

   6 ml 100 mM sodium pyruvate (Invitrogen, final concentration 1 mM).

   6 ml 10 mM B-mercaptoethanol (Sigma, final concentration 100 μM) (14.3 M BME use 35 μl in 50 ml to make 100× stock).

   1,000× FGF4 (R&D Systems).

   1,000× FGF Basic (R&D Systems).

   1,000× Heparin (1mg/ml).

   6 ml 200 mM L-glutamine (Sigma, final conc 2 mM).

   FCS to final volume of 15% (Hyclone ES Serum Fisher Scientific).

2. Freezing Medium (50 ml).

   25 ml FBS.

   20 ml of TS cell Medium.

   5 ml of DMSO.

3. Cryo-Vials.
4. −80 °C Freezer.
5. −160 °C Freezer or Liquid Nitrogen.
6. Cell Freezer.

3. Methods

3.1. Production of Embryonic Feeders: Mitomycin C Treatment of Mouse Embryonic Fibroblasts

Trophectoderm Stem Cells require several soluble, secreted factors, including Activin and TGF-Beta, in order to maintain an undifferentiated state (19). Moreover, TS cell lines grown on MEF-feeder layers are easier to maintain than those grown in MEF-conditioned medium. Here we will describe the production of growth-arrested mouse embryonic fibroblasts to be used as feeder layers in the culture of trophectoderm stem cells. Using the powerful chemotherapeutic Mitomycin-C to irreversibly inhibit DNA replication, treated MEFs can be plated and although their growth has been arrested they still continue to secrete factors necessary for TS cell maintenance. It is best to prepare multiple vials of feeder cells at once to ensure uniformity.

1. Thaw a frozen vial of MEFs in a 37 °C water bath and transfer entire contents into a 1.5-ml tube and centrifuge at 400 ×*g* for 4 min.
2. Remove the supernatant and gently resuspend the cells in 1 ml of DMEM/10% FBS.
3. Split cells (500 μl each) into two 15 cm dishes, each containing 25 ml DMEM/10% FBS.
4. Culture cells at 37 °C for 3–4 days or until cells reach ~90% confluence. Do not let the cells become confluent.
5. Passage the cells by removing the growth medium and rinse twice with 10 ml PBS per dish.
6. Add 2.5 ml 0.1% trypsin to each dish and incubate for ~2 min at 37 °C. Tap each dish to dissociate the cell monolayer.
7. Add 10 ml of DMEM/10% FBS to the dish and gently pipette to break cell aggregates.
8. Transfer cells to a 15 ml tube and centrifuge at 200 ×*g* for 4 min. Resuspend cell pellet in 15 ml of DMEM/10% FBS.
9. Split (3 ml each) into five new 15 cm plates containing 22 ml DMEM/10% FMS.
10. Culture cells at 37 °C for 3–4 days or until cells reach ~90% confluence. Do not let the cells become confluent.

Caution—Mitomycin C is extremely toxic. Please exercise caution when handling.

11. When cells are ready prepare 200 ml of DMEM/10%FBS containing 10 μg/ml Mitomycin-C. Typically this compound is sold in 2 mg aliquots. Resuspend the entire contents of the vial in 4 ml of DMEM/10% FBS and add this to a final volume of 200 ml of DMEM/10% FBS. This will produce enough Mitomycin-C medium to treat ten 15 cm plates. Add Mitomycin-C medium and incubate cells for 2 h at 37 °C.

12. Remove the medium and rinse cells twice, with 20 ml of PBS.
13. Add 5 ml of Trypsin as described above and place in incubator for ~2 min.
14. Tap each dish to dissociate the cell monolayer. Add 10 ml of DMEM/10% FBS to the dish and gently pipette to break cell aggregates.
15. Transfer the cells from each plate into a 15 ml tube and centrifuge at 200 × *g* for 4 min.
16. Resuspend cell pellet in 5 ml of Freezing Medium—60% FBS, 30% DMEM, and 10% DMSO—and aliquot into freezing vials. Typically, one vial will contain enough cells to cover one 10 cm dish or two 12-well plates.
17. Place vials in the −80 °C freezer overnight and transfer to liquid nitrogen or a −160 °C freezer for long-term storage.

3.2. Derivation of Trophectoderm Stem Cells

TS Cells are derived by plating blastocyst stage embryos on an MEF-feeder layer and allowing outgrowths to form. When these outgrowths are dissociated, culture of the derivative cells in TS cell medium will promote the growth of trophectoderm and XEN stem cell colonies. These colonies can be picked and stable TS cell lines derived.

3.2.1. Preparation for TS Cell Medium

1. FGF Basic (R&D Systems) and FGF4 (R&D Systems) need to be suspended in 1 ml of PBS/0.1% BSA (we use the NEB FBS that comes with Restriction Enzymes and filter sterilize) and make 50 μl aliquots and freeze at −80 °C.
2. 1000 × Heparin is made by diluting 1 mg/ml of Heparin (Sigma catalogue # H3393) in PBS. Make 1 ml aliquots and store at −80 °C.

3.2.2. Derivation of Mouse TS Cells

1. One or two days before day before blastocyst collection, plate Mitomycin C-treated MEFs in low-wall 4-well plates in a final volume of 0.5 ml DMEM/10%FBS per well.
2. On the day of blastocyst collection replace the DMEM/10% FBS on the feeders with TS cell medium adding fresh FGF basic and FGF4 to the medium.
3. Sacrifice mated females using methods approved by your institution's animal use and care committee at a time that will allow the collection of late morula or early blastocyst stage embryos (Day 3.0 to Day 3.5 days post coitus).
4. Isolate blastocysts by dissection of the uterine horns and utilizing a 1 ml syringe, M2 medium, and a 26 gauge needle to flush the embryos into a petri dish. Detailed protocols describing mouse blastocyst collection have been described elsewhere (20).

5. Wash the embryos through PBS–PVP and add a single embryo to each well of the 4-well plate containing the feeder MEFs.
6. Return the plates to a standard tissue culture incubator (37 °C, 5% CO_2) and allow blastocyst outgrowths to form.
7. After 2 days of culture replace medium with fresh TS cell medium containing FGF basic and FGF4.
8. Plate a second group of mouse feeders on low-wall 4-well plates.
9. After a 4 or 5 days of growth blastocyst outgrowths should be easily visible and have reached a size of greater than 750 μm. Do not let outgrowths become too large as the efficiency of stem cell isolation will rapidly begin to diminish (see Note 1).
10. Carefully aspirate the TS cell medium with a pipette and wash with PBS. The outgrowths will be very loosely attached, so take great care not to knock them loose.
11. To dissociate these structures add 0.1 ml 0.1% trypsin/1 mM EDTA and incubate for 5 min at 37 °C.
12. After incubation, use a 200 μl pipette (set to a volume of 75 μl) to dissociate the outgrowths by pipetting up and down.
13. Add 600 μl of Fesh TS cell medium containing fresh FGF basic and FGF4 to the dissociated outgrowths and transfer to a well in the new 4-well plates.
14. After 12 h, replace medium with fresh TS cell medium (+ heparin, FGF basic, and FGF4).
15. Culture TS cells in the above medium in a standard tissue culture incubator (37 °C, 5% CO_2 incubator) replacing TS cell medium every 2 days. After approximately 1 week, TS cell colonies will begin to form. Often, XEN cells will also begin to grow. These cells tend to grow as single cells in clumps that begin to branch out. It is important to pick TS cell colonies before the plate becomes too overgrown so as to minimize the chance of XEN contamination (see Note 2) (Fig. 1).
16. Plate mouse feeders in either a flat-bottom 96-well plate or 24-well plate depending on the number of colonies.
17. Aspirate the medium from the 4-well plates and wash TS cell colonies twice with PBS.
18. After the last wash, cover the cells with 150 μl of PBS.
19. Using a dissecting scope use a 20 μl pipette to add approximately 10 μl of 0.1% trypsin/1 mM EDTA directly onto the target colony, then mechanically break the colony away from the tissue culture surface, and pick up with the pipette.
20. Eject colony in a 0.5 ml tube on a 37 °C heating block.
21. After ~5 min add 100 μl of fresh TS cell medium containing heparin, FGF basic, and FGF4.

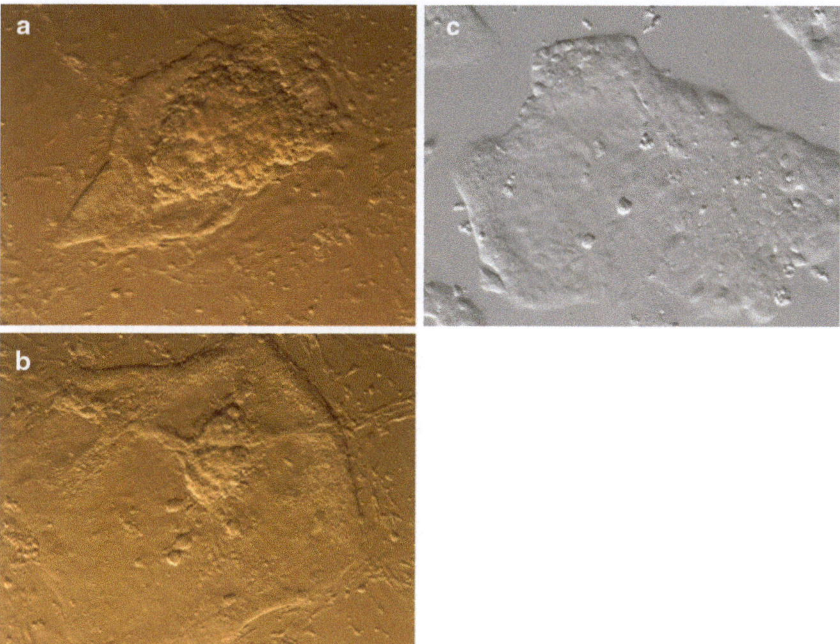

Fig. 1. Light micrographs depicting sequential stages in the isolation and culture of trophectoderm stem cells. (**a**) Early passage TS cells 4–6 days after dissociation of the initial embryonic body. Note the "smooth" cells along the expanding margins and "rough" cells in the center. (**b**) TS cells after the third passage. Colonies at this point will be primarily composed of "smooth" cells growing in individual colonies. (**c**) TS cells after ten passages in culture, growing on gelatin-coated plastic, in conditioned medium. Cells grow in relatively homogenous colonies with actively proliferating, expanding margins of smooth cells. The colonies shown here are confluent and need to be split within 12 h.

22. Pipette up and down to dissociate the cells and transfer to a single well of the tissue culture plate containing the newly plated feeders.
23. After 12–24 h replace medium with fresh.
24. Culture TS cells in the above medium in a standard tissue culture incubator (37 °C, 5% CO_2 incubator) replacing TS cell medium every 2 days.
25. TS cells may now be cultured in the above medium in a standard tissue culture incubator (37 °C, 5% CO_2 incubator). Cells are typically passaged (1:20) every 2–3 days. If they are split 1:20, they may become confluent in 2–3 days. TS cell media is changed every second day; however, when they reach >60% confluence, the media should be changed daily.

3.3. Maintenance and Passage of TS Cells

1. To passage TS cells wash twice with PBS, and dissociate colonies with enough 0.1% trypsin/1 mM EDTA to cover the bottom of the plate.
2. Cells should begin to lift off the plate after ~1–2 min at which point colonies should be dissociated by gentle pipetting up and down.

3. Add TS cell medium containing heparin, FGF basic, and FGF4 to stop the reaction. Split cells 1:20 and transfer into a new well of feeder cells.
4. After 12 h replace the medium on the newly plated cells with fresh TS cell medium containing heparin, FGF basic, and FGF4.

3.4. Culture and Passage of TS Cells in the Absence of Feeders

In certain experimental situations it is necessary to culture TS cells in the absence of MEF-feeders. For example, when diagnosing imprinted gene expression it is best not to have feeder lines contributing to the analysis. To this end, TS cells may be grown in gelatin-treated plastic (see Note 3).

MEFs and differentiated trophoblast cells adhere to the tissue culture dish more quickly than TS cells. This differential plating time can be used to recover floating TS cells in the medium after the MEFs and other cell types have adhered to the tissue culture plastic. TS cells can be maintained in the absence of MMC-MEFs in medium supplemented with 70% MEF-conditioned medium. The example below is for a 100-mm cell culture dish. Adjust volumes accordingly for different sizes of dishes or flasks.

3.4.1. Production of MEF-Feeder-Conditioned Medium

1. Thaw a frozen vial of Mitomycin-C-treated MEFs in a 37 °C water bath, transfer entire contents into a 1.5-ml tube, and centrifuge at 400 × *g* for 4 min.
2. Remove the supernatant and gently resuspend the cells in 1 ml of DMEM/10% FBS.
3. Transfer cells into a 15 cm dish containing 25 ml DMEM/10% FBS.
4. Culture cells at 37 °C for 24–48 h to let cells settle.
5. Replace medium with 25 ml TS cell medium that does not contain heparin, FGF basic, and FGF4.
6. Incubate cells in a standard tissue culture incubator (37 °C, 5% CO_2 incubator) for 3 days.
7. On the third day pre-wet a 0.45 μm syringe filer with fresh TS cell medium that does not contain FGF basic and FGF4.
8. Draw the medium on the 15 cm plate up into the 15 ml syringe.
9. Filter medium into a 50 ml conical tube. It may take a few pulls from each plate.
10. The medium is now feeder conditioned and may be stored at −80 °C.
11. Replace aspirated medium with a second 25 ml volume of TS cell medium that does not contain heparin, FGF basic, and FGF4.
12. A second collection of filtered medium is possible by repeating steps 5–10.

3.4.2. Culture of TS Cells in MEF-Feeder-Conditioned Medium

1. Grow TS cells on MEF-Feeders as described above.
2. Two to 3 h before passaging the cells, the tissue culture dishes that will be used to culture the MEF-feeder-free TS cells need to be treated with 1% Gelatin. Add enough 1% Gelatin in PBS to completely cover the bottom of the tissue culture well and place in a standard tissue culture incubator for 2–3 h.
3. Wash TS cells twice with PBS, and dissociate colonies with enough 0.1% trypsin/1 mM EDTA to cover the bottom of the plate.
4. Cells should begin to lift off the plate after ~1–2 min at which point colonies should be dissociated by gentle pipetting up and down.
5. Add conditioned TS cell medium containing FGF basic and FGF4 to stop the reaction.
6. Transfer the cells to a new tissue culture plate and allow feeder cells to settle our for 1 h.
7. After 1 h a large proportion of the MEF-feeders will have attached to the tissue culture dish. Pick up the remaining cells and place in a suitable tube. Discard the tissue culture well/dish with the attached feeders.
8. Remove the 1% gelatin solution from the tissue culture plates and split the TS cells 1:7.5 or 1:10 and transfer into a well of the gelatin-coated dish.
9. After 12 h replace the medium on the newly plated cells with fresh conditioned TS cell medium containing heparin, FGF basic, and FGF4.
10. Cells may be passaged in conditioned TS cell medium containing FGF basic and FGF4 as described above.

3.5. Freezing TS Cell Cultures

1. Prepare the TS Cell Freezing Medium by combining the components listed above. Place Medium on ice.
2. Wash confluent TS cell cultures twice with PBS, and dissociate colonies with enough 0.1% trypsin/1 mM EDTA to cover the bottom of the plate.
3. Cells should begin to lift off the plate after ~1–2 min at which point colonies should be dissociated by gentle pipetting up and down.
4. Add TS cell medium containing FGF basic and FGF4 to stop the reaction.
5. Remove an aliquot of cells to passage as necessary.
6. Transfer the remaining cells to a 1.5 ml tube and spin at 4,000 × g for 4 min.
7. Remove media and resuspend cells in 1 ml of TS cell freezing medium.

8. Transfer cells to a 15 ml tube containing 4 ml of TS Cell Freezing Medium.
9. Mix the cells by inversion and aliquot the cells in 1 ml volumes into five cryovials.
10. Place cells in cell freezer and place in the −80 °C freezer overnight.
11. The next day transfer frozen cells to either liquid nitrogen or −160 °C freezer for long-term storage (see Note 4).

4. Notes

1. TS cells are a challenging cell type to derive and maintain in culture. TS cells grow very slowly when plated at low density but upon reaching a critical mass begin to grow very quickly. This property is likely due to an as yet unidentified secreted factor. Given these observations we have always found that derivation of TS cells is more efficient when multiple blastocysts are plated and dissociated in a single culture dish. Once multiple TS cell colonies begin to emerge clonal populations are picked.
2. When deriving TS cells, care must be taken not to allow XEN stem cells to take over the culture dish. In contrast to the smooth morphology of TS cell colonies, XEN cells grow in clumps of individual cells but will quickly spread throughout the dish.
3. TS cells typically take a long time to recover after being frozen. Be sure to plate a large number of cells in a dish to ensure a rapid recovery. Furthermore, after freezing a significant number of cells will spontaneously begin to differentiate. Allow two to three passages for the stem cell population to stabilize before proceeding with your experiments.
4. Culture of TS cells in the absence of feeders is very challenging. A significant number of TS cell colonies will have subpopulations that may differentiate when plated on plastic or glass. Again, allow cells to passage two to three times in conditioned medium before beginning experiments.

Acknowledgement

This work was supported by the NIH grant AA020129-02.

References

1. Verona RI, Mann MR, Bartolomei MS (2003) Genomic imprinting: intricacies of epigenetic regulation in clusters. Annu Rev Cell Dev Biol 19:237–259
2. Odom LN, Segars J (2010) Imprinting disorders and assisted reproductive technology. Curr Opin Endocrinol Diabetes Obes 17: 517–522
3. Uribe-Lewis S, Woodfine K, Stojic L, Murrell A (2011) Molecular mechanisms of genomic imprinting and clinical implications for cancer. Expert Rev Mol Med 13:e2
4. Eggan K, Akutsu H, Hochedlinger K, Rideout W, Yanagimachi R, Jaenisch R (2000) X-Chromosome inactivation in cloned mouse embryos. Science 290:1578–1581
5. Hill JR, Burghardt RC, Jones K, Long CR, Looney CR et al (2000) Evidence for placental abnormality as the major cause of mortality in first-trimester somatic cell cloned bovine fetuses. Biol Reprod 63:1787–1794
6. Bourc'his D, Le Bourhis D, Patin D, Niveleau A, Comizzoli P et al (2001) Delayed and incomplete reprogramming of chromosome methylation patterns in bovine cloned embryos. Curr Biol 11:1542–1546
7. Xue F, Tian XC, Du F, Kubota C, Taneja M et al (2002) Aberrant patterns of X chromosome inactivation in bovine clones. Nat Genet 31:216–220
8. Santos F, Zakhartchenko V, Stojkovic M, Peters A, Jenuwein T et al (2003) Epigenetic marking correlates with developmental potential in cloned bovine preimplantation embryos. Curr Biol 13:1116–1121
9. Mann MR, Lee SS, Doherty AS, Verona RI, Nolen LD et al (2004) Selective loss of imprinting in the placenta following preimplantation development in culture. Development 131:3727–3735
10. Lin J, Shi L, Zhang M, Yang H, Qin Y et al (2011) Defects in trophoblast cell lineage account for the impaired in vivo development of cloned embryos generated by somatic nuclear transfer. Cell Stem Cell 8:371–375
11. Martin GR (1981) Isolation of a pluripotent cell line from early mouse embryos cultured in medium conditioned by teratocarcinoma stem cells. Proc Natl Acad Sci U S A 78:7634–7638
12. Nagy A, Rossant J, Nagy R, Abramow-Newerly W, Roder JC (1993) Derivation of completely cell culture-derived mice from early-passage embryonic stem cells. Proc Natl Acad Sci U S A 90:8424–8428
13. Tanaka S, Kunath T, Hadjantonakis AK, Nagy A, Rossant J (1998) Promotion of trophoblast stem cell proliferation by FGF4. Science 282:2072–2075
14. Kunath T, Arnaud D, Uy GD, Okamoto I, Chureau C et al (2005) Imprinted X-inactivation in extra-embryonic endoderm cell lines from mouse blastocysts. Development 132: 1649–1661
15. Lewis A, Mitsuya K, Umlauf D, Smith P, Dean W et al (2004) Imprinting on distal chromosome 7 in the placenta involves repressive histone methylation independent of DNA methylation. Nat Genet 36:1291–1295
16. Terranova R, Yokobayashi S, Stadler MB, Otte AP, van Lohuizen M et al (2008) Polycomb group proteins Ezh2 and Rnf2 direct genomic contraction and imprinted repression in early mouse embryos. Dev Cell 15:668–679
17. Latos PA, Stricker SH, Steenpass L, Pauler FM, Huang R et al (2009) An in vitro ES cell imprinting model shows that imprinted expression of the Igf2r gene arises from an allele-specific expression bias. Development 136:437–448
18. Market-Velker BA, Zhang L, Magri LS, Bonvissuto AC, Mann MR (2010) Dual effects of superovulation: loss of maternal and paternal imprinted methylation in a dose-dependent manner. Hum Mol Genet 19:36–51
19. Erlebacher A, Price KA, Glimcher LH (2004) Maintenance of mouse trophoblast stem cell proliferation by TGF-beta/activin. Dev Biol 275:158–169
20. Nagy A (2003) Manipulating the mouse embryo: a laboratory manual. CSHL Press, Cold Spring Harbor, NY

Chapter 4

Immunomagnetic Purification of Murine Primordial Germ Cells

Emily Y. Smith and James L. Resnick

Abstract

Primordial germ cells (PGCs) play essential roles in both reproduction and development. In this chapter, we describe a method used in our laboratory for the immunopurification of PGCs from the mouse embryo. After dissection and disruption of the fetal gonad, PGCs are identified by a monoclonal antibody recognizing an epitope characteristic of pluripotent stem cells. After reaction with a paramagnetic bead-linked secondary antibody, the cell mixture is applied to a strong magnetic field. PGCs are recovered by release from the magnetic field. Purity is assessed by the alkaline phosphatase activity inherent to PGCs.

Key words: Primordial germ cells, Immunomagnetic purification

1. Introduction

Primordial germ cells (PGCs) are the embryonic progenitors of eggs and sperm and are thus vital to both reproduction and development. In addition to ensuring transgenerational continuity of the germ line, murine PGCs implement mitotic amplification of the germ lineage, colonization of the gonad, epigenetic reprogramming of both imprinted and non-imprinted genes, and reactivation of the X chromosome, and set the stage for entry into meiosis. After a complex pattern of specification in the extraembryonic mesoderm during gastrulation, PGCs perform these tasks while migrating toward the genital ridge, the gonadal anlage. PGCs undergo sexual differentiation shortly after arrival in the genital ridge. In male embryos the PGCs arrest mitotically as prospermatogonia. In female embryos PGCs enter meiotic prophase (1).

Given their diverse roles and functions, it is not surprising that investigators would seek to morphologically, physiologically,

Nora Engel (ed.), *Genomic Imprinting: Methods and Protocols*, Methods in Molecular Biology, vol. 925, DOI 10.1007/978-1-62703-011-3_4, © Springer Science+Business Media, LLC 2012

and molecularly characterize PGCs. During the migratory phase PGCs are not a uniform tissue, but instead are present either singly or in small aggregates (2). This has naturally led to the development of several methods for PGC isolation and purification. McLaren and colleagues were among the first to purify PGCs. They developed a method in which isolated genital ridges were mashed to release the germ cells into the media. PGCs were then identified by their unique morphology and isolated manually (3). Shortly thereafter, De Felici and McLaren described a method for larger scale PGC fractionation on Percoll gradients (4). Mayanagi et al. have more recently described improvements in this method (5). Over the past 15 years, however, FACS sorting and immunomagnetic sorting have become the most widely used methods. Both techniques yield high purity with reasonable effort but require either special mouse strains or immunological reagents. FACS sorting to purify PGCs often takes advantage of monoclonal antibodies that recognize SSEA-1, a PGC cell surface antigen (6, 7). The use of fluorescent substrates of beta-galactosidase in combination with a mouse engineered to express LacZ in PGCs has also been described (8). Currently, however, the most widely used FACS method employs sorting of PGCs engineered to express green fluorescent protein (GFP) (9–11).

Our lab has had reliable success using immunomagnetic sorting based on the SSEA1 surface antigen present on pluripotent mouse cells. Compared to GFP-based FACS sorting, immunomagnetic sorting has the disadvantages of requiring more "hands-on" effort and is dependent upon both primary and secondary antibodies. An advantage of immunomagnetic purification is that PGCs can be obtained from mouse strains lacking the GFP expression marker. We have found the method to be compatible with PGC culture assays (12), RNA expression analysis, and DNA methylation analysis. It can be used to isolate PGCs from embryos between 10.5 and 14.5 dpc. Beyond 14.5 dpc the expression of the SSEA-1 epitope is reduced (13).

The method used in our lab closely resembles that first described by Pesce and DeFelice (14).

2. Materials

2.1. Dissection Requirement

1. Two jewelers forceps (such as Dumont #5).
2. PBS (1× without Ca^{2+} or Mg^{2+}).
3. Bacterial petri dishes.
4. Stereomicroscope.

2.2. Immunomagnetic Purification

1. 0.05% Trypsin–EDTA. Warm to 37 °C before use. Store at 4 °C.
2. PBS–DNase buffer: 1× PBS, pH 7.2, 5 mM EDTA, 0.5% BSA, 20 μg/ml DNase. Prepare 5 ml per purification, making fresh each time.
3. Equilibration buffer: 1× PBS, pH 7.2, 3% BSA. Prepare fresh, 500 μl for each purification.
4. TG-1 mAb or equivalent as hybridoma cell culture supernatant (see Note 1).
5. MiniMACS starting kit (Miltenyi Biotech). Kit includes integral components for the purification procedure: Anti-Mouse IgM MicroBeads, MiniMACS MS separation columns, MACS MultiStand, and MiniMACS Separation Unit.

2.3. Alkaline Phosphatase Staining Components

1. Cytospin.
2. CSA-100 silanated slides.
3. Poly-D-lysine dissolved at 50 mg/ml in H_2O.
4. 4% Paraformaldehyde.
5. Fast Red TR Salt (Sigma) dissolved at 1 mg/ml in H_2O.
6. Naphthol AS-MX phosphate (Sigma).

3. Methods

3.1. Recovery of Urogenital Ridges

1. Euthanize gravid female in accordance with institutional policy and recover reproductive tract bearing embryos. Submerge uterine horns in PBS in bacterial petri dishes for further dissection.
2. Working under the buffer and with the aid of a dissecting microscope, separate embryos from decidual tissue.
3. Use a "scissor" action of two jewelers forceps to decapitate the embryo at a position just posterior to the forelimb buds.
4. Remove the viscera. This is most easily accomplished as follows. With the ventral surface facing up steady the embryo by inserting the forceps tips at the base of each hindlimb bud. The viscera may then be removed using forceps held with the other hand. The urogenital ridges remain attached to the dorsal body wall.
5. Dissect the urogenital ridges away from the body wall.
6. Yield per column can be increased by teasing the genital ridge away from the more dorsally located mesenephros with the aid of a 28 gauge syringe needle, but for many applications we find this to be unnecessary. If desired, 12.5 dpc or older male and

female genital ridges may be pooled separately based on their morphological differences. We find it simplest to transfer genital ridges from the dissection medium to a microfuge tube using a 1,000 μl pipette tip.

3.2. Immunomagnetic Purification of PGCs

1. Digest approximately 8–16 urogenital ridge pairs in 0.5 ml trypsin–EDTA at 37 °C for 5 min in a microcentrifuge tube. Use about 15 pairs for 10.5 dpc and 10 pairs 13.5. Do not overload the column (see Note 2).
2. Triturate (see Note 3) and then centrifuge at 250×*g* for 2 min in a microcentrifuge. Carefully remove most of trypsin–EDTA leaving tissue clumps behind in about 50–100 μl. Triturate again.
3. Add 1 ml PBS–DNase buffer, triturate, and centrifuge.
4. Remove the majority of the PBS–DNase, leaving approximately 160 μl behind. Triturate thoroughly to generate a single cell suspension.
5. Place on ice and add 40 μl TG-1 mAb. Incubate on ice with shaking for 30 min.
6. Centrifuge, aspirate supernatant, and then triturate. Add 100 μl PBS–DNase buffer and wash two additional times, triturating each time.
7. After final wash, resuspend pellet in 180 μl ice-cold PBS–DNase. Add 20 μl of Anti-Mouse IgM MicroBeads. Incubate on ice for 30 min with shaking.
8. Place MiniMacs separation column in magnetic holder and prewash the column with 500 μl equilibration buffer (see Note 4). Take care to avoid bubble formation on the column. Allow column to empty by gravity.
9. Apply the cell suspension and collect flow through into a 1.5 ml microfuge tube while the column is in the magnet. Allow column to empty by gravity.
10. Reapply the flow through to the magnet two additional times. The third flow through is the immunodepleted fraction, consisting of the somatic cells of the urogenital ridge (see Note 5).
11. With the column still in the magnet, apply 500 μl PBS–DNase buffer and allow column to empty by gravity. Wash this way three additional times.
12. Elute column to obtain purified PGCs. Remove column from magnet, apply 500 μl PBS–DNase buffer, and allow column to empty by gravity. Collect this purified fraction.
13. Apply 1 ml PBS–DNase buffer to the column and gently force through using the plunger supplied with the column. Collect

and add to original purified fraction obtained in previous step. The cells can be concentrated by centrifugation. If assessing purity, remove 150 µl before centrifugation.

14. Follow alkaline phosphatase staining to assess purity.

3.3. Alkaline Phosphatase Staining

1. Cytospin a 10% aliquot (150 µl) onto silanated slides for 10 min at 55 × *g*.
2. Fix in 4% paraformaldehyde for 10–20 min at room temperature. Wash by gently immersing the slide into water two to four times.
3. Blot away excess water and overlay with Fast Red TR/Napthol AS-MX. PGCs will stain red (see Note 6).

4. Notes

1. TG-1 is a mouse IgM that was originally intended to recognize human thymocytes (15); however, it also recognizes SSEA-1 on the surface of mouse pluripotent teratocarcinoma cells and PGCs. If this monoclonal is not available, substitute MC-480, available from the Developmental Studies Hybridoma Bank.
2. Overloading the column will result in very slow column clearing times and a reduction in the level of purity. As a general rule, one column is sufficient for about fifteen 10.5 dpc genital ridges or eight to ten 12.5–14.5 dpc genital ridges.
3. Trituration steps should be performed by 50–100 passages through a 200 µl pipet tip. Be careful not to suck air into the tip and thereby introduce bubbles into the solution.
4. We use two magnets and are able to run two samples (often male and female) simultaneously. Samples may be processed in sequence if only one magnet is available. Samples awaiting the magnet should be held on ice after addition of the anti-IgM microbeads.
5. We regularly observe a very small number of PGCs in the immunodepleted fraction.
6. Add 40 µl of 1 mg/ml Fast Red TR to 1 ml of Napthol AS-MX phosphate. Gently layer mixture over the deposited cells at room temperature. The stain may be very slow under these conditions, but may be sped up by placing on a slide warmer. If stain still does not appear after 25 min, blot away old staining solution and apply fresh.

References

1. Ewen KA, Koopman P (2010) Mouse germ cell development: from specification to sex determination. Mol Cell Endocrinol 323(1): 76–93. doi:S0303-7207(09)00621-2 [pii] 10.1016/j.mce.2009.12.013
2. Gomperts M, Garcia-Castro M, Wylie C, Heasman J (1994) Interactions between primordial germ cells play a role in their migration in mouse embryos. Development 120(1):135–141
3. Monk M, McLaren A (1981) X-chromosome activity in foetal germ cells of the mouse. J Embryol Exp Morphol 63:75–84
4. De Felici M, McLaren A (1982) Isolation of mouse primordial germ cells. Exp Cell Res 142(2):476–482. doi:0014-4827(82)90393-7 [pii]
5. Mayanagi T, Kurosawa R, Ohnuma K, Ueyama A, Ito K, Takahashi J (2003) Purification of mouse primordial germ cells by Nycodenz. Reproduction 125(5):667–675
6. McCarrey JR, Hsu KC, Eddy EM, Klevecz RR, Bolen JL (1987) Isolation of viable mouse primordial germ cells by antibody-directed flow sorting. J Exp Zool 242(1):107–111. doi:10.1002/jez.1402420116
7. Yamazaki Y, Mann MR, Lee SS, Marh J, McCarrey JR, Yanagimachi R, Bartolomei MS (2003) Reprogramming of primordial germ cells begins before migration into the genital ridge, making these cells inadequate donors for reproductive cloning. Proc Natl Acad Sci U S A 100(21):12207–12212. doi:10.1073/pnas.2035 119100 2035119100 [pii]
8. Abe K, Hashiyama M, Macgregor G, Yamamura K (1996) Purification of primordial germ cells from TNAPbeta-geo mouse embryos using FACS-gal. Dev Biol 180(2): 468–472. doi:S0012160696903206 [pii]
9. Szabo PE, Hubner K, Scholer H, Mann JR (2002) Allele-specific expression of imprinted genes in mouse migratory primordial germ cells. Mech Dev 115(1–2):157–160. doi:S0925477302000874 [pii]
10. Yoshimizu T, Sugiyama N, De Felice M, Yeom YI, Ohbo K, Masuko K, Obinata M, Abe K, Scholer HR, Matsui Y (1999) Germline-specific expression of the Oct-4/green fluorescent protein (GFP) transgene in mice. Dev Growth Differ 41(6):675–684
11. Yeom YI, Fuhrmann G, Ovitt CE, Brehm A, Ohbo K, Gross M, Hubner K, Scholer HR (1996) Germline regulatory element of Oct-4 specific for the totipotent cycle of embryonal cells. Development 122(3):881–894
12. Cooke JE, Godin I, Ffrench-Constant C, Heasman J, Wylie CC (1993) Culture and manipulation of primordial germ cells. Methods Enzymol 225:37–58
13. Donovan PJ, Stott D, Cairns LA, Heasman J, Wylie CC (1986) Migratory and postmigratory mouse primordial germ cells behave differently in culture. Cell 44(6):831–838. doi:0092-8674(86)90005-X [pii]
14. Pesce M, De Felici M (1995) Purification of mouse primordial germ cells by MiniMACS magnetic separation system. Dev Biol 170(2):722–725. doi:S0012-1606(85)71250-X [pii] 10.1006/dbio.1995.1250
15. Beverley PC, Linch D, Delia D (1980) Isolation of human haematopoietic progenitor cells using monoclonal antibodies. Nature 287(5780): 332–333

Part II

Identifying Imprinted Genes

Chapter 5

Whole Genome Methylation Profiling by Immunoprecipitation of Methylated DNA

Andrew J. Sharp

Abstract

I provide a protocol for DNA methylation profiling based on immunoprecipitation of methylated DNA using commercially available monoclonal antibodies that specifically recognize 5-methylcytosine. Quantification of the level of enrichment of the resulting DNA enables DNA methylation to be assayed for any genomic locus, including entire chromosomes or genomes if appropriate microarray or high-throughput sequencing platforms are used. In previous studies (1, 2), I have used hybridization to oligonucleotide arrays from Roche Nimblegen Inc, which allow any genomic region of interest to be interrogated, dependent on the array design. For example, using modern tiling arrays comprising millions of oligonucleotide probes, several complete human chromosomes can be assayed at densities of one probe per 100 bp or greater, sufficient to yield high-quality data. However, other methods such as quantitative real-time PCR or high-throughput sequencing can be used, giving either measurement of methylation at a single locus or across the entire genome, respectively. While the data produced by single locus assays is relatively simple to analyze and interpret, global assays such as microarrays or high-throughput sequencing require more complex statistical approaches in order to effectively identify regions of differential methylation, and a brief outline of some approaches is given.

Key words: Immunoprecipitation, DNA methylation, Imprinting

1. Introduction

One feature that has been associated with many imprinted genes is the presence of parent-of-origin-specific Differentially Methylated Regions (DMRs). Thus, the maternal and paternal genomes possess distinct epigenetic marks which distinguish them at imprinted loci. Here we describe a DNA immunoprecipitation method to perform comparative DNA methylation profiling between the two parental genomes that can detect DMRs associated with imprinted genes. This methodology takes advantage of the fact that patients

Nora Engel (ed.), *Genomic Imprinting: Methods and Protocols*, Methods in Molecular Biology, vol. 925, DOI 10.1007/978-1-62703-011-3_5, © Springer Science+Business Media, LLC 2012

with uniparental disomy contain chromosomes inherited from a single parent, representing a system that allows the independent study of a paternally or a maternally derived epigenome. Systematic comparison of the two parental epigenomes in this way represents a powerful method for detecting epigenetic differences between the two parental epigenomes associated with imprinted loci (2).

This described protocol has been optimized for hybridization of the immunoprecipitated DNA to Nimblegen microarrays. However, the level of enrichment for methylated DNA can also be assayed using other methodologies such as qPCR or high-throughput sequencing. In this case, it may be appropriate to modify the amount of starting DNA accordingly.

2. Materials

2.1. Methylated DNA Immunoprecipitation Components

1. Antibody: Mouse monoclonal anti 5-methyl cytidine (Diagenode, Liege, Belgium).
2. Beads: Protein A Agarose Beads (Invitrogen, Carlsbad, CA).
3. 5× IP buffer: 50 mM Sodium Phosphate (pH 7), 0.7 M NaCl, 0.25 % Triton X-100. Total volume 100 ml. Mix 50 ml 100 mM Na-Phosphate (pH 7), 14 ml 5 M NaCl, 2.5 ml 10 % Triton X-100, 33.5 ml distilled H_2O.
4. Digestion buffer: 50 mM Tris–HCl (pH 8), 10 mM EDTA, 0.5 % SDS. Total volume 100 ml. Mix 5 ml 1 M Tris–HCl (pH 8), 2 ml 0.5 M EDTA, 5 ml 10 % SDS, 88 ml distilled H_2O. Filter using a 0.2 μm filter and store at 4 °C.
5. Phosphate-Buffered Saline (PBS), pH 7. Store at 4 °C.
6. Proteinase K solution (10 mg/ml). Store at −20 °C.
7. 25:24:1 Phenol:chloroform:isoamyl alcohol. Store at 4 °C.
8. 1× TE Buffer (pH 8). Store at room temperature.
9. 24:1 Chloroform:isoamyl alcohol. Store at 4 °C.

3. Methods

3.1 Methylated DNA Immunoprecipitation

Due to the use of overnight incubations at two points, this protocol is best performed over a period of 3 days.

Day 1

1. Dilute 15 μg genomic DNA (see Note 1) in 440 μl sterile H_2O in a 1.5 ml screw-top tube (see Note 2).

2. Fragment DNA to a size range of approximately 200–800 bp. Fragmentation can be achieved by sonication of DNA with a Branson 450 sonifier using a standard tapered probe (see Note 3). Program the sonifier as follows: Time = 70 s; Amplitude = 10%; Pulse on 0.5 s; Pulse off 0.5 s. Suspend the tube of DNA in a polystyrene float in ice/water bath during sonication to keep the DNA solution cool, as significant heating of the solution occurs during sonication which can cause denaturation of the DNA and result in a nonrandom fragmentation pattern. Clean the sonication probe with 70 % ethanol before and after fragmenting each DNA sample to avoid contamination between samples.
3. Check the size of the DNA fragments produced by sonication by loading 15 μl of the sonicated DNA (equivalent to ~300 ng) on 1.5 % agarose gel with a 100 bp ladder alongside. Most fragments should be between 200 and 800 bp, with an average size of ~500 bp (see Note 4).
4. Denature the DNA for 5 min at 95 °C in a hot block, and immediately place the samples on ice.
5. Remove 75 μl (~1.5 μg) of each DNA sample into a new tube labeled "*sample name*_sonicated" and store at +4 °C. This sonicated DNA will be used later as the reference (input) DNA, and will be precipitated on Day 3 of the protocol.
6. Add 88 μl of 5× IP buffer to the remaining 350 μl denatured/sonicated DNA and mix by briefly vortexing.
7. Add 10 μg of 5meth-C antibody (1 μg/μl) to each tube of DNA/1× IP buffer (see Note 1). Ensure that the lids are screwed on each tube tightly, and secure in a rack.
8. Incubate the tubes of DNA/IP buffer/antibody at 4 °C with gentle rotation/rocking overnight.

Day 2

1. Prechill a micro-centrifuge to 4 °C.
2. Resuspend the Protein A Agarose beads by gently shaking the bottle. Remove 80 μl of beads per IP reaction into an Eppendorf tube and place on ice.
3. Make 3 ml of PBS:0.1 % BSA (2,970 μl cold PBS + 30 μl 10 mg/ml BSA) per IP reaction. Mix by vortexing and chill on ice.
4. Wash beads twice with 1 ml cold PBS:0.1 % BSA, as follows:
   (i) Centrifuge the beads for 2 min at 3,824 × g, 4 °C.
   (ii) Remove the majority of the supernatant from each tube with a 1 ml pipette, taking care not to disturb the beads. The remaining supernatant can be removed using a BD

UltraFine needle and syringe, which has a needle bore smaller than the beads, such that they cannot be pipetted.

(iii) Add 1 ml of PBS:0.1 % BSA and mix well by inversion.

(iv) Repeat steps (i) and (ii), and place the washed beads on ice.

5. Make 1× IP Buffer by diluting one part 5× IP buffer with four parts cold sterile water, and place on ice to chill.
6. Resuspend each aliquot of the washed agarose beads in 80 μl of chilled 1× IP Buffer.
7. Add one tube of the agarose bead/1× IP Buffer slurry to each tube of the DNA–antibody mixture. Flick each tube gently to ensure that the agarose beads are fully resuspended in the 1× IP Buffer before pipetting, as the beads settle in solution.
8. Ensure that the lids are screwed on tightly to each tube of DNA/1× IP Buffer/antibody/agarose beads, secure in a rack, and incubate the mix for 2 h at 4 °C with gentle rotation/rocking.
9. Transfer the mix of DNA/1× IP Buffer/antibody/agarose beads into a new labeled 1.5 ml screw-top tube to avoid carryover of contaminating un-precipitated DNA from sides of the first tube.
10. Wash beads twice with 1 ml cold 1× IP Buffer, as follows:

    (i) Centrifuge the beads for 2 min at 3,824 × *g*, 4 °C.

    (ii) Remove the majority of the supernatant from each tube with a 1 ml pipette, taking care not to disturb the beads. Remove the remaining supernatant using a BD UltraFine needle and syringe.

    (iii) Add 1 ml cold 1× IP Buffer and mix well by inversion.

    (iv) Repeat steps (i) and (ii), and place the washed beads on ice.

11. Resuspend the DNA/antibody/beads in 250 μl digestion buffer.
12. Add 7 μl Proteinase K solution (10 mg/ml) to each tube, ensure that the lids are screwed on tightly, and incubate overnight at 55 °C with rotation.

Day 3

1. Add 250 μl of sterile H_2O to each tube.
2. Working in a fume hood, add 500 μl of 25:24:1 phenol:chloroform:isoamyl alcohol to each tube.
3. Ensure that lids are screwed on tightly and vortex each tube thoroughly for ~30 s.
4. Centrifuge tubes for 3 min at 13,000 rpm.

5. Working in a fume hood, remove the majority of the upper aqueous phase from each tube into a new labeled 2 ml screw-top tube, taking care not to disturb the precipitates at the interface or the lower organic phase.
6. To maximize the recovery of DNA, working in a fume hood add 300 μl 1× TE Buffer to each tube containing the remaining upper aqueous phase and phenol:chloroform.
7. Ensure that the lids are screwed on tightly, vortex thoroughly, and centrifuge for 3 min at 13,000 rpm.
8. Working in a fume hood, remove the upper aqueous phase (avoiding the interface/lower organic phase) and add it to the first aliquot of the aqueous phase that was removed in step 5 into a labeled 2 ml screw-top tube.
9. Perform a second phenol:chloroform extraction on the DNA solution by repeating steps 2–8 above to ensure removal of any residual protein from the DNA.
10. Working in a fume hood, add an equal volume of 24:1 chloroform:isoamyl alcohol to each tube containing the DNA solution.
11. Ensure that lids are screwed on tightly, and vortex thoroughly.
12. Centrifuge tubes for 3 min at 13,000 rpm.
13. For each sample, label two 1.5 ml Eppendorf tubes with "*sample name*_meDIP." Remove the aqueous (upper) phase, avoiding the interface and lower organic phase, and divide equally between the two new labeled tubes.
14. Precipitate both the tubes of IP DNA and the 75 μl sonicated input DNA from Day 1 as follows:
    (i) Add 0.7 μl glycogen (20 mg/ml), 1 ml of ice-cold 100 % ethanol, and 50 μl of 5 M NaCl to each tube. Mix by vortexing.
    (ii) Incubate tubes at −20 °C for 1 h.
    (iii) Centrifuge tubes at 13,000 rpm for 15 min at 4 °C (see Note 5).
    (iv) Carefully remove the supernatant from each tube using a pipette, taking care not to remove the DNA pellet.
15. Wash each DNA pellet by adding 300 μl cold 70 % ethanol, mixing by briefly vortexing.
16. Centrifuge tubes for 5 min at 13,000 rpm, 4 °C.
17. Carefully remove all the supernatant, taking care not to remove the DNA pellet.
18. Air-dry the pellets for ~30 min until all liquid has evaporated.
19. Resuspend each DNA pellet by adding 15 μl 1× TE buffer per tube and incubating at 65 °C for 30 min.

20. Vortex each tube well, pulse spin to collect all the droplets, and combine together the two tubes of IP DNA per sample into a single tube.
21. Measure the DNA concentration of each sample using a Nanodrop spectrophotometer or a similar method. The amount of IP DNA recovered is usually ~10–20 % of the amount of input DNA.

3.2. Measurement of Enrichment for Methylated DNA

Immunoprecipitated DNA and the corresponding input DNA are labeled with cy3 and cy5 fluorescent dyes, hybridized to tiling oligonucleotide arrays, scanned, and the images analyzed to extract $\log_2$ ratios representing the relative quantity of methylated:unmethylated DNA at each probe locus. All steps are performed according to manufacturer's protocols. Alternatively, relative amounts of IP and input DNA can be quantified by alternative technologies, such as real-time PCR or high-throughput sequencing.

3.3. Processing of Microarray Data to Identify Regions of Differential Methylation Between Samples

Rigorous analysis and interpretation of large datasets produced by hybridization of IP and input DNA to oligonucleotide arrays require some programming and/or statistical knowledge, such as use of the Bioconductor project (3). Below I outline a framework for the analysis of data from Nimblegen microarrays with a median density of one probe per 100 bp across entire chromosomes that can be used to identify regions of differential methylation between samples. The exact thresholds and analysis approaches used in any such global analysis will vary depending on the nature of the underlying data, specifics of the technical platform used, number and nature of the samples assayed, underlying biological question, and potentially many other factors specific to each experiment. As a result, the outline below should be treated as a set of guidelines that should be modified to suit each specific situation.

1. Due to technical variations between samples that may be caused by a variety of factors, including (but not limited to) variable efficiency of immunoprecipitation reactions, sample labeling, or hybridization, it is usually necessary to perform normalization across all samples within an experiment to try and remove systematic sample-to-sample biases that might otherwise result in significant artifacts when comparing different individuals. A variety of normalization approaches are available, but I have found quantile normalization (4) to be effective (Fig. 1). Following quantile normalization, the mean and standard deviation of the data distributions in each sample are identical, allowing relatively unbiased comparison of $\log_2$ ratios across different samples.
2. A small fraction of probes on any microarray perform poorly, yielding highly unreliable data. These low-quality data points may represent probes that are inherently unreliable due to their sequence characteristics (5), or, for example, may be located

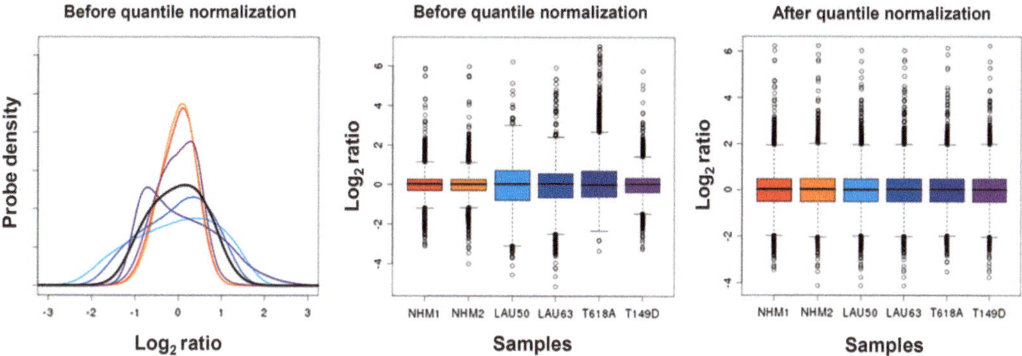

Fig. 1. Transformation of microarray hybridization data by quantile normalization allows unbiased comparison across arrays. (*Left panel*) Density plot showing the varying distribution of raw $\log_2$ ratios in six individual hybridizations. Due to these differing distributions, comparisons across samples using these raw data would result in the detection of many differences that are likely artifacts resulting from the inherently different underlying data distributions. (*Middle panel*) Raw data from six individual hybridizations was transformed by quantile normalization to remove sample-specific biases resulting from differences in antibody enrichment, labeling, or hybridization. (*Right panel*) After quantile normalization, the six datasets show identical distributions, allowing unbiased comparison across samples to identify differentially methylated regions.

on a section of the microarray surface containing hybridization or scanning artifacts (e.g., dust). Given that with high-density tiling arrays the probe spacing throughout the genome (~1 per 100 bp) is significantly smaller than the size of DNA fragments being hybridized to the array (~500 bp), it is expected that closely spaced probes will show somewhat correlated $\log_2$ values. Based on this expectation, low-quality data points can be identified by implementing a sliding window analysis that identifies probes that deviate significantly from the $\log_2$ value of their immediate neighbors (outlier data points). I have found that an effective approach is to use a sliding window to identify all clusters of five consecutive probes which span a physical distance of 1 kb or less. For each group of probes, if the difference in $\log_2$ values between the central probe and the median value of the five probes in that cluster exceeds the interquartile range of $\log_2$ values on that entire chromosome, it is flagged as an outlier. Rather than completely removing outliers, which would result in loss of data, an alternative approach is to replace them with the median $\log_2$ ratio of the remaining four probes in the group. Based on these criteria, the $\log_2$ values of approximately 2–4 % of probes per array are replaced.

Overall, these normalization and filtering steps resulted in significant noise reduction and improvements to data quality. For example, in one prior study in which six samples were tested in duplicate, the mean correlation between $\log_2$ ratios in technical replicate hybridizations for the six individuals tested increased from 0.83 in the raw data to 0.93 after quantile normalization and outlier replacement (2) (Fig. 2).

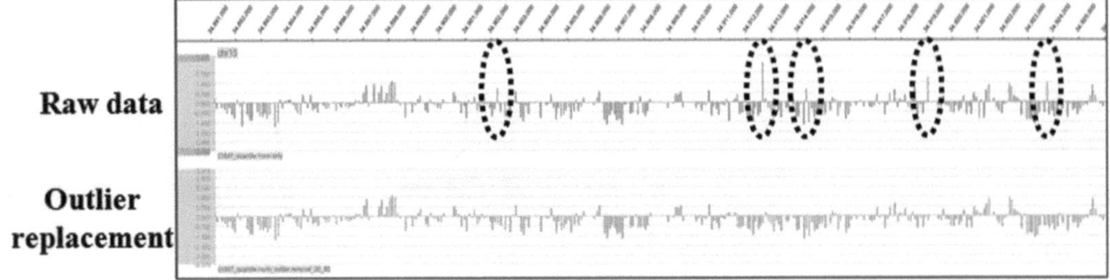

Fig. 2. Effects of outlier probe replacement on methylation profiles. The image shows a screenshot of probe $\log_2$ ratios in a 25 kb region of chromosome 15 from one array hybridization. The *top track* shows the raw data, while the *lower panel* shows the same data after replacement of outlier probes (*dotted ellipses*). This step can significantly reduce noise caused by poor-performing probes on the array.

3. Data can be further treated by application of a linear smoothing function (6), which acts to reduce probe-to-probe variation.
4. Before performing any analysis to compare between samples to detect regions of differential methylation, it is often useful to remove probes that show very low variance between samples. Probes that are inherently invariant have very low power to detect differences between samples, and are not therefore useful to include in downstream analysis. In fact, removal of invariant probes from the dataset prior to performing formal statistical testing can actually increase the overall statistical power of an analysis by reducing the burden of multiple testing correction that needs to be performed in any microarray analysis. Generally a simple filter to remove any probe that shows low variance across a population (e.g., standard deviation <0.2) is appropriate.
5. To identify probes that show potential differential methylation between samples, significance testing can be performed, using, e.g., a Student's *t*-test. Given that most microarray analyses will involve large numbers of probes, it is also necessary to apply a multiple testing correction to any p values generated in these testes to avoid large numbers of false positive associations (e.g., False discovery rate correction) (7).
6. Given that data from single probes is often unreliable and consideration of single probe events will often result in a high false positive rate, differentially methylated regions of higher confidence can be identified by searching for clusters of >1 closely spaced probes that exceed a given threshold of significance. In a previous study (2), use of a sliding window analysis to identify clusters of five probes separated by <1 kb which exceeded a relatively low-stringency statistical threshold was successful in identifying regions of differential methylation (the central probe shows an FDR-adjusted $p<0.1$, and at least two of the four neighboring probes show an unadjusted $p<0.05$).

4. Notes

1. The amount of starting DNA can be varied depending on the amount of immunoprecipitated material that is required for later quantification by array hybridization, sequencing, or qPCR. However in this case it is necessary to also vary the amount of antibody used in step 7 of Day 1 of the protocol in order to retain the same 3:2 ratio of DNA:antibody. As a rule, generally the final yield of the immunoprecipitation reaction is approximately 10–15 % of the starting amount.
2. The use of screw-top tubes is strongly recommended where indicated, as there are multiple steps in this protocol where standard snap-top Eppendorf tubes can leak, resulting in the loss of DNA and the release of organic solvents.
3. Other random DNA fragmentation methods can be used, such as Covaris DNA shearing or chemical cleavage methods.
4. In some cases, after sonication the DNA sample may have a larger size range than anticipated. In this case, additional fragmentation can be performed to reduce the size to an acceptable range (200–500 bp).
5. The final DNA pellets are generally very small, and are sometimes barely visible to the naked eye. It is recommended to rotate each Eppendorf tube so that the hinges are on the outer edge of the centrifuge rotor. If this is done, even if the precipitated DNA pellet is invisible after centrifugation, its location in the tube will always be below the hinge, making it easier to subsequently remove the supernatant without losing the DNA pellet.

Acknowledgements

This work was supported by funding from the Fondation Jerome LeJeune.

References

1. Sharp AJ, Stathaki E, Migliavacca E, Brahmachary M, Montgomery S, Dupre Y, Antonarakis SE (2011) DNA methylation profiles of human active and inactive X chromosomes. Genome Res 21(10):1592–1600
2. Sharp AJ, Migliavacca E, Dupre Y, Stathaki E, Sailani MR, Mackay D, Robinson DO, Cobellis G, Cobellis L, Brunner H, Steiner B, Antonarakis SE (2010) Methylation profiling in cases with uniparental disomy identifies novel differentially methylated regions on chromosome 15. Genome Res 20:1271–1278
3. Gentleman RC, Carey VJ, Bates DM, Bolstad B, Dettling M, Dudoit S, Ellis B, Gautier L, Ge Y, Gentry J, Hornik K, Hothorn T, Huber W, Iacus S, Irizarry R, Leisch F, Li C, Maechler M,

Rossini AJ, Sawitzki G, Smith C, Smyth G, Tierney L, Yang JY, Zhang J (2004) Bioconductor: open software development for computational biology and bioinformatics. Genome Biol 5:R80

4. Bolstad BM, Irizarry RA, Astrand M, Speed TP (2003) A comparison of normalization methods for high density oligonucleotide array data based on bias and variance. Bioinformatics 19:185–193
5. Sharp AJ, Itsara A, Cheng Z, Alkan C, Schwartz S, Eichler EE (2007) Optimal design of oligonucleotide microarrays for measurement of DNA copy-number. Hum Mol Genet 16: 2770–2779
6. Pelizzola M, Koga Y, Urban AE, Krauthammer M, Weissman S, Halaban R, Molinaro AM (2008) MEDME: an experimental and analytical methodology for the estimation of DNA methylation levels based on microarray derived MeDIP-enrichment. Genome Res 18: 1652–1659
7. Benjamini Y, Hochberg Y (1995) Controlling the false discovery rate: a practical and powerful approach to multiple testing. J R Stat Soc B 57:289–300

Chapter 6

Identification of Imprinted Loci by Transcriptome Sequencing

Tomas Babak

Abstract

Enabled by high-throughput technologies that are capable of generating millions of sequencing reads, transcriptome sequencing is emerging as an important approach for mapping allelic imbalance (AI), where transcription is biased toward one allele in a diploid system. AI is identified by counting sequencing reads that map to genomic regions containing heterozygous SNPs, where the base identity of the SNP is used to distinguish allelic origin. Genomic imprinting is a special case of AI where bias is toward parental sex and can be identified by transcriptome sequencing of systems that represent reciprocally inherited loci. The focus of this protocol is on experimental design, analysis, and interpretation of genomic imprint discovery using whole transcriptome sequencing.

Key words: Imprinting, Transcriptome sequencing, Allelic imbalance

1. Introduction

While a comprehensive map of imprinted loci in all cell/tissue types of all mammals facilitates the evolutionary characterization of genomic imprinting, inherent challenges of traditional discovery approaches have mostly limited their application to developing mouse stages. Classical genetic screens based on uniparental disomies and reciprocal translocations (1, 2) and genetic mapping of parent-of-origin phenotypes in humans (e.g., ref. 3) revealed the initial imprinted loci. Most imprinted genes emerged from fine mapping of these initial regions, typically by RFLP analysis of cDNA. Microarray profiling of embryos with uniparental disomies or entire genomes (4, 5) extended the map, but embryonic lethality induced by these genetic perturbations has limited their discovery potential. A two-dimensional RFLP approach (6) and genotyping microarrays (7) have been applied to imprint discovery in adult/wild-type tissues but require extensive analysis to rule out signal

Nora Engel (ed.), *Genomic Imprinting: Methods and Protocols*, Methods in Molecular Biology, vol. 925,
DOI 10.1007/978-1-62703-011-3_6, © Springer Science+Business Media, LLC 2012

from noise. Mapping imprinting by transcriptome sequencing, which has only recently become possible, is advantageous in that it does not require a priori knowledge of the expected genomic localization or phenotype, and can be applied to any diploid progeny of genetically diverged parents.

NextGen sequencing (NGS) has rapidly changed the landscape of nucleic acid-driven research. In addition to standard sequence determination, the ability to generate millions of sequencing reads has also enabled quantification of input abundance by simply counting the reads. RNA-Seq (8, 9), the most commonly utilized form of transcriptome sequencing, is based on random priming of polyA + purified RNA to generate cDNA, which is again random primed to generate double-stranded DNA that is then sequenced. Biological inference typically involves mapping of the reads to a reference genome and quantifying events (e.g., splicing or gene expression) by counting. Similarly, when a mapped sequencing read overlaps a heterozygous SNP, allelic origin of that read can be discerned using the identity of the polymorphic base. AI can be inferred by comparing the two sums of allelic reads that map to that SNP. Since AI is widespread in mammalian cells (10) and genomic imprinting is likely to be causal for only a small proportion (11), demonstrating AI in a reciprocally inherited fashion is imperative for imprinting discovery. The most straightforward experimental design consists of two crosses of inbred strains, where each parental sex is represented by both genetic backgrounds. In this simple system each progeny is heterozygous at all SNPs and both maternal and paternal copies of each allele are represented (Fig. 1). RNA-Seq is then applied to both crosses and genomic imprinting distinguished as AI that is consistently biased toward parental sex (vs. AI biased toward same parental strain which is much more common). Although the classical definition of genomic imprinting is based on monoallelic expression, it has been expanded

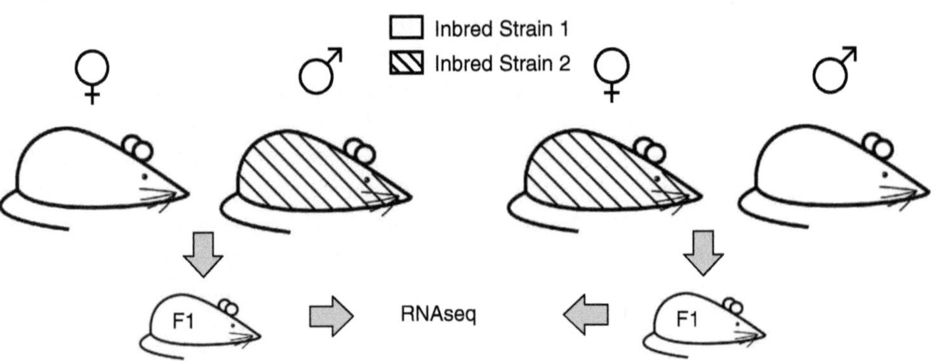

Fig. 1. Schematic of reciprocal cross. Inbred parental strains are crossed in reciprocal to produce F1 progeny that are sequenced and analyzed in pairs.

to include incomplete parent-of-origin expression biases (11) and enables identification of cell/tissue-specific imprinting that is diluted by tissue heterogeneity.

Several groups have successfully used transcriptome sequencing for imprint discovery (12–14) and the approach has outstanding promise, but unresolved challenges exist. For example, all published studies using RNA-Seq to map AI have thus far not addressed systematic bias. It has been shown that AI reproduces very well across technical and biological replicates when considering the same SNP (15) but little has been done to assess concordance across different SNPs. Systematic priming biases induced by SNPs could lead to incorrect AI calls. One would expect strong agreement on AI among two SNPs within the same exon, for example, and these could be used to empirically estimate systematic bias. Until AI is robustly modeled, mock reciprocal crosses (i.e., biological replicates) can be used to gauge false-discovery and I expand on this below. A second challenge is application outside the somewhat artificial setting of inbred mouse crosses. Transcriptome sequencing has not yet been applied for imprint discovery in species where inbred strains or controlled crosses are not available (such as humans). Two experimental designs could be employed to make this feasible. The first is a controlled screen in a family where parents and offspring are genotyped; AI is determined at all heterozygous SNPs in offspring, and parent-of-origin inheritance determined from phased haplotypes. A large family with distantly related parents would be ideal. The second approach is to screen an outbred population and identify imprinting as AI with no sequence dependence (i.e., AI is always observed but biased for either allele with equal likelihood) since consistent bias toward one allele suggests a genetic mechanism and this is generally the case (10).

The focus of this protocol is on identification of genomic imprinting in reciprocally crossed F1 mouse strains using standard RNA-Seq with emphasis on analysis practices. The assumption is that the reader has access to total RNA from reciprocally crossed mouse tissues and a reasonable (>5 million) map of SNPs for these strains. The scope of the approach could be expanded to any diploid species and additional recommendations for application outside inbred mouse strains are also included.

2. Materials

2.1. Constructed RNA-Seq Libraries

Construction of RNA-Seq libraries was first described in yeast and mouse (8, 16) and is now available in kit format from several manufacturers. In my experience the standard RNA-Seq kit sold by Illumina works very well and the end result is a library of high complexity (measured by the proportion of sequencing reads that align to unique genomic locations). The TruSeq RNA kit (Illumina), which is designed for higher sample throughput, works well and I have seen great data from as little as 100 ng total RNA input. A few practical changes involving deoxy-uridine triphosphate (dUTP) and uracil-*N*-glycosylase (UNG) (17, 18) effectively introduce strand specificity and are recommended since imprinted antisense transcription is known to exist. Library complexity and even coverage are essential for measuring AI and some of the low-input kits suffer in this regard because they have multiple series of amplification. NSR-seq (not-so-random primer sequencing) (19) is one of the earlier approaches that was applied for identifying AI (12), and has the advantage of capturing non-polyadenylated transcripts and is also strand specific, but personal experience and a recent evaluation (17) have revealed undesirable evenness of coverage (i.e., coverage is "spiky"). In summary, any approach that quantitatively captures input transcript abundance and yields a library of high complexity will work and I have seen excellent data suitable for mapping AI from libraries made with mRNAseq/TruSeq RNA kits purchased from Illumina modified with dUTP/UNG treatment (18) to achieve strand specificity.

2.2. NextGen Sequencing Capacity

454, SOLiD, and Illumina are currently the major suppliers of NGS sequencers. Any of these platforms and likely many other emerging platforms will work, although Illumina and SOLiD are currently the only commercially available RNA-Seq platforms for generating tens to hundreds of millions of reads. Overall sequencing depth is dependent on the length and number of sequencing reads and the heterozygous SNP density of the system. Methods exist to estimate the minimum required sequencing (20) and more will always improve sensitivity. In practice, 4 Gb of single-end RNA-Seq data from reciprocally crossed C57BlxCAST samples (i.e., 8 Gb total data, 4 Gb from each cross) is sufficient to confidently identify >90% of previously validated imprints in that tissue. 2 Gb will result in slightly lower performance (70–80% sensitivity at the same detection threshold) and even 1 Gb will yield acceptable results (~60% sensitivity). The ideal read length is a trade-off between molecular complexity (long reads and PE reads limit the number of molecules represented in the library) and sequencing of SNPs. The ideal read length would on average capture 1 SNP/read and can be estimated using a published model (20). Considering practical challenges I recommend using single-end 75–100 bp reads. Paired-end (PE) data improves mapping performance but only marginally. With a mean RNA-Seq insert size of ~200 bp, the 3′ ends of pairs can overlap which leads to diminishing returns. Reads shorter than 50 bp are not recommended since this will lead to significant mapping bias (see Note 9).

2.3. Computational Resources

1. Access to Linux/Unix working environment with at least 6 Gb RAM.
2. Installation of Novoalign v2.07.11 or newer (21) or equivalent short read sequence aligner. V2.07.11 has capability of reporting mismatches to masked bases (Ns).
3. Reference genome. Most sequenced mammalian genomes can be downloaded from UCSC (22). If the genome for the species is not available, reads could be assembled into transcript models that then serve as a reference, but this is beyond the scope of this protocol.
4. Map of SNPs. 15 mouse strains were recently sequenced by the Sanger Institute and SNP maps are available for download (23). If working with a system that has a reference genome but where SNPs are unknown, genotyping arrays or genome sequencing can be used to map heterozygous SNPs. In humans additional SNPs can be imputed and phased using MaCH (24) to improve sensitivity of AI mapping. SNPs can also be inferred from the RNA-Seq data. This does not work well for mapping AI since discovery favors SNPs biased in expression toward the non-reference allele and thus the resulting AI profile becomes artificially skewed toward non-reference alleles. However, the approach can be effective for imprint discovery by calling SNPs on pooled (in equal amount) sequencing data from the reciprocal crosses where imprinted SNPs are supported by near 50:50 proportions. SAMtools (25), GATK (26), or soapSNP (27) can all be used to identify SNPs from mapped RNA-Seq data and are comparable in performance.
5. Perl/Python installation or equivalent for custom manipulation of data.
6. Matlab, R, Excel, or equivalent for visualizing results.

3. Methods

1. Generate a masked version of genome where known SNPs are replaced with Ns (see Note 1).
2. Index genome by running Novoindex (default options and -k 14 -s 3) on a single fasta file of masked genome (see Note 2).
3. Align fastq-formatted raw reads files (paired-end or single-end) against masked genome using Novoalign (default options and -a -o IUBMatch -r None, and -i 0 1000 if aligning PE reads) (see Note 3). A summary of the alignment approach is shown in Fig. 2.

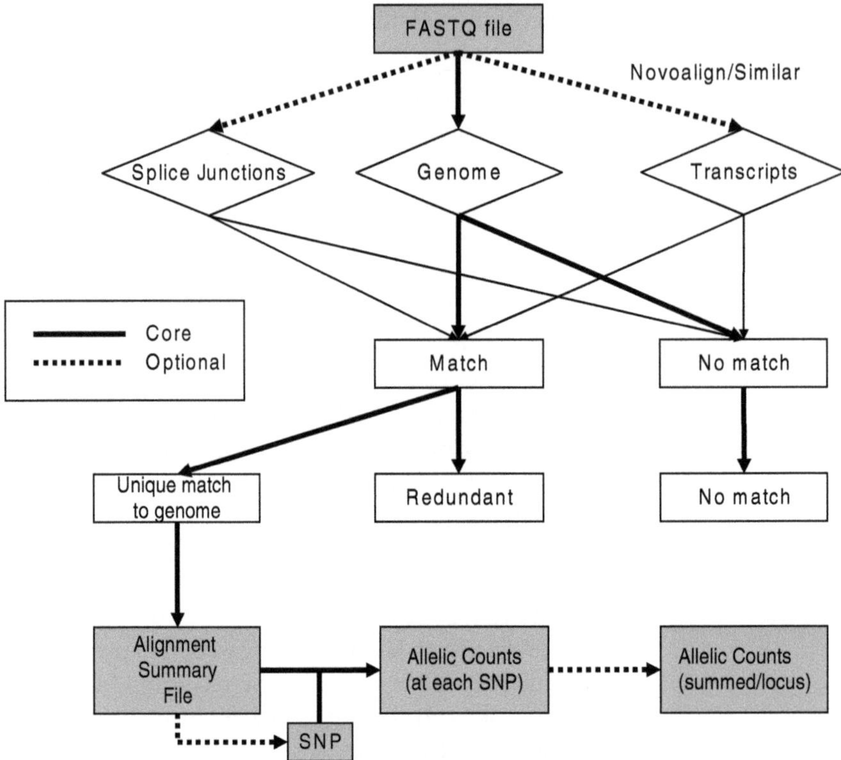

Fig. 2. Alignment, SNP-identification, AI-quantification pipeline. Alignment is accomplished with an independent algorithm (e.g., Novoalign (21)) against the genome, and optionally splice junctions and full-length transcripts. Unique matches (in the genome) are retained and used for SNP prediction and quantification of ASE.

4. If improved alignment sensitivity is desired, also align reads to splice junctions and full-length transcripts (especially useful if aligning paired-end reads) using Novoalign (same settings as above except -r All 50) (see Note 4). Convert transcript alignment coordinates to genomic coordinates using the transcript genomic coordinates (see Note 5).
5. Discard redundant alignments (where reads map to more than one genomic location) and generate report files that store alignment coordinates and genomic mismatches for each read (see Note 6).
6. For each SNP, tally the number of reads that support the reference and alternate bases (see Note 6).
7. At each SNP, let A represent the number of reference-specific reads and B the number of alternate allele-specific reads. Quantify the degree of AI as $A/(A+B)$. The probability of AI can be estimated using the cumulative binomial distribution. This can also be done in Excel where binomial-p (probability of no AI) = binomdist(min(A, B), $A+B$, 0.5, 1). In Matlab the binomcdf function from the statistics package can be called.

The same principles can also be used on all allele-specific reads summed across a transcript (i.e., sum over all SNPs within the transcript) (see Note 7).

8. Genomic imprinting requires AI to be measured in tissue-matched reciprocally crossed samples. If s1 = sample 1 and s2 = reciprocal sample, Genomic imprinting may exist if ($AI_{s1} > 0.5$ and $AI_{s2} < 0.5$) or if ($AI_{s1} < 0.5$ and $AI_{s2} > 0.5$), i.e., reciprocal bias exists. The probability of imprinting can be estimated as the less significant binomial estimate of the two samples (see Note 8).
9. Select a suitable threshold of significance for calling imprinting by using a mock reciprocal cross as a negative control (see Note 9).

4. Notes

1. SNP maps can be downloaded from Sanger (23) and masking greatly reduces alignment biases (28).
2. This step creates an .idx file that is used as input for genome alignment. Junction and transcript indices (step 4) can be made with the same settings.
3. Over a dozen short-read alignment algorithms are currently available. BWA (29), SOAP2 (30), and Bowtie (31) are based on the Burrows–Wheeler Transform (BWT) algorithm and are by far the fastest aligners with sensitivity and specificity comparable or better than most. However, in testing these and eight other popular aligners on simulated single-end and paired-end data (with imputed mismatches representative of quality scores and expected variation), Novoalign (21) attained the highest sensitivity (7–8% higher sensitivity than BWT approaches with avg. alignment rate of 87% vs. 79–80%) and comparable specificity (<0.1% erroneous alignments) to all aligners. Not surprisingly, the AI profile was less biased toward reference alleles than for other approaches tested owing to a better ability to align over SNPs. -a will trim adapter sequences, -o IUBMatch will report N > (ACGT) base changes, -r None will not report reads that align in more than one genomic regions, and -i 0 1000 will allow pairs to match up to 1,000 bp apart.
4. Extensive custom scripting will be required to perform this step and there is more than one way to compile a reference transcriptome. I made a splice junction coordinate file from all possible exon skipping events (up to two exons skipped) from RefSeq, ENSEMBL, UCSC known gene, and Genbank mRNA BED files downloaded from UCSC (22). I then sorted to

remove duplicate entries (sort -k 6,6 -k 1,1 -k 2,2n -k 3,3n -u unsorted_with_6_columns.bed > sorted_unique_junctions.bed) and retrieved the fasta equivalent from UCSC Table Browser and indexed using Novoindex (-k 14 -s 3). Aligning paired-end reads to transcripts will considerably improve the number of reads that align as pairs. I again recommend RefSeq, ENSEMBL, UCSC KG, and Genbank mRNAs as a comprehensive transcript set. It is important to allow reporting of all matches since -r None will ignore matches to multiple isoforms which will be most of them (i.e., use -r All 50). Redundant filtering (step 5) done in genomic space removes truly redundant matches. BED files for junctions and transcripts can be used to convert alignments back into genomic coordinates.

5. If mapping paired-end reads, a paired match takes precedence over single matches (i.e., if maps as a pair once take that alignment and disregard all others). At this point reads that do not contain N > (ACGT) changes can be discarded if further SNP discovery will not be done.

6. Reads mapping to opposite strands should be tallied independently if strand-specific RNA-Seq was used (i.e., each SNP may have up to two sets of counts).

7. The cumulative binomial distribution models the maximum number of successes in a sequence of independent binary events, each of which yields success with some probability. For example, the chance of getting three or fewer heads when flipping a fair coin ten times is 17.2%. Summing reads across SNPs violates the binomial assumption when a single sequencing read spans more than one SNP since it expects all counts to be independent. Ideally, a read (whether single-end or paired-end) should only be counted once. An ad hoc approach to ensure that this is the case is to only consider SNPs that are further apart than the read length (fragment length if using paired-reads). In practice, the extent of systematic error in measuring AI with RNA-Seq contributes significantly more uncertainty in the binomial calculation than violating counting independence as described. Negative controls are imperative for estimating false-discovery (see step 9).

8. A suitable threshold for making an imprint call depends on the extent of acceptable false-discovery (i.e., proportion of calls that are not truly imprinted; see step 9) and will vary from sample to sample and with the selected RNA-Seq protocol. In practice, a binomial p-value of 0.001 results in a false-discovery rate (FDR) of ~10% using standard Illumina mRNAseq.

9. For this control to be valid, samples need to be prepared completely in parallel, they must be sequenced to equivalent depths, and all must pass quality control criteria. The FDR can be

estimated by plotting the number of imprinted sites as a function of binomial-p cutoff from data generated from biological replicates. Since there is no genuine reciprocal inheritance of any allele in this scheme, all calls are false-positives and their rate will translate to a genuine cross if all samples are sequenced to an equal depth. Random removal of reads should be done to ensure that all samples have the same number of input reads. A similar plot for a genuine reciprocal cross can be used to estimate sensitivity (number of known imprinted sites detected) and by combining the data into a plot of FDR vs. sensitivity a useful threshold for making imprinting calls can be selected. The same criteria can be applied to AI inferred from reads summed across SNPs in the same transcript.

References

1. Cattanach BM, Kirk M (1985) Differential activity of maternally and paternally derived chromosome regions in mice. Nature 315: 496–498
2. Surani MA, Reik W, Allen ND (1988) Transgenes as molecular probes for genomic imprinting. Trends Genet 4:59–62
3. Nicholls RD, Knoll JH, Butler MG, Karam S, Lalande M (1989) Genetic imprinting suggested by maternal heterodisomy in nondeletion Prader-Willi syndrome. Nature 342: 281–285
4. Choi JD, Underkoffler LA, Collins JN, Marchegiani SM, Terry NA, Beechey CV, Oakey RJ (2001) Microarray expression profiling of tissues from mice with uniparental duplications of chromosomes 7 and 11 to identify imprinted genes. Mamm Genome 12: 758–764
5. Mizuno Y, Sotomaru Y, Katsuzawa Y, Kono T, Meguro M, Oshimura M, Kawai J, Tomaru Y, Kiyosawa H, Nikaido I, Amanuma H, Hayashizaki Y, Okazaki Y (2002) Asb4, Ata3, and Dcn are novel imprinted genes identified by high-throughput screening using RIKEN cDNA microarray. Biochem Biophys Res Commun 290:1499–1505
6. Plass C, Shibata H, Kalcheva I, Mullins L, Kotelevtseva N, Mullins J, Kato R, Sasaki H, Hirotsune S, Okazaki Y, Held WA, Hayashizaki Y, Chapman VM (1996) Identification of Grf1 on mouse chromosome 9 as an imprinted gene by RLGS-M. Nat Genet 14:106–109
7. Morcos L, Ge B, Koka V, Lam KC, Pokholok DK, Gunderson KL, Montpetit A, Verlaan DJ, Pastinen T (2011) Genome-wide assessment of imprinted expression in human cells. Genome Biol 12:R25
8. Mortazavi A, Williams BA, McCue K, Schaeffer L, Wold B (2008) Mapping and quantifying mammalian transcriptomes by RNA-Seq. Nat Methods 5:621–628
9. http://www.illumina.com
10. Pickrell JK, Marioni JC, Pai AA, Degner JF, Engelhardt BE, Nkadori E, Veyrieras JB, Stephens M, Gilad Y, Pritchard JK (2010) Understanding mechanisms underlying human gene expression variation with RNA sequencing. Nature 464:768–772
11. Morison IM, Ramsay JP, Spencer HG (2005) A census of mammalian imprinting. Trends Genet 21:457–465
12. Babak T, Deveale B, Armour C, Raymond C, Cleary MA, van der Kooy D, Johnson JM, Lim LP (2008) Global survey of genomic imprinting by transcriptome sequencing. Curr Biol 18:1735–1741
13. Gregg C, Zhang J, Weissbourd B, Luo S, Schroth GP, Haig D, Dulac C (2010) High-resolution analysis of parent-of-origin allelic expression in the mouse brain. Science (New York, NY) 329:643–648
14. Wang X, Sun Q, McGrath SD, Mardis ER, Soloway PD, Clark AG (2008) Transcriptome-wide identification of novel imprinted genes in neonatal mouse brain. PLoS One 3:e3839
15. Babak T, Garrett-Engele P, Armour CD, Raymond CK, Keller MP, Chen R, Rohl CA, Johnson JM, Attie AD, Fraser HB, Schadt EE (2010) Genetic validation of whole-transcriptome sequencing for mapping expression affected by cis-regulatory variation. BMC Genomics 11:473
16. Nagalakshmi U, Wang Z, Waern K, Shou C, Raha D, Gerstein M, Snyder M (2008) The transcriptional landscape of the yeast genome

defined by RNA sequencing. Science (New York, NY) 320:1344–1349

17. Levin JZ, Yassour M, Adiconis X, Nusbaum C, Thompson DA, Friedman N, Gnirke A, Regev A (2010) Comprehensive comparative analysis of strand-specific RNA sequencing methods. Nat Methods 7:709–715
18. Parkhomchuk D, Borodina T, Amstislavskiy V, Banaru M, Hallen L, Krobitsch S, Lehrach H, Soldatov A (2009) Transcriptome analysis by strand-specific sequencing of complementary DNA. Nucleic Acids Res 37:e123
19. Armour CD, Castle JC, Chen R, Babak T, Loerch P, Jackson S, Shah JK, Dey J, Rohl CA, Johnson JM, Raymond CK (2009) Digital transcriptome profiling using selective hexamer priming for cDNA synthesis. Nat Methods 6:647–649
20. Fontanillas P, Landry CR, Wittkopp PJ, Russ C, Gruber JD, Nusbaum C, Hartl DL (2010) Key considerations for measuring allelic expression on a genomic scale using high-throughput sequencing. Mol Ecol 19(Suppl 1):212–227
21. http://www.novocraft.com/main/index.php
22. Rhead B, Karolchik D, Kuhn RM, Hinrichs AS, Zweig AS, Fujita PA, Diekhans M, Smith KE, Rosenbloom KR, Raney BJ, Pohl A, Pheasant M, Meyer LR, Learned K, Hsu F, Hillman-Jackson J, Harte RA, Giardine B, Dreszer TR, Clawson H, Barber GP, Haussler D, Kent WJ (2010) The UCSC Genome Browser database: update. Nucleic Acids Res 38:D613–D619
23. http://www.sanger.ac.uk/resources/mouse/genomes/
24. Li Y, Willer CJ, Ding J, Scheet P, Abecasis GR (2010) MaCH: using sequence and genotype data to estimate haplotypes and unobserved genotypes. Genet Epidemiol 34:816–834
25. Li H, Handsaker B, Wysoker A, Fennell T, Ruan J, Homer N, Marth G, Abecasis G, Durbin R (2009) The Sequence Alignment/Map format and SAMtools. Bioinformatics (Oxford, England) 25:2078–2079
26. McKenna A, Hanna M, Banks E, Sivachenko A, Cibulskis K, Kernytsky A, Garimella K, Altshuler D, Gabriel S, Daly M, DePristo MA (2010) The genome analysis toolkit: a MapReduce framework for analyzing next-generation DNA sequencing data. Genome Res 20:1297–1303
27. Li R, Li Y, Fang X, Yang H, Wang J, Kristiansen K, Wang J (2009) SNP detection for massively parallel whole-genome resequencing. Genome Res 19:1124–1132
28. Degner JF, Marioni JC, Pai AA, Pickrell JK, Nkadori E, Gilad Y, Pritchard JK (2009) Effect of read-mapping biases on detecting allele-specific expression from RNA-sequencing data. Bioinformatics (Oxford, England) 25: 3207–3212
29. Li H, Durbin R (2009) Fast and accurate short read alignment with Burrows-Wheeler transform. Bioinformatics (Oxford, England) 25:1754–1760
30. Li R, Yu C, Li Y, Lam TW, Yiu SM, Kristiansen K, Wang J (2009) SOAP2: an improved ultrafast tool for short read alignment. Bioinformatics (Oxford, England) 25:1966–1967
31. Langmead B, Trapnell C, Pop M, Salzberg SL (2009) Ultrafast and memory-efficient alignment of short DNA sequences to the human genome. Genome Biol 10:R25

Chapter 7

Data Mining as a Discovery Tool for Imprinted Genes

Chelsea Brideau and Paul Soloway

Abstract

This chapter serves as an introduction to the collection of genome-wide sequence and epigenomic data, as well as the use of these data in training generalized linear models (glm) to predicted imprinted status. This is meant to be an introduction to the method, so only the most straightforward examples will be covered. For instance, the examples given below refer to 11 classes of genomic regions (the entire gene body, introns, exons, 5′ UTR, 3′ UTR, and 1, 10, and 100 kb upstream and downstream of each gene). One could also build models based on combinations of these regions. Likewise, models could be built on combinations of epigenetic features, or on combinations of both genomic regions and epigenetic features.

This chapter relies heavily on computational methods, including basic programming. However, this chapter is not meant to be an introduction to programming. Throughout the chapter, the reader will be provided with example code in the Perl programming language.

Key words: Epigenetics, Epigenomics, Imprinting, Data mining, Bioinformatics, Generalized linear model

1. Introduction

Genomic imprinting refers to genes that are expressed from one of the two parental alleles in a parent-of-origin-specific manner. Until recently, about 100 mouse imprinted genes had been identified, with many more genes predicted to be imprinted (http://igc.otago.ac.nz/home.html) (1, 2). However, application of new methods, such as whole transcriptome sequencing and computational prediction, has identified additional imprinted genes (3–7).

The identification of novel imprinted genes has become increasingly important with the realization that imprinting defects are associated with a variety of complex disorders, such as obesity, diabetes, and schizophrenia (8–11). Given the importance that imprinted genes play in human health, several studies have attempted genome-wide identification of imprinted genes

Nora Engel (ed.), *Genomic Imprinting: Methods and Protocols*, Methods in Molecular Biology, vol. 925,
DOI 10.1007/978-1-62703-011-3_7,

(1–6, 12–24). These have done so mainly using experimental methods, with some success.

The first studies identified loci with allele-specific DNA methylation, one hallmark of imprinted genes, applying methods such as Restriction Landmark Genome Scanning (RLGS) to DNAs from progeny of interstrain reciprocal crosses (18, 25–28). In this method, DNA is cut with a methylation-sensitive restriction enzyme, followed by radioactive end-labeling. Then, the radioactive DNA fragments are digested with a second restriction enzyme and run on the first dimension of a two-dimensional agarose gel. The DNA fragments are then digested in the gel with a third restriction enzyme, and the second dimension of the two-dimensional gel is run. After exposing the gel to film, a pattern of spots is visible and can be compared between reciprocal crosses to determine whether there are any potential differences in methylation between the two alleles. One of the major drawbacks of the RLGS method was the low throughput, as the genomic region associated with each spot needed to be cloned and identified. This issue has been resolved, since all spots have been cloned and identified. However, this is a low-resolution method, as there are a limited number of sites that can be queried using this method. Reduced representational bisulfite sequencing (29) or whole genome bisulfite sequencing (30) of DNAs from progeny of reciprocal crosses can also be used to comprehensively identify sites of allele-specific DNA methylation.

In the past decade, methods for genome-wide identification of imprinted genes have increased dramatically in terms of throughput. One large-scale study identified candidate imprinted transcripts in the mouse genome by expression profiling of cDNA clones. cDNA microarrays were used to detect differential expression by comparing mRNA levels in the P9.5 gynogenetic and androgenetic mouse embryos (1). Of the ~28,000 FANTOM2 transcripts analyzed, ~2,000 were identified as imprinted candidates. Interestingly, 39 of the 2,000 transcripts mapped to known imprinted regions of the mouse genome, while 56 were ncRNAs, and 159 were antisense transcripts. Experimental validation of two transcripts located in the Prader–Willi syndrome region identified these transcripts as imprinted, indicating that allele-specific array-based methods are useful for large-scale identification of novel imprinted genes.

Additionally, four recent papers have successfully identified novel imprinted genes using massively parallel sequencing approaches (3–6). In each case, RNA and cDNA were prepared from reciprocal F1 mouse tissues and the cDNA subjected to massively parallel sequencing. By using polymorphic strains for the reciprocal crosses, SNPs in the sequenced material can be used to identify the expressed alleles, as well as those genes that express only, or predominantly, one allele. Use of reciprocal crosses is important, as it allows one to distinguish strain-specific expression effects from parent-of-origin expression effects. (For a more detailed description of F1 hybrid studies, please refer to Chapter 6.) The first of the four studies used neonatal mouse brain and successfully identified three novel imprinted genes in this tissue. Imprinting of each gene was confirmed by Sanger and pyrosequencing of PCR products spanning allele-specific SNPs (3). The second used e9.5 mouse embryos from reciprocal F1 mouse crosses and identified six novel imprinted genes (4). This study also suggests that many ncRNAs are subject to imprinted expression, as more than half of all imprinted single-nucleotide polymorphisms did not overlap previously discovered imprinted transcripts and a large fraction of these represent novel ncRNAs within known imprinted loci. The two most recent studies examined patterns of imprinted expression in brains of adult mice from reciprocal crosses (5, 6). The authors found elevated expression from the maternal X chromosome, indicating a bias in X-chromosome activation. Furthermore, over 1,300 candidate autosomal imprinted genes were identified. Two of the candidate genes were examined further and found to be imprinted in female, but not in male, adult mouse brain. These studies demonstrate the feasibility of unbiased, transcriptome-wide analysis for the identification of novel imprinted genes.

A third method has also been used in imprinted gene identification: computational prediction. This method will be discussed in detail throughout the remainder of the chapter. Briefly, this method involves identifying features of interest (for example, DNA sequence features, transcription factor or chromatin-remodeling protein-binding sites, or epigenetic status), demonstrating enrichment for those features at known imprinted loci, identifying loci of unknown imprinting status that carry those features, and experimentally testing imprinted status. Characteristic epigenetic features have been identified at gene regulatory elements of both nonimprinted and imprinted genes (31–33). Furthermore, epigenetic mechanisms are known to regulate genomic imprinting at several well-studied imprinted loci (34–51). With the application of genome-wide sequencing technologies to chromatin immunoprecipitation experiments (52), epigenomic data sets have become widely available for a variety of epigenetic marks, allowing the importance of epigenetic marks in the control of imprinted expression to be used as a tool to predict which additional genes in a given genome may be imprinted.

In this method, species-specific epigenomic data on a variety of features of interest are collected and, using training sets of genes, a computer learning approach is used to identify those features of interest, which are most important for the prediction of imprinted status. Once trained models have been created, they are used to search genome-wide for predicted novel imprinted genes. By identifying the epigenetic features that serve as the strongest predictors of imprinting, this approach can also identify those epigenetic mechanisms that are most likely to control imprinted states. This is something that identification of imprinted genes by transcriptome sequencing cannot do. Genome-wide identification of novel imprinted genes based on sequence features alone was pioneered in a series of two studies, which used a two-tiered machine-learning program to predict novel mouse and human imprinted genes genome-wide (2, 16). The first tier used a training set of known imprinted genes and presumed non-imprinted control genes to train the prediction program based on data on a variety of sequence features, but focusing on repetitive elements and transcription factor binding sites. The second tier was where the resulting model was run on the genome-wide data to predict novel imprinted genes. Although they did not experimentally verify any candidate imprinted genes in the mouse genome, they predicted a total of 600 imprinted mouse genes. A similar approach was used for the human genome and successfully verified two new imprinted genes on a chromosome that was not previously known to contain any imprinted genes. However, as a cautionary note, the imprinting status of genes was not verified using reciprocal F1 crosses, so false positives due to genetic background, but not parent of origin expression bias, cannot be ruled out. Subsequent studies using computational methods have resulted in experimental validation of twelve imprinted genes in the mouse genome, in addition to the two candidate genes from analysis of the human genome mentioned above (2, 7, 16, 20, 24).

As data become available that describe placement of additional epigenetic marks in other tissues, or other features altogether, such as sequences that physically interact, these methods can complement experimental methods such as transcriptome sequencing to identify imprinted genes and to provide insights into mechanisms controlling imprinting (3, 4, 53).

2. Materials

2.1. Hardware

A computer connected to the Internet.

A multi-CPU cluster will be helpful, although not strictly necessary. For some steps dealing with large data sets (e.g., ChIP-Seq data sets), processing on a laptop or a desktop may not be possible.

2.2. Software

Perl: http://www.perl.org/get.html.

A text editing program, such as Notepad++: http://notepad-plus-plus.org/download.

[R]: http://www.r-project.org/.

Microsoft Excel, or similar spreadsheet program.

Microsoft Word, or similar word processing program.

Unzipping program capable of handling tar.gz files.

3. Methods

3.1. Data Mining: Extracting Genomic Regions of Interest

In this section, you will extract genomic regions you wish to include in any analysis planned. This is done using UCSC and Galaxy and saving .txt files containing those genomic regions.

1. Direct your Web browser of choice to the UCSC Genome Browser Web site: http://genome.ucsc.edu/cgi-bin/hgGateway (54). From the menu at the top of the page, click on the "Tables" link. Once the page has loaded, select the relevant "clade" and "genome" from the drop-down list next to each. Next to the "assembly" option, the most recent assembly will be sufficient for most purposes.
2. To download genomic coordinates for all known genes, select "Genes and Gene Prediction Tracks" from the "group" menu, select the desired track from the "track" menu (see Note 1), ensure that "knownGene" is selected next to the "table" option, and that "selected fields from primary and related tables" is selected next to the "output format" option. To download coordinates for all known genes, leave everything else as is. However, to download coordinates for only a subset of all known genes, click on either the "paste list" or the "upload list" button next to "identifiers (names/accessions)." Then, if you have selected "paste," paste the names of the genes you wish to work with into the box and press the "submit" button. If you have selected "upload," click "browse," select the appropriate file from your computer, and then press the "submit" button. Once you have done this, enter the file name to which you will save your downloaded genomic coordinates in the box next to "output file" (e.g., All Gene Coordinates.xls). Then, press the "get output" button. You will now be taken to a different Webpage where you will be given options to select. This page will be divided into separate sections. In the very top section, click the "select all" button. Then search (control + F) for "Gene Symbol" and make sure that the box next to this option is selected. Finally, scroll back up to the top of the page and press the "Get output" button in the very top section of the page. Once the file has downloaded, check that you can open it using Excel, or a similar spreadsheet management program. You will notice that the first column does not contain conventional gene names. If you would like, you can replace this column with the last column, which will contain more familiar looking gene names. Make sure that you have saved any changes.
3. Next, filter this file to remove all duplicate entries. Under the data tab, select "Remove Duplicates." In the pop-up menu, unselect all except "Chrom," "txStart," and "txEnd," which

should be in columns B, D, and E. Click OK. Once your duplicates have been removed, open a new Excel file, and name it accordingly (e.g., Genes Filtered). Copy the data for "chrom" from column B of your filtered file into the first column of your new file. Also copy the "txStart" and "txEnd" data from columns D and E of your filtered file into the second and third columns of your new file. Delete the header row, by right-clicking on the "1" to the left of the header row and selecting delete. Save this file with an appropriate name (e.g., Genes.txt, see Note 2) and close this file.

4. To calculate genomic coordinates for upstream genic regions, first sort your "Genes Filtered" Excel spreadsheet by the data in the "strand" column. To do this, select "sort" from the "data" menu. Make sure to indicate that your data contain headers and that you want to sort by the "strand" column, which should be column "C." Then, select sort ascending (see Note 3). You should now have all of the genes on the "–" strand at the top of your Excel sheet and all of the genes on the "+" strand at the bottom of your Excel sheet. Now, insert two columns by right-clicking on the "C" above "strand." Select "Insert" from the menu that pops up. Repeat to add a second column. You should now have two empty columns between "chrom" and "strand." Label your new columns with an appropriate name. For this example, we will be calculating interval 100 kb upstream of your transcription start site for all genes, so we will label column C as "100 kb up Start" and column D as "100 kb up End." To perform this calculation for all genes on the "–" strand, type =G2 + 100,000 into cell D2 and press enter. Copy the formula from cell D2. Then, go to the last cell in column D that still has a "–" in the "strand" column. Select this cell and paste the formula. Now, go back to the top of your sheet and select cell D2. The cell should be outlined with a thick black border and a small square should be apparent at the lower right-hand corner of the cell. Position your mouse cursor over this square. A cross should form. When you see the cross form, double click on the cross with the left-hand mouse button. The formula should fill in for all of the genes on the "–" strand. To calculate the interval 100 kb upstream of your transcription start site for all genes on the "+" strand, locate the first cell under column C that has a "–" in the "strand" column. This cell should be empty. Note the row number you are on and then type =F followed immediately by the row number you are on and –100,000 (e.g., =F22485–100,000). Press enter. Select this cell and double click on the cross with the left-hand mouse button. The formula should fill in for all of the genes on the "+" strand. Now, you should have two columns partially filled with numbers. Next, type =G2 into cell C2. Select this cell and double click on the cross with the

left-hand mouse button. Then, find the first empty cell in column E and note the row number you are on. Into this empty cell, type =F, followed immediately by the row number you are on (e.g., =F22485). Press enter. Select this cell and double click on the cross with the left-hand mouse button. You should now have two columns filled with numbers. Open a new Excel file and name it accordingly (e.g., 100kbUP.txt). Highlight the data in columns B, C, and D, copy, and paste these into your new spreadsheet using the "paste special" function (see Note 4). Delete the header row, by right-clicking on the "1" to the left of the header row and selecting delete. Save as a .txt file and close the file. Repeat this procedure for each upstream interval you wish to examine, but change the addition and subtraction of 100,000 accordingly (e.g., 10,000 and 1,000 for 10 kb and 1 kb upstream, respectively).

5. To calculate genomic coordinates for downstream genic regions, the procedure is very similar, but with a couple of subtle, but important, differences. Label your newly emptied columns, which previously held the genomic coordinates for upstream genic regions with an appropriate name. For this example, we will be calculating interval 100 kb downstream of your transcription start site for all genes, so we will label column C as "100 kb dn Start" and column D as "100 kb dn End." To perform this calculation for all genes on the "–" strand, type =F2–100,000 into cell D2 and press enter. Copy the formula from cell D2. Then, go to the last cell in column D that still has a "–" in the "strand" column. Select this cell and paste the formula. Now, go back to the top of your sheet and select cell D2. The cell should be outlined with a thick black border and a small square should be apparent at the lower right-hand corner of the cell. Position your mouse cursor over this square. A cross should form. When you see the cross form, double click on the cross with the left-hand mouse button. The formula should fill in for all of the genes on the "–" strand. To calculate the interval 100 kb upstream of your transcription start site for all genes on the "+" strand, locate the first cell under column C that has a "–" in the "strand" column. This cell should be empty. Note the row number you are on and then type =G followed immediately by the row number you are on and +100,000 (e.g., =G22485 + 100,000). Press enter. Select this cell and double click on the cross with the left-hand mouse button. The formula should fill in for all of the genes on the "+" strand. Now, you should have two columns partially filled with numbers. Next, type =I2 into cell C2. Select this cell and double click on the cross with the left-hand mouse button. Then, find the first empty cell in column E and note the row number you are on. Into this empty cell, type =H, followed immediately by the row number you are on (e.g., H22485). Press enter. Select this cell

and double click on the cross with the left-hand mouse button. You should now have two columns filled with numbers. Open a new Excel file and name it accordingly (e.g., 100kbDN.txt). Highlight the data in columns B, C, and D, copy, and paste these into your new spreadsheet using the "paste special" function (see Note 4). Delete the header row, by right-clicking on the "1" to the left of the header row and selecting delete. Save as a .txt file and close the file. Repeat this procedure for each downstream interval you wish to examine, but change the addition and subtraction of 100,000 accordingly (e.g., 10,000 and 1,000 for 10 kb and 1 kb downstream, respectively).

6. Download and install Perl to C:\Perl from http://www.perl.org/get.html.
7. Download and install a text editing program (e.g., Notepad++, gedit, Aquamacs, etc.).
8. To download genomic coordinates for exons, go back to the Table browser at the UCSC Genome Browser Web site. Make sure that the correct "clade," "genome," and "assembly" are still selected. Select "Genes and Gene Prediction Tracks" from the "group" menu and select the desired track from the "track" menu, as above in step 2. However, this time, next to "output format," select "BED—Browser Extensible Data" and tick the box next to "Send to Galaxy." Click "get output." On the next screen, under "Create one BED record per," make sure that "Coding Exons" is ticked. Click "send query to Galaxy." Your browser should be redirected to the Galaxy Web site and your data will appear under "History" on the right-hand side of the browser screen. In the left-hand "Tools" menu, click on "Text Manipulation," and then "Cut columns from a table." Now, cut all columns except those containing "Chrom," "Start," and "End." To do this, type the columns you wish to keep into the box next to "Cut columns" (e.g., c1, c2, c3). You can view the existing column order by clicking on the data set in the "History" pane on the right-hand side of the screen. The option under "Delimited by" should set to "Tab." Press execute and wait for your job to finish. When it has finished running, click on the name of the job. This should expand the file window and you should see several icons at the top right of the file window. Click on the pencil icon to "edit attributes." Once the new window has opened, scroll down to the "change data type" heading and select txt from the drop-down menu. Click the save button. This will change the format of the file. Once the file has finished updating, click on the name of the job to expand the file window, if needed, and save the file by clicking on the disk icon. Save the file under an appropriate name ("ExonsAll.txt"). Now, move this file to the same directory where you have installed Perl (see Note 5).

9. Now, open Notepad++ or a similar text editing program, and copy or type the text below into a new file (see Note 6).

```
#!/usr/local/bin/perl -w
# removes duplicates
my $usage = 'duplicates.pl
Removes duplicate entries from a file.
USAGE:
./duplicates.pl input.txt output.txt';
# the next 2 lines tell the program that the user will enter the input and output files to use
my $input = shift @ARGV || die "$usage\n";
my $output = shift @ARGV || die "$usage\n";
open (INPUT, "<$input")
or die "can't open INPUT FILE"; #opens the input file or dies trying
open (OUT, ">$output")
or die "can't open OUTPUT FILE"; #opens the output file or dies trying
print "Running...\n";
my @INPUT = <INPUT>;
chomp @INPUT;
 my %hash = map {$_, 1} @INPUT;
 # enters data into a hash
my @unique = keys %hash;
foreach (@unique) {
print (OUT $_, "\n");
}
print "Done!\n";
close (INPUT); #closes input file
close (OUT); #closes output file
exit 0; #closes program
```

Type or paste the text above into a new file and name this file Duplicates. To save as a Perl file, select "Perl source file (*.pl, *.pm, *.plx)" from the drop-down menu next to "Save as type." Make sure that the file is saved in the same directory as Perl. You will also need to check that the file called ExonsAll.txt, which you created in Subheading 3.1, step 8 above, is saved in this directory as a .txt file. Open Command Prompt, or similar command line program. In Windows, Command Prompt can be found by following Start ->All Programs ->Accessories ->Command Prompt. Change directory from the current directory to the folder where you have installed Perl (see Note 5). To change a directory, type cd followed by the new directory (e.g., cd C:\Perl). Your Command Prompt should now read C:\Perl>, or something very similar. Type perl Duplicates.pl ExonsAll.txt Exons.txt and press enter. This will create a file called "Exons.txt," which contains a list of all the genomic coordinates of all exons, filtered for duplicates.

10. To download genomic coordinates for introns, go back to the Table browser at the UCSC Genome Browser Web site. Make sure that the correct "clade," "genome," and "assembly" are still selected. Select "Genes and Gene Prediction Tracks" from the "group" menu and select the desired track from the "track" menu, as above in step 2. However, this time, next to "output format," select "BED—Browser Extensible Data" and tick the box next to "Send to Galaxy." Click "get output." On the next screen, under "Create one BED record per," make sure that "Introns" is ticked. Click "send query to Galaxy." Your browser should be redirected to the Galaxy Web site and your data will appear under "History" on the right-hand side of the browser screen. In the left-hand "Tools" menu, click on "Text Manipulation," and then "Cut columns from a table." Now, cut all columns except those containing "Chrom," "Start," and "End." To do this, type the columns you wish to keep into the box next to "Cut columns" (e.g., c1, c2, c3). Press execute and wait for your job to finish. When it has finished running, click on the name of the job. This should expand the file window and you should see several icons at the top right of the file window. Click on the pencil icon to "edit attributes." Once the new window has opened, scroll down to the "change data type" heading and select txt from the drop-down menu. Click the save button. This will change the format of the file. Once the file has finished updating, click on the name of the job to expand the file window, if needed, and save the file by clicking on the disk icon. Save the file under an appropriate name ("IntronsAll.txt"). Now, move this file to the same directory where you have installed Perl (see Note 5).

11. Open Command Prompt, or similar command line program. In Windows, Command Prompt can be found by following Start ->All Programs ->Accessories ->Command Prompt. Change directory from the current directory to the folder where you have installed Perl. To change a directory, type cd followed by the new directory (e.g., cd C:\Perl); see Note 5. Your Command Prompt should now read C:\Perl>, or something very similar. Type perl Duplicates.pl IntronsAll.txt Introns.txt and press enter. This will create a file called "Introns.txt," which contains a list of all the genomic coordinates of all introns, filtered for duplicates.

12. To download genomic coordinates for 5′ UTRs, go back to the Table browser at the UCSC Genome Browser Web site. Make sure that the correct "clade," "genome," and "assembly" are still selected. Select "Genes and Gene Prediction Tracks" from the "group" menu and select the desired track from the "track" menu, as above in step 2. However, this time, next to "output format," select "BED—Browser Extensible Data" and tick the box next to "Send to Galaxy." Click "get output." On the next screen, under "Create one BED record per," make sure that "5′ UTRs" is ticked. Click "send query to Galaxy." Your browser should be redirected to the Galaxy Web site and your data will appear under "History" on the right-hand side of the browser screen. In the left-hand "Tools" menu, click on "Text Manipulation," and then "Cut columns from a table." Now, cut all columns except those containing "Chrom," "Start," and "End." To do this, type the columns you wish to keep into the box next to "Cut columns" (e.g., c1, c2, c3). Press execute and wait for your job to finish. When it has finished running, click on the name of the job. This should expand the file window and you should see several icons at the top right of the file window. Click on the pencil icon to "edit attributes." Once the new window has opened, scroll down to the "change data type" heading and select txt from the drop-down menu. Click the save button. This will change the format of the file. Once the file has finished updating, click on the name of the job to expand the file window, if needed, and save the file by clicking on the disk icon. Save the file under an appropriate name ("5UTRAll.txt"). Now, move this file to the same directory where you have installed Perl (see Note 5).
13. Open Command Prompt, or similar command line program. In Windows, Command Prompt can be found by following Start ->All Programs ->Accessories ->Command Prompt. Change directory from the current directory to the folder where you have installed Perl (see Note 5). To change a directory, type cd followed by the new directory (e.g., cd C:\Perl). Your Command Prompt should now read C:\Perl>, or something very similar. Type perl Duplicates.pl 5UTRAll.txt 5UTR.txt and press enter. This will create a file called "5UTR.txt," which contains a list of all the genomic coordinates of all introns, filtered for duplicates.
14. To download genomic coordinates for 3′ UTRs, go back to the Table browser at the UCSC Genome Browser Web site. Make sure that the correct "clade," "genome," and "assembly" are still selected. Select "Genes and Gene Prediction Tracks" from the "group" menu and select the desired track from the "track" menu, as above in step 2. However, this time, next to "output format," select "BED—Browser Extensible Data" and tick the box next to "Send to Galaxy." Click "get

output." On the next screen, under "Create one BED record per," make sure that "3′ UTRs" is ticked. Click "send query to Galaxy." Your browser should be redirected to the Galaxy Web site and your data will appear under "History" on the right-hand side of the browser screen. In the left-hand "Tools" menu, click on "Text Manipulation," and then "Cut columns from a table." Now, cut all columns except those containing "Chrom," "Start," and "End." To do this, type the columns you wish to keep into the box next to "Cut columns" (e.g., c1, c2, c3). Press execute and wait for your job to finish. When it has finished running, click on the name of the job. This should expand the file window and you should see several icons at the top right of the file window. Click on the pencil icon to "edit attributes." Once the new window has opened, scroll down to the "change data type" heading and select txt from the drop-down menu. Click the save button. This will change the format of the file. Once the file has finished updating, click on the name of the job to expand the file window, if needed, and save the file by clicking on the disk icon. Save the file under an appropriate name ("3UTRAll.txt"). Now, move this file to the same directory where you have installed Perl (see Note 5).

15. Open Command Prompt, or similar command line program. In Windows, Command Prompt can be found by following Start ->All Programs ->Accessories ->Command Prompt. Change directory from the current directory to the folder where you have installed Perl (see Note 5). To change a directory, type cd followed by the new directory (e.g., cd C:\Perl). Your Command Prompt should now read C:\Perl>, or something very similar. Type perl Duplicates.pl 3UTRAll.txt 3UTR.txt and press enter. This will create a file called "3UTR.txt," which contains a list of all the genomic coordinates of all introns, filtered for duplicates.

3.2. Data Mining: Extracting Additional Features

In this section, you will identify locations of additional features you wish to correlate with imprinting status. The examples below cover miRNAs, CpG islands, G-quartets, CTCF sites, a variety of epigenetic features, and GC percent. Identifying miRNAs and CpG islands requires the use of UCSC and Galaxy, in a way that is very similar to what was already described for extracting the genic regions. For additional features, other databases are used (insulatorDB, Broad, Quadruplex). You may want to consider other features (for example, transcription factor binding sites) and additional databases exist where those locations can be captured (http://ecrbase.dcode.org/).

1. To download genomic coordinates for CpG islands, go back to the Table browser at the UCSC Genome Browser Web site. Make sure that the correct "clade," "genome," and "assembly" are still selected. Select "Expression and Regulation" from the

"group" menu and select the "CpG Islands" track from the "track" menu. However, this time, next to "output format," select "all fields from selected table" and tick the box next to "Send to Galaxy." Click "get output." On the next screen, click "send query to Galaxy." Your browser should be redirected to the Galaxy Web site and your data will appear under "History" on the right-hand side of the browser screen. In the left-hand "Tools" menu, click on "Text Manipulation," and then "Cut columns from a table." Now, cut all columns except those containing "Chrom," "ChromStart," and "ChromEnd." To do this, type the columns you wish to keep into the box next to "Cut columns" (e.g., c2, c3, c4). Press execute and wait for your job to finish. When it has finished running, click on the name of the job. This should expand the file window and allow you to save the file by clicking on the disk icon. Make sure that you can open your file in Excel. Delete the header row, by right-clicking on the "1" to the left of the header row and selecting delete. Name the file accordingly (e.g., CpG.txt) and close the file.

2. To download genomic coordinates for micro-RNA (miRNA) clusters, go back to the Table browser at the UCSC Genome Browser Web site. Make sure that the correct "clade," "genome," and "assembly" are still selected. Select "Genes and Gene Prediction Tracks" from the "group" menu and select the "miRNA" track from the "track" menu. However, this time, next to "output format," select "all fields from selected table" and tick the box next to "Send to Galaxy." Click "get output." On the next screen, click "send query to Galaxy." Your browser should be redirected to the Galaxy Web site and your data will appear under "History" on the right-hand side of the browser screen. In the left-hand "Tools" menu, click on "Text Manipulation," and then "Cut columns from a table." Now, cut all columns except those containing "Chrom," "ChromStart," and "ChromEnd." To do this, type the columns you wish to keep into the box next to "Cut columns" (e.g., c2, c3, c4). Press execute and wait for your job to finish. When it has finished running, click on the name of the job. This should expand the file window and allow you to save the file by clicking on the disk icon. Open the file with Excel. Delete the header row, by right-clicking on the "1" to the left of the header row and selecting delete. Then, save the file as a .txt file, making sure to give it an appropriate name (e.g., miRNA.txt), and close the file.

3. To obtain genomic coordinates for CTCF sites for human or mouse, direct your Web browser to the Insulator Database Web site at http://insulatordb.uthsc.edu/help.php#download (55). On this page, you will have two choices of download: experimentally verified CTCF binding sites and computationally

predicted CTCF binding sites. Once you have downloaded the file(s), open it first in Word or a similar text editing program and save as a .txt (Plain Text File) file; see Note 2. When prompted in the next Window, leave all options as they are, except choose “CR only” from the drop-down menu next to “End lines with:”. Open the saved .txt file with Excel. Sort the file based on species. To do this, click on the “Data” tab and select “Sort.” Sort by the “Species” column, which should be column B. Make sure to indicate that your data contain headers and click OK. Now, erase any data that do not apply to the species in which you are interested. The easiest way to do this is to search (press control and f at the same time) for your species name. This should take you to the first row containing data applying to your species of interest. Delete all rows above this, except for the header, by clicking on the row number to the left of the first row you wish to delete, scrolling to the top of the page, and holding down shift while clicking on the last row you wish to delete. Then, right-click within the highlighted area and select “delete.” To delete any unwanted species following your species of interest, find the last row containing data you wish to keep, and delete click on the row number to the left of the first row you wish to delete, scroll to the bottom of the page, and hold down shift while clicking on the last row you wish to delete.

4. Then, delete all columns, except the “Chromosome Location” file. To do this, click on the letter above each column you wish to eliminate, and hold down the control key while clicking. Once you have selected all of the columns you wish to delete, right-click and select delete. You should be left with a single column labeled “Chromosome Location.” Click on the A above “Chromosome Location” and select “Text-to-Columns” from the “Data” tab. Make sure that “Delimited” is selected and then click “Next.” In the next window, deselect “Tab” and select “Other.” In the box next to “Other,” type a colon (:) and click “Finish.” Part of the data formerly contained in column A will have moved to column B. Click on the B above your new column and select “Text-to-Columns” from the “Data” tab. Make sure that “Delimited” is selected and then click “Next.” In the next window, deselect “Tab” and select “Other.” In the box next to “Other,” type a dash (-) and click “Finish.” You should now have your chromosomal locations split between three columns. Delete the header row, by right-clicking on the “1” to the left of the header row and selecting delete. Save your file under an appropriate name (e.g., CTCFcompOLD.txt or CTCFexpOLD.txt) and close the file.

5. The data from the Insulator Database are mapped to the hg18 assembly for human, the mm8 assembly for mouse, the rn3 assembly for rat, and the galGal2 assembly for chicken. If this

is not the option you have been selecting next to the "assembly" option on the UCSC Genome Browser, you will need to convert the genomic coordinates from February 2006, mm8 to the assembly you are working with. To convert CTCF site genomic coordinates from one genome assembly to another, direct your Web browser to the Galaxy Web site at http://main.g2.bx.psu.edu/ (56, 57). Upload your file, making sure to choose the version without column names, by clicking on "Get Data" and selecting "Upload File" from the left-hand menu. Click "Browse" and select the file you wish to upload. At the top of the window will be a drop-down menu under "File Format." Select "bed" from this menu. Above the "Execute" button, there should be a drop-down menu which allows you to select the genome. Make sure to select both the correct species and genome assembly (hg18 assembly for human, the mm8 assembly for mouse, the rn3 assembly for rat and the galGal2 assembly for chicken) from this menu. Then, click "Execute." When your file has finished loading to Galaxy, click on "Lift Over" and select "Convert genome coordinates" from the menu on the left. Select your newly uploaded file from the drop-down menu under "Convert coordinates of" and select the genome assembly you wish to convert to from the drop-down menu under "To:". Click execute. Two new files will appear in the menu to the right. Once the jobs have finished running, click on the header of the file with "MAPPED COORDINATES" in the name. Click on the disk icon to save the file. Locate the file and open it with Excel. Then, save the file as a .txt file, making sure to give it an appropriate name (e.g., CTCFcomp.txt or CTCFexp.txt).

6. To obtain genomic coordinates for histone modification data, direct your Web browser to ftp://ftp.broad.mit.edu/pub/papers/chipseq/ at the Broad Institute. The data are arranged by first author and publication date of each of the papers in which the data are published. Select the data set you are most interested in by double-clicking on the relevant folder. In each folder is a file called "Readme.txt." If you click on this file, you will be able to read a key indicating the type of data found in each subfolder. The subfolder of most interest is likely to be the "Alignments" folder, which contains both the sequences and the coordinates of uniquely aligned ChIP-Seq reads. However, some data sets contain files called "WindowIntervals.tar.gz" and "HMMIntervals.tar.gz," which may be of interest as well. These, respectively, contain intervals enriched for certain histone modifications inferred by fixed-size windows and intervals Enriched for certain histone modifications inferred by a Hidden Markov Model (HMM). Within the "Alignments" subfolder, you are able to choose those histone ChIP-Seq data sets which most interest you. All of the downloadable files are

in tar.gz format, and will require an unzipping program capable of handling this type of file. Also, keep in mind that the genomic coordinates for the histone-modification data might be from a different assembly than the genomic coordinates you are using. The ReadMe file will tell you which assembly was used, and you can use Galaxy, as in step 14 above, to convert coordinates between different assemblies; see Note 7.

7. To obtain genomic coordinates for predicted G-quartet sites in your genome of interest, direct your Web browser to http://www.quadruplex.org/?view=quadbaseDownload and click on the link for your organism and genome assembly of interest (58). If your particular genome assembly is not available, it is possible to convert coordinates between genome assemblies using Galaxy, so choose a different assembly, and make a note of which version you have selected. Click on the "All Files" link to download all predicted G-quartet sites in your genome of interest. Once the file has downloaded, rename it appropriately (e.g., GQraw) and upload the file to Galaxy by clicking on "Get Data" and selecting "Upload File" from the left-hand menu. Click "Browse" and select the file you wish to upload. At the top of the window will be a drop-down menu under "File Format." Select "bed" from this menu. Above the "Execute" button, there should be a drop-down menu which allows you to select the genome. Make sure to select both the correct species and genome assembly. Now, cut all columns except the first three, which contain information regarding genomic location. In the left-hand "Tools" menu, click on "Text Manipulation," and then "Cut columns from a table." Type the columns you wish to keep into the box next to "Cut columns" (e.g., c1, c2, c3). Press execute and wait for your job to finish. When it has finished running, click on the name of the job. This should expand the file window and allow you to save the file by clicking on the disk icon. Locate the file and open it with Excel. Delete the header row, by right-clicking on the "1" to the left of the header row and selecting delete. Then, save the file as a .txt file, making sure to give it an appropriate name (e.g., GQs.txt), and close the file. Genomic coordinates for predicted G-quartet sites in your genome of interest can also be obtained in the method described in Note 8.

8. To calculate the GC% of each gene body, introns, exons, 5′ UTRs, 3′ UTRs, and any upstream and downstream regions, direct your Web browser to the Galaxy Web site at http://main.g2.bx.psu.edu/. Upload your file containing the filtered genomic coordinates for each of your regions of interest, making sure to choose the version without column names, by clicking on "Get Data" and selecting "Upload File" from the left-hand menu. Click "Browse" and select the file you wish to upload. At the top of the window will be a drop-down menu

under “File Format.” Select “bed” from this menu. Above the “Execute” button, there should be a drop-down menu which allows you to select the genome. Make sure to select both the correct species and genome assembly. Once your file has loaded to Galaxy, select “Fetch Sequences” and “Extract Genomic DNA” from the “Tools” menu on the left-hand side of the page. Make sure that the file you just uploaded is selected in the drop-down menu under “Fetch sequences corresponding to Query:” and change “Output” data type to “Interval.” Click “Execute.” When it has finished running, click on the name of the job. This should expand the file window and allow you to save the file by clicking on the disk icon. Locate the file and open it with Excel. Delete the header row, by right-clicking on the “1” to the left of the header row and selecting delete. Then, save the files as a .txt file, making sure to give it an appropriate name (e.g., CGintron.txt), and close the file. Do this for each of the genomic regions you are interested in.

9. Now, open Notepad++ or a similar text editing program, and copy or type the text below into a new file (see Note 6).

```
#!/usr/local/bin/perl -w
# calculate GC%
my $usage = ‘gc-count.pl
Compute GC content in a set of sequences.
USAGE:
./gc-count.pl input.txt output.txt
‘;
# the next 2 lines tell the program that the user will enter the input and output files to use
my $input = shift @ARGV || die “$usage\n”;
my $output = shift @ARGV || die “$usage\n”;
open (INPUT, “<$input”)
or die “can’t open INPUT FILE”; #opens the input file or dies trying
open (GCOUT, “ > $output”)
or die “can’t open OUTPUT FILE”; #opens the output file or dies trying
print “Running…\n”;
while (<INPUT>) {#tells the program what to do while the input file is open
 chomp; #removes any new line symbols from the end of each line
(@INPUT) = split/\t/; #splits each line on tabs
$SEQ = $INPUT[3]; #defines the variable $SEQ
```

```
my @seqarray = split '', $SEQ; #splits the variable $SEQ at each
    character
my $GC = 0; # counter for G's and C's
foreach my $char (@seqarray) {#tells the program what to do
    with each character
if ($char =~ m/[GgCc]/) {#regular expression to search for
    matches to any of G g C or c
$GC++; # increment the GC counter
}
}
$percentGC = (($GC/(length($SEQ)-1))*100);   #calculates
    GC%
print (GCOUT $GC, "\t", length($SEQ)-1, "\t", $percentGC,
    "\n"); #prints GC%
}
print "Done!\n";
close (INPUT); #closes input file
close (GCOUT); #closes outout file
exit 0; #closes program
```

Type or paste the text above into a new file and name this file gc-count. To save as a Perl file, select "Perl source file (*.pl, *.pm, *.plx)" from the drop-down menu next to "Save as type." Make sure that the file is saved in the same directory as Perl and click "Save." Open Command Prompt, or similar command line program. Command Prompt can be found by following Start ->All Programs ->Accessories ->Command Prompt. Change directory from the current directory to the folder where you have installed Perl (see Note 5). To change a directory, type cd followed by the new directory (e.g., cd C:\Perl). Your Command Prompt should now read C:\Perl>, or something very similar. Type perl gc-count.pl followed by your input and output file names (e.g., perl gc-count.pl CGintron.bed CGintronOUT.txt); see Note 9. Repeat this process for every region you are interested in (gene body, introns, exons, 5′ UTRs, 3′ UTRs, and any upstream and downstream regions). Make sure to change the name of the input and output file names each time.

3.3. Data Mining: Identifying Known Imprinted Genes for Model Training

1. To obtain genomic coordinates of all known imprinted genes, direct your Web browser to the Otago Catalogue of Imprinted Genes: http://igc.otago.ac.nz/home.html. Click on the "Summary Tables" link. Select your organism of interest from the drop-down menu next to "Taxon" and select "Imprinted Genes" from the drop-down menu next to "Category." Click "Search." Copy and paste the resulting list into Excel and save

under an appropriate name (e.g., Allimprinted.xls). Remove the extra data from the "Gene" column by clicking on the C above the column. Next, select "Text-to-Columns" from the "data" tab. Make sure that "delimited" is selected and click "Next." In the new window, tick both "comma" and "space" and click "Finish." Copy the gene names from the "Gene" column. Direct your Web browser of choice to the Table browser at the UCSC Genome Browser Web site. Make sure to select the relevant "clade," "genome," and "assembly." Then, click on the "paste list" button next to "identifiers names/ associations:". A new window will open. Paste the names of the genes you have copied from your Excel sheet into the box and press the "submit" button. Once you have done this, choose the "selected fields from primary and related tables" option from the drop-down list next to "output format." Enter the file name to which you will save your downloaded genomic coordinates in the box next to "output file" (e.g., KnownImprinted.txt). Then, press the "get output" button. In the next window, tick the boxes next to "chrom," "txStart," and "txEnd." Once the file has downloaded, check that you can open it using Excel, or a similar spreadsheet management program. Next, filter this file to remove all duplicate entries. Under the data tab, select "Remove Duplicates." In the pop-up menu, unselect all boxes. Click OK. Delete the header row, by right-clicking on the "1" to the left of the header row and selecting delete. Save and close the file.

2. To obtain a list of whether each gene in your filtered list of all known genes (e.g., Genes.txt) is imprinted or not, install Perl from http://www.perl.org/get.html by following the instructions on the Webpage. Open Notepad++ or a similar text editing program, and copy the text below into a new file:

```
#!/usr/local/bin/perl -w
# Determines whether each known gene is imprinted.
my $usage = 'Imprinted.pl
Determines whether each known gene in your filtered list of all known genes is imprinted.
USAGE:
./Imprinted.pl within.txt find.txt output.txt
';
# the next 3 lines tell the program that the user will enter the
    input and output files to use
my $within = shift @ARGV || die "$usage\n";
my $find = shift @ARGV || die "$usage\n";
my $output = shift @ARGV || die "$usage\n";
open (WITHIN, "<$within")
```

```
or die "can't open file to search within"; #opens the first input
    file or dies trying
open (OUT, "> $output")
or die "can't open OUTPUT file"; #opens the output file or dies
    trying
print "Running...\n";
while (<WITHIN>) {#tells the program what to do while the
    input file is open
chomp; #removes any new line symbols from the end of each line
(@WITHIN) = split/\t/; # splits each line on tabs
# the next 3 lines define variables
$chrI = $WITHIN [0];
$startI = $WITHIN [1];
$endI = $WITHIN [2];
my $NUM = 0; #sets the variable $NUM equal to 0
open (FIND, "<$find")
or die "can't open file to search with"; #opens the second input
file or dies trying
foreach $chrI (@WITHIN){#loops through 1st file
while (<FIND>) {#loops through 2nd file
chomp; #removes any new line symbols from the end of each line
(@FIND) = split/\t/; # splits each line on tabs
 # the next 3 lines define variables
$chrS = $FIND [0];
$startS = $FIND [1];
$endS = $FIND [2];
if ((($chrI = $chrS) && ($startI<= $startS) && ($endI >=
$endS))) {#finds matches
 $NUM++; #increases count if match is found
}
}
}
#the lines below tell the program to print a 0 if no matches were
found and a 1 if any #match was found.
if ($NUM > 0){
print (OUT $NUM/$NUM, "\n");
}else{
print (OUT 0, "\n");
}
}
```

```
print "Done!\n";
close (WITHIN); #closes first input file
close (FIND); #closes second input file
close (OUT); #closes output file
exit 0; #closes program
```

Type or paste the text above into a new file and name this file Imprinted. To save as a Perl file, select "Perl source file (*.pl, *.pm, *.plx)" from the drop-down menu next to "Save as type." Make sure that the file is saved in the same directory as Perl and click "Save." Open Command Prompt, or similar command line program. In Windows, Command Prompt can be found by following Start ->All Programs ->Accessories ->Command Prompt. Change directory from the current directory to the folder where you have installed Perl (see Note 5). To change a directory, type cd followed by the new directory (e.g., cd C:\Perl). Your Command Prompt should now read C:\Perl>, or something very similar. Type perl Imprinted.pl followed by your input and output file names (e.g., perl Imprinted.pl Genes.txt KnownImprinted.txt Imprinted.txt, see Note 10). Now you will have a list containing a column of 1's and 0's. A value of 1 means that the gene is imprinted and a value of 0 means that the gene is not imprinted. This file will not make much sense on its own, but it will once we combine it with another file, which we will do in Subheading 3.4, step 3.

3.4. Data Manipulation: Counting Occurrences of Features of Interest

Now that you have collected data on each of your features of interest, you need to count the number of times each of those features occurs within each of your genomic regions of interest, for every gene in the genome. To do this, you will use some basic Perl scripts.

1. To tabulate the number of times each of the features you have collected occurs within each of these regions of interest: Introns, Exons, 5′ UTRs, or 3′ UTRs, open Notepad++, or a similar text editing program, and copy the text below into a new file (see Note 6):

```
#!/usr/local/bin/perl -w
# tabulate number of occurrences of one file within another file
my $usage = 'countIN.pl
```

Tabulates the number of occurrences of one file within another file. Meant to be a 2 round counting program.

E.g. the first round fiunds the exons which contain CpG islands. The 2nd round will find the exons within each known gene in the genome, effectively counting the number of CpG islands within exons, but classified by gene.

```
USAGE:
./countIN.pl within.txt find.txt output.txt
';
# the next 3 lines tell the program that the user will enter the input and output files to use
my $within = shift @ARGV || die "$usage\n";
my $find = shift @ARGV || die "$usage\n";
my $output = shift @ARGV || die "$usage\n";
open (WITHIN, "<$within")
or die "can't open file to search within"; #opens the first input file or dies trying
open (OUT, "> $output")
or die "can't open OUTPUT file"; #opens the output file or dies trying
print "Running...\n";
while (<WITHIN>) {#tells the program what to do while the first file is open
chomp; #removes any new line symbols from the end of each line
(@WITHIN) = split/\t/; #splits each line on tabs
# the next 3 lines define variables
$chrI = $WITHIN [0];
$startI = $WITHIN [1];
$endI = $WITHIN [2];
my $NUM = 0;
open (FIND, "<$find")
or die "can't open file to search with"; #opens the second input file or dies trying
foreach $chrI (@WITHIN){#loops through 1st file
while (<FIND>) {#loops through 2nd file
chomp; #removes any new line symbols from the end of each line
(@FIND) = split/\t/; #splits each line on tabs
# the next 3 lines define variables
$chrS = $FIND [0];
$startS = $FIND [1];
$endS = $FIND [2];
if ((($chrI = $chrS) && ($startI<= $startS) && ($endI >= $endS))) {#finds matches
print (OUT $chrS, "\t", $startS, "\t", $endS, "\n"); #prints matches if found
```

```
}
}
}
}
print "Done!\n";
close (WITHIN); #closes first input file
close (FIND); #closes second input file
close (OUT); #closes output file
exit 0; #closes program
```

Type or paste the text above into a new file and name this file countIN. To save as a Perl file, select "Perl source file (*.pl, *.pm, *.plx)" from the drop-down menu next to "Save as type." Make sure that the file is saved in the same directory as Perl and click "Save." You will use a different process for the remaining regions; make sure to use this program only for introns, exons, 5′ UTRs, and 3′ UTRs. The aim of this program is to identify which introns, exons, 5′ UTRs, or 3′ UTRs contain any of your marks of interest. The second program, which we will run below, will identify the gene to which each of the introns, exons, 5′ UTRs, and 3′ UTRs belongs. To run the countIN program, open Command Prompt, or similar command line program. In Windows, Command Prompt can be found by following Start ->All Programs ->Accessories ->Command Prompt. Change directory from the current directory to the folder where you have installed Perl (see Note 5). To change a directory, type cd followed by the new directory (e.g., cd C:\Perl). Your Command Prompt should now read C:\Perl>, or something very similar. Type perl count.pl followed by your first input file containing one of the following genomic intervals, introns, exons, 5′ UTRs, or 3′ UTRs, then your second input file containing the marks of interest you have collected (e.g., your list of CpG Islands, miRNA clusters, CTCF binding sites, etc.), and, finally, the output file (e.g., perl countIN.pl Exons.txt miRNA.txt miRNAexonsOUT_1.txt). For very large data sets, like ChIP-Seq data sets, it may not be possible to perform this analysis on a laptop or a desktop computer. In this case, it may be useful to have access to a multi-CPU cluster. See Notes 11 and 12. Repeat this process for each of the following regions you are interested in: introns, exons, 5′ UTRs, or 3′ UTRs. Make sure to change the name of both the input file as well as the output file, each time.

2. Once you have run the above program on introns, exons, 5′ UTRs, and 3′ UTRs, you will want to run the program below which will both tabulate the number of times each of the features you have collected occurs within each of these regions of

interest: gene body, introns, exons, 5′ UTRs, 3′ UTRs, and any upstream and downstream regions. And, importantly, this program will also determine in which gene each of the counts occurs. To run this program, open Notepad++ or a similar text editing program, and copy the text below into a new file (see Note 6):

```
#!/usr/local/bin/perl -w
# tabulate number of occurrences of one file within another file
my $usage = ‘count.pl
Tabulates the number of occurrences of one file within another file. E.g. number of CpG Islands found within each known gene in the genome.
USAGE:
./count.pl within.txt find.txt output.txt
‘;
# the next 3 lines tell the program that the user will enter the input and output files to use
my $within = shift @ARGV || die “$usage\n”;
my $find = shift @ARGV || die “$usage\n”;
my $output = shift @ARGV || die “$usage\n”;
open (WITHIN, “<$within”)
or die “can’t open file to search within”; #opens the first input file or dies trying
open (OUT, “> $output”)
or die “can’t open OUTPUT file”; #opens the output file or dies trying
print “Running…\n”;
while (<WITHIN>) {#tells the program what to do while the first file is open
chomp; #removes any new line symbols from the end of each line
(@WITHIN) = split/\t/; #splits each line on tabs
# the next 3 lines define variables
$chrI = $WITHIN [0];
$startI = $WITHIN [1];
$endI = $WITHIN [2];
my $NUM = 0;
open (FIND, “<$find”)
or die “can’t open file to search with”; #opens the second input file or dies trying
foreach $chrI (@WITHIN){#loops through 1st file
```

```
while (<FIND>) {#loops through 2nd file
chomp; #removes any new line symbols from the end of each line
(@FIND) = split/\t/; #splits each line on tabs
# the next 3 lines define variables
$chrS = $FIND [0];
$startS = $FIND [1];
$endS = $FIND [2];
if ((($chrI = $chrS) && ($startI<= $startS) && ($endI >=
$endS))) {#finds matches
$NUM++; #increases count if match is found
}
}
}
print (OUT $NUM, "\n");
}
print "Done!\n";
close (WITHIN); #closes first input file
close (FIND); #closes second input file
close (OUT); #closes output file
exit 0; #closes program
```

Type or paste the text above into a new file and name this file count. To save as a Perl file, select "Perl source file (*.pl, *.pm, *.plx)" from the drop-down menu next to "Save as type." Make sure that the file is saved in the same directory as Perl and click "Save." Open Command Prompt, or similar command line program. In Windows, Command Prompt can be found by following Start ->All Programs ->Accessories ->Command Prompt. Change directory from the current directory to the folder where you have installed Perl (see Note 5). To change a directory, type cd followed by the new directory (e.g., cd C:\Perl). Your Command Prompt should now read C:\Perl>, or something very similar. Type perl count.pl followed by your first input file containing the genomic intervals you wish to examine (e.g., your known gene locations or upstream/downstream genomic intervals), your second input file containing the marks of interest you have collected (e.g., your list of CpG Islands, miRNA clusters, CTCF binding sites, etc.), and output file names (e.g., perl count.pl Genes.txt miRNA.txt miRNAgenesOUT.txt). Important: For introns, exons, 3′ UTRs, and 5′ UTRs, you will follow a slightly different format. First of all, the first input file will remain as Genes.txt independent of whether you are analyzing introns, exons, 5′ UTRs, or 3′ UTRs. Second, the second input file will be different than the example given above. In these cases only, you will want to use the output file from Subheading 3.4, step 1, above (e.g., miRNAexonsOUT_1.txt). Therefore, the full command for these regions will look something like perl count.pl Genes.txt miRNAexonsOUT_1.txt miRNAgenesOUT.txt. See Notes 12 and 13. Repeat this process for every region you are interested in (gene body, introns, exons, 5′ UTRs, 3′ UTRs, and any upstream and downstream regions). Make sure to check that you have entered the correct name for both the input files as well as the output file, each time.

3. Next, we need to get all of the data you have collected into a format that we can use for model training. To do this, open your file containing your filtered list of all known genes (e.g., Genes.txt), as well as the file containing the information regarding whether each known gene is imprinted or not (e.g., Imprinted.txt, from Subheading 3.3 above). The file containing the information regarding whether each known gene is imprinted or not should contain a single column of 1's and 0's. The first empty column in your file containing your filtered list of all known genes should be column D. Copy the column of numbers you have in the file containing the information regarding whether each known gene is imprinted (e.g., Imprinted.txt) and paste it into column D of your file containing the genomic coordinates of all know genes (e.g., Genes.txt). Now, open the file containing data on one of your features of interest within that genomic region (e.g., miRNAgenesOUT.txt). The only thing present in this file should be a column of numbers. The next empty column in your file containing the filtered list of all known genes should be column E. Copy the column of numbers into column E of the file containing your filtered list of all known genes. Repeat this copy and paste process for each of the features of interest you have collected data for. Once you have done this, save your file with an appropriate name, and as a .csv file (e.g., Genes.csv, see Note 14). Then, repeat this entire process for each of the genomic intervals you are interested in examining, making sure to add your features of interest to the columns in the *same order* as you did in this first file. For example, if in your first file, you copied GC% to column E, miRNA clusters to column F, and predicted CTCF binding sites to column G, make sure you copy GC% to column E, miRNA clusters to column F, and predicted CTCF binding sites to column G in all of the other files you create for introns, exons, 5′ UTRs, 3′ UTRs, and upstream and downstream genomic regions. Please note that you will reuse the file containing the information regarding whether each known gene is imprinted (e.g., Imprinted.txt). Therefore, each of the new files you create will contain the exact same information in column D. Save each file with an appropriate name as a .csv file.

3.5. Data Manipulation: Creating Data Sets for Model Training and Model Testing

In this section, you will need to create two separate data sets. One data set will be used to train your models. This data set will contain a mix of known imprinted genes and non-imprinted control genes. The program will be told which genes are imprinted and which are not. This will allow the computer to identify those features which distinguish imprinted genes from non-imprinted genes. The second data set, which you will use for model testing will also contain a mixture of imprinted and non-imprinted genes. The difference here is that the computer will not be told which genes are imprinted and which are not, although you will have a record of this information. You will be able to use the results of this analysis to determine the specificity and sensitivity of your trained models. This is a very important step, as it will allow you to evaluate the performance of your models before running your programs on the genome-wide data. Again, you will use some basic Perl scripts to manipulate your data.

1. To obtain genomic coordinates of a random mix of known imprinted genes and non-imprinted control genes to use in the future as training and test data sets, paste or type the following program into Notepad++ (see Note 6):

   #!/usr/bin/perl -w

   open (OUT, ">Random Numbers.txt") or die "can't open output file"; *#opens output file or dies trying*

   $i=0; *#sets variable $i equal to 0*

   while ($i<100,000) {*#tells the program what to do while $i is less than 100,000*

   my $numbers=100000; *#sets the variable $numbers to 100000*

   my $random_number=int(rand($numbers)); *#tells the program to generate a random number*

   print (OUT $random_number, "\n"); *#prints the random number*

   $i++; *#increases the value of $i by 1*

   }

   close (OUT); *#closes the program*

   exit 0; *#exits the program*

   Save the file as Numbers. To save as a Perl file, select "Perl source file (*.pl, *.pm, *.plx)" from the drop-down menu next to "Save as type." Make sure that the file is saved in the same directory as Perl. Open Command Prompt, or similar command line program. In Windows, Command Prompt can be found by following Start ->All Programs ->Accessories ->Command Prompt. Change directory from the current directory to the folder where you have installed Perl (see Note 5). To change a directory, type cd followed by the new directory (e.g., cd C:\Perl). Your Command Prompt should now read C:\Perl>, or something very similar. Type perl Numbers.pl and press enter. This will create a file called "Random Numbers.txt."

2. Open a file containing data for one of your genomic regions of interest (e.g., Genes.csv). *Do not save any of the following changes to this file.* It is important to keep all rows in the exact same order between all of your files, because eventually all of the data will be put together into one file. To avoid saving any changes accidentally, open the file and immediately save it under a different name (e.g., GenesCOPY.csv). Sort your spreadsheet in descending order, based on the column that contains information on whether each gene is imprinted. This should be column D. Once you have done this, those genes with a 1 will be on top. Copy the genes with a 1 to a new file. Open the file named "Random Numbers.txt." You should see a single column of numbers. Paste the column of numbers from "Random Numbers.txt" into the first empty column of your new file. Delete any numbers that extend past the rows containing your data. To do this, click on the first cell which contains a random number beyond your rows of data. While holding down the shift key, double click on the bottom boundary of the cell. You should now have highlighted to the bottom of the column of random numbers. Right-click and select delete. Now, sort your spreadsheet in ascending order. To do this, select "sort" from the "data" menu. Copy approximately three-quarters of the rows from this sheet to a new Excel spreadsheet. This will become part of the data set you will use to train your computational prediction models. It is important to reserve a subset of genes that were not used for training so that you can test how well your trained models perform before running them on your genome-wide data. The three-quarter to one-quarter split is somewhat arbitrary, and can therefore be changed to suit your needs, but it is important to strike a balance between having more genes in the training set (which gives better opportunities for training) and having an independent set of genes for testing (which is essential). Now, delete the column containing the random numbers and save your new sheet with an appropriate name (e.g., Genes Training.csv). Copy the remaining one-quarter of your known imprinted genes to a new Excel spreadsheet. This will become part of the data set you will use to test the performance of your computational prediction models. Then, delete the column containing the random numbers and save your new sheet with an appropriate name (e.g., Genes Test.csv). Now, go back to your GenesCOPY.csv file and copy the genes with a 0 to a new file. Again, paste the column of numbers from "Random Numbers.txt" into the first empty column of your new file. Delete any numbers that extend past the rows containing your data. To do

this, click on the first cell which contains a random number beyond you rows of data. While holding down the shift key, double click on the bottom boundary of the cell. You should now have highlighted to the bottom of the column of random numbers. Right-click and select delete. Sort your spreadsheet in ascending order. To do this, select "sort" from the "data" menu. Copy the first three-quarters of the rows from this sheet, add it to the bottom of the Excel spreadsheet containing your list of training genes (e.g., Genes Training.csv), delete the column containing the random numbers, and save the file. Copy the remaining one-quarter of your known imprinted genes, add it to the bottom of the Excel spreadsheet containing your list of test genes (e.g., Genes Test.csv), delete the column containing the random numbers, and save the file. If your original file is still open (e.g., Genes.csv), close your original file containing data for one of your genomic regions of interest genes *without saving any changes.* Repeat this process for each genomic regions you are interested in (e.g., gene body, introns, exons, 5′ UTRs, 3′ UTRs, and any upstream or downstream genomic intervals). Now, you have collected all of the epigenomic and sequence data you need to start model training.

3.6. Data Analysis: Calculation of Correlation Coefficients

This step uses R to determine which of your features of interest are correlated with known imprinted genes.

1. Download and install [R] from http://www.r-project.org/ (59).
2. Often, you will want to know how well each feature you have collected data on correlates with Imprinted status in each of the genomic intervals. The examples discussed below deal only with the most straightforward correlation calculations. For instance, the examples given below refer to 11 classes of genomic regions (the entire gene body, introns, exons, 5′ UTR, 3′ UTR, and 1, 10, and 100 kb upstream and downstream of each gene). One could also calculate correlation with imprinted status based on combinations of these regions. Correlation could also be calculated for combinations of epigenetic features, or combinations of both genomic regions and epigenetic features, with imprinted status. To determine the degree of correlation between a given feature in a given genomic region and imprinting, you can calculate both a correlation coefficient and a corresponding p-value using R. First, open R. Go to File, and select New Script. Then, paste or type the text below into the script editing box that opens up (the example shown is for the gene body, see Note 6):

   setwd("c:/Perl") # *where I want my working directory. Tells the #program where to look #for files*

   dataG<- read.csv("Genes.csv", header=TRUE)[1:xxx,] #*tells the #program which file to load* t((cor(dataG$Imprinted, dataG[,4:yyy]))) *#calculates correlation #coefficients and prints #them in an easy to use format*

Before running this script, you will need to make a few changes. You may need to change the first line, where you set your working directory. To do this, see Note 15. At the end of the second line, where it reads [1:xxx,] the xxx needs to be changed to a number. This number will depend on how many rows are there in your input file (e.g., Genes.csv). Once you have determined how many rows are there in this file, simply substitute this number for xxx. Likewise, at the end of the script, where it reads [,4:yyy], the yyy needs to be changed to a number. This number depends on how many features of interest you are working with. The simplest way to find this number is to type

setwd("c:/Perl") # *where I want my working directory. Tells the #program where to look for files*

dataG<- read.csv("Genes.csv", header=TRUE)[1:xxx,] *#tells the #program which file to load*

length(dataG)

into your R script editing box, and then run the script. To run a script in Windows, highlight all of the text in your script and press both control and the r key at the same time. A number should appear on the screen, and this will tell you how many columns are there in your file. You can also open the file in Excel and manually count the number of columns. Replace yyy with this number. Then, run the script. To run a script in Windows, highlight all of the text in your script and press both control and the r key at the same time. To determine how to run in a script using other operating systems, right click within the script editing box. When the script has finished running, you should see two columns, one with the names of all of your features of interest, and the next with numbers, which are your correlation coefficients. Copy the columns and open a new Excel file. Label the first column "Features" and the second column "Genes," or whatever genomic interval you are examining. Click on the cell below "Features" and then select "Paste Special" by right-clicking. In the box that pops up, select "Text" and click OK. When you repeat this process, you can delete the additional "Features" columns leaving only the correlation coefficients.

3. To calculate p-values for your correlation coefficients, paste or type the text below into your script editing box in R (the example shown is a continuation of the example in step 1 and deals with the gene body, see Note 6). The script below is taken from http://tolstoy.newcastle.edu.au/R/help/01c/2272.html.

cors<-cor(dataG[,4:yyy]) *#saves columns 4-yyy in a variable #called cors*

```
#the lines below calculate the correlation coefficient p-values
cor.prob<- function(X, dfr=nrow(X) - 2) {
R<- cor(X)
above<-row(R)<col(R)
r2<- R[above]^2
Fstat<-r2 * dfr/(1 - r2)
R[above]<-1 - pf(Fstat, 1, dfr)
R}
probs<-cor.prob(dataG[,4:33]) #stores correlation coefficient
    p-#values for columns 4-yyy in a variable named probs
as.matrix(probs[1,]) #prints the correlation coefficient p-values
    #in an easily useable format
```

Do not forget to change xxx to the number of lines contained in your training file. And, do not forget to change yyy to the number of columns contained in your file. Run the script. When the script has finished running, you should see two columns, one with the names of all of your features of interest, and the next with numbers, which are your correlation coefficients. Copy the columns and open a new Excel file. To do this, label the first column "Features" and the second column "Genes," or whatever genomic interval you are examining. Click on the cell below "Features," and then select "Paste Special" by right-clicking. In the box that pops up, select "Text" and click OK. When you repeat this process, you can delete the additional "Features" columns, leaving only the p-values of the correlation coefficients.

4. Repeat steps 1 and 2 for all of your genomic regions of interest (e.g., gene body, introns, exons, 5′ UTRs, 3′ UTRs, and any upstream or downstream genomic intervals). For each run, you will need to make a few changes. First, do not forget to change xxx to the number of lines contained in your input file. Second, do not forget to change yyy to the number of columns contained in your file. Also, in the script for step 1, you will need to change the input file (e.g., Genes.csv to 100kbUP.csv), as well as the variable called "dataG." If you wanted to calculate correlation coefficients for the region 100 kb upstream of all known genes, this might be changed to data100u. For example (the example shown is for the region 100 kb upstream of each gene, see Note 6):

```
setwd("c:/Perl") #tells the program where to look for files
data100u<- read.csv("100kbUP.csv", header=TRUE)
    [1:xxx,] #tells the program which file to load
t((cor(data100u$Imprinted, data100u[,4:yyy]))) #calculates
    #correlation coefficients and prints them in an easy to use
    format
```

In the script for step 2, you will have to make some similar changes. Again, you will need to change the variable called "dataG." Do not forget to change xxx to the number of lines contained in your training file. And, do not forget to change yyy to the number of columns contained in your file. Finally, you may need to change the first line where you set your working directory. To do this, see Note 15. Using the same example, if you wanted to calculate p-values for the correlation coefficients for the region 100 kb upstream of all known genes, this might be changed to data100u. For example:

```
cors<-cor(data100u[,4:yyy]) #saves columns 4-yyy in a variable
    #called cors
#the lines below calculate the correlation coefficient p-values
cor.prob<- function(X, dfr=nrow(X) - 2) {
R<- cor(X)
above<-row(R)<col(R)
r2<- R[above]^2
Fstat<-r2 * dfr/(1 - r2)
R[above]<-1 - pf(Fstat, 1, dfr)
R}
probs<-cor.prob(data100u[,4:yyy]) #stores correlation coefficient
    #p-values for columns 4-#yyy in a variable named probs
as.matrix(probs[1,]) #prints the correlation coefficient p-values
    #in an easily useable format
```

3.7. Model Training

In this step, you will use R to train computational prediction models for each of your genomic regions of interest. At this point, the computer is told which genes within the data set are imprinted and which genes are not. This allows the computer to pick out features that are important for distinguishing imprinted from non-imprinted genes. Again, only the most straightforward examples of model training will be discussed below. For instance, the following examples train models using 11 classes of genomic regions (the entire gene body, introns, exons, 5′ UTR, 3′ UTR, and 1, 10, and 100 kb upstream and downstream of each gene). One could also build models using combinations of these genomic regions. Furthermore, models could be built using combinations of epigenetic features, or combinations of both genomic regions and epigenetic features.

1. The next step is to train the model using the training set files you created in Subheading 3.5 above. To do this, paste or type the text below into your R script editing box (the example shown is for the region 100 kb upstream of each gene, see Note 6):

```
setwd("c:/Perl") # where I want my working directory
```

```
Data100u<- read.csv("Genes Training.csv", header=TRUE)
    [1:xxx,],] #tells the program which file to load

Variables100u<- data100[,5:yyy] #stores data contained in
#columns 4-yyy in a variable named Variables100u

#the following lines specify your models

Base100u<- glm(data100$Imprinted~1, family=binomial)

Full100u<- glm(data100u$Imprinted~variables100u[,1]+
variables100u[,2] + variables100u[,3] + variables
100u[,4]+variables100u[,5]+variables100u[,6]+variables
100u[,7]+variables100u[,8]+variables100u[,9]+variables
100u[,10]+variables100u[,11]+variables100u[,12]+variables
100u[,13]+variables100u[,14]+variables100u[,15]+variables
100u[,16]+variables100u[,17]+variables100u[,18]+variables
100u[,19]+variables100u[,20]+variables100u[,21]+variables
100u[,22]+variables100u[,23]+variables100u[,24]+variables
100u[,25]+variables100u[,26]+variables100u[,27]+variables
100u[,28]+variables100u[,29], family=binomial)

#removes unnecessary variables from model

Res100u<-step(full100u,scope=list(upper=100u,lower=~1),
direction="both", trace=FALSE)

#prints a summary of which variables are included in the model
#and the p-value of each variable

summary(res100u)
```

Do not forget to change xxx to the number of lines contained in your training file. And, do not forget to change yyy to the number of columns contained in your file. Finally, you may need to change the first line where you set your working directory. To do this, see Note 15. Also, you will notice that variables100u[,x] range from 1 through 29 in the example above. If you have more features of interest or fewer features of interest, you will need to adjust the number of variables100u entries. For example, if you have 8 features of interest, you will have variables100u 1 through 8. Likewise, if you have 40 features of interest, you will have variables100u 1 through 40. To determine how many features of interest you have, type length(data100) into your R script editing box and run the script. The output will be the number of columns in your input file. However, remember that the first three columns contain information about Chromosome, Start, and End locations. Also, the fourth column contains information regarding whether each gene is imprinted. Therefore, the number of features of interest will be the number returned by length(data100) minus 4. Once you have made the necessary adjustments, run the script. When the script has finished running, you will have a trained model for this genomic region.

2. Repeat step 4 for all of your genomic regions of interest (e.g., gene body, introns, exons, 5′ UTRs, 3′ UTRs, and any upstream or downstream genomic intervals). For each run, you will need to make a few alterations. To change between regions, alter the text in bold accordingly. For example, to move from 100 kb up to 3′ UTR (see Note 6):

setwd("c:/Perl") # *where I want my working directory*

data3utr <- read.csv("**3UTR Training.csv**", header = TRUE) [1:xxx,] *#tells the program which file to load*

variables3utr <- data**3utr**[,5:yyy] *#stores data contained in #columns 4-yyy in a variable #named Variables100u*

#the following lines are specify your models

base3utr <- glm(**data3utr**$Imprinted ~ 1, family = binomial)

full3utr <- glm(**data3utr**$Imprinted ~ **variables3utr**[,1] + **variables3utr**[,2] + **variables3utr**[,3] + **variables3utr**[,4] + **variables3utr**[,5] + **variables3utr**[,6] + **variables3utr**[,7] + **variables3utr**[,8] + **variables3utr**[,9] + **variables3utr**[,10] + **variables3utr**[,11] + **variables3utr**[,12] + **variables3utr**[,13] + **variables3utr**[,14] + **variables3utr**[,15] + **variables3utr**[,16] + **variables3utr**[,17] + **variables3utr**[,18] + **variables3utr**[,19] + **variables3utr**[,20] + **variables3utr**[,21] + **variables3utr**[,22] + **variables3utr**[,23] + **variables3utr**[,24] + **variables3utr**[,25] + **variables3utr**[,26] + **variables3utr**[,27] + **variables3utr**[,28] + **variables3utr**[,29], family = binomial)

#removes unnecessary variables from model

res3utr <- step(**full3utr**, scope = list(upper = **full3utr**, lower = ~1), direction = "both", trace = FALSE)

#prints a summary of which variables are included in the model #and the p-value of each variable

summary(**res3utr**)

Also, do not forget to change xxx to the number of lines contained in your training file. Likewise, do not forget to change yyy to the number of columns contained in your file. And, you may need to change the first line where you set your working directory. To do this, see Note 15. Finally, do not forget to adjust the number of variables3utr to match the number of features of interest you have.

3.8. Model Testing

Here you will use R and the second data set of genes, which were not used for training your models, to test how well your trained models perform before running them on your genome-wide data.

1. Once your models have been trained, you will want to test how well the trained models perform in predicting imprinted status.

To do this, paste or type the following script into your R script editing box (the example shown is for introns, see Note 6):

setwd("c:/Perl") # *where I want my working directory*

dataI <- read.csv("Introns Test.csv", header = TRUE)[1:xxx,] *#tells the program which file to load*

predictedI <-predict.glm(resI,newdata = dataI,type = "response") *#predicts whether each #gene in your test file is imprinted and #saves the information in the variable predictedI*

names(which(predictedI > 0.8)) *#prints the index of each gene #predicted to be imptinted at 80% confidence*

Do not forget to change xxx to the number of lines contained in your test file. Also, you may need to change the first line where you set your working directory. To do this, see Note 15. Run the script. In this example, we are testing how well the model for introns performs. The output from this script will tell you the index of any genes that are predicted as imprinted. To determine which gene the index corresponds to, simply find that row number in your test file. The gene name can be obtained by searching the UCSC Genome Browser using the genomic coordinates of the gene. At this point, you will want to record whether each gene in your test set was called as imprinted or not imprinted, and whether that call was accurate. If the results are not satisfactory, you can change the stringency of the calls. To do this, locate the line above which reads names(which(predictedI > 0.8)). To make calls more stringent, increase the number at the end of the line from 0.8. To make the calls less stringent, decrease the number at the end of the line from 0.8. Repeat this process for each genomic region you are interested in. For each run, you will need to make a few alterations. To change between regions, alter the text in bold accordingly. For example, to move from Introns to Exons (see Note 6):

setwd("c:/Perl") # *where I want my working directory*

dataE <- read.csv(**"Exons Test.csv"**, header = TRUE)[1:xxx,] *#tells #the program which file to load*

predictedE <-predict.glm(**resE**, newdata = **dataE**, type = "response") *#predicts whether #each gene in your test file is imprinted and #saves the information in the variable #predictedE*

names(which(**predictedE** > 0.8)) *#prints the index of each gene #predicted to be imptinted at 80% confidence*

If your models are not performing as well as you had hoped, you may want to try adding additional epigenetic or sequence features of interest. You can also try increasing the number of genes in your training set, if you have enough genes to do so.

If not, you can increase the ratio of non-imprinted to imprinted genes in your training set. Finally, you can adjust the confidence level (in the line **predictedE > 0.8**) to see whether a different cutoff gives better results for your particular data set.

3.9. Genome-Wide Prediction of Imprinted Status

Here you will use R to run your trained and tested computational prediction models on the genome-wide data you have gathered on each of your features of interest across each of your genomic regions of interest.

1. Once you have an idea of how well your models are performing, you can then run your models on your genome-wide data sets to predict which genes in your genome of interest might be imprinted. To run your models on your genome-wide data, paste or type the following script into your R script editing box (the example shown is for exons, see Note 6):

   setwd("c:/Perl") # *where I want my working directory*

   dataE <- read.csv("Exons.csv", header = TRUE)[1:xxx,] #*tells the #program which file to load*

   predictedE <-predict.glm(resE, newdata = dataE, type = "response") #*predicts whether #each gene in your test file is imprinted and #saves the information in the variable predictedE*

   ImpE <-names(which(predictedE > 0.8)) #*saves the index of each gene #predicted to be imprinted at 80% confidence to a variable #called ImpE*

   ImpE #*prints the data contained in ImpE to the screen*

   Do not forget to change xxx to the number of lines contained in your test file. Also, you may need to change the first line where you set your working directory. To do this, see Note 15. Run the script. The output from this script will tell you the index of any genes that are predicted as imprinted. Remember, you can change the stringency of the calls by locating the line above which reads ImpE<-names(which(predictedI > 0.8)). To make calls more stringent, increase the number at the end of the line from 0.8. To make the calls less stringent, decrease the number at the end of the line from 0.8. There may be several hundred genes that are predicted as imprinted in each genomic region, so you may want to wait until the next step to identify individual genes. However, if you would like to determine which gene the index corresponds to, simply find that row number in your test file. The gene name can be obtained by searching the UCSC Genome Browser using the genomic coordinates of the gene. At this point, you will want to record whether each gene in your test set was called as imprinted or not imprinted. Repeat this process for each genomic region you are interested in. For each run, you will need to make a few alterations. To change between regions, alter the text in bold accordingly. For example, to move from Introns to Exons (see Note 6):

setwd("c:/Perl") # *where I want my working directory*

dataI <- read.csv("**Introns**.csv", header = TRUE)[1:xxx,] *#tells #the program which file to load*

predictedI <-predict.glm(**resI**, newdata = **dataI**, type = "response") *#predicts whether #each gene in your test file is imprinted and #saves the information in the variable predictedI*

ImpI <-names(which(**predictedI** > 0.8)) *#saves the index of each gene predicted to be imprinted at 80% confidence to a variable called ImpI*

ImpI *#prints the data contained in ImpI to the screen*

3.10. Compiling a Candidate Gene List

In this step, you will use R to compile a list of all predicted imprinted genes, genome-wide.

1. Now you are ready to compile a candidate list of imprinted genes.

   In the following example,

   imp100u corresponds to 100 kb upstream.

   imp10u corresponds to 10 kb upstream.

   imp1u corresponds to 1 kb upstream.

   imp5utr corresponds to 5′ UTRs.

   impG corresponds to the gene body.

   impE corresponds to exons.

   impI corresponds to introns.

   imp3utr corresponds to 3′ UTRs.

   imp1d corresponds to 1 kb downstream.

   imp10d corresponds to 1 kb downstream.

   imp100d corresponds to 1 kb downstream.

   To compile a single list of all genes predicted as imprinted across all genomic regions of interest, paste or type this script into your R script editing box (refer to the key above to determine which genomic interval corresponds to which shorthand notation, see Note 6):

   setwd("c:/Perl") # *where I want my working directory*

   imp100u <-which(predicted100 > 0.8) *#saves the genes which are #predicted as imprinted at 80% confidence using data from the #region 100 kb upstream of all genes in a variable called imp100u*

   imp10u <-which(predicted10 > 0.8) *#saves the genes which are #predicted as imprinted at 80% confidence using data from the #region 10 kb upstream of all genes in a variable called imp10u* imp1u<-which(predicted1>0.8) *#saves the genes which are predicted #as imprinted at 80% confidence using data from the region 10 kb #upstream of all genes in a variable called imp1u*

imp5utr<-which(predicted5>0.8) *#saves the genes which are #predicted as imprinted at 80% confidence using data from the #5′ UTRs of all genes in a variable called imp5utr*

impG<-which(predictedG>0.8) *#saves the genes which are predicted #as imprinted at 80% confidence using data from within each gene #body in a variable called impG*

impE<-which(predictedE>0.8) *#saves the genes which are predicted #as imprinted at 80% confidence using data from the exons of all #genes in a variable called impE*

impI<-which(predictedI>0.8) *#saves the genes which are predicted #as imprinted at 80% confidence using data from the introns of all #genes in a variable called impI*

imp3utr<-which(predicted3utr>0.8) *#saves the genes which are #predicted as imprinted at 80% confidence using data from the #3′ UTRs of all genes in a variable called imp3utr*

imp1d<-which(predicted1d>0.8) *#saves the genes which are predicted #as imprinted at 80% confidence using data from the region 1 kb #downstream of all genes in a variable called imp1d*

imp10d<-which(predicted10d>0.8) *#saves the genes which are #predicted as imprinted at 80% confidence using data from the #region 10 kb downstream of all genes in a variable called imp10d*

imp100d<-which(predicted100d>0.8) *#saves the genes which are #predicted as imprinted at 80% confidence using data from the #region 100 kb downstream of all genes in a variable #called imp100d*

impALL<-c(imp100u, imp10u, imp1u, imp5utr, impG, impE, impI, imp3utr, imp1d, imp10d, imp100d) *#stores all predicted imprinted #genes from all of the regions above in a single variable called #impALL*

TableImpALL<-table(impALL) *#changes the way the data in impALL are #stored*

names(which(TableImpALL>8) *#prints to the screen those genes which #are predicted as imprinted by 8 or more of the genomic regions of #interest*

Do not forget that you may need to change the first line where you set your working directory. To do this, see Note 15. If you do not have one of the genomic regions included above, or if you have a genomic region in addition to those listed above, simply add or delete these regions as necessary. The output from this script will list those genes that are predicted as imprinted in 8 or more of your genomic regions of interest. To change this threshold, simply replace the 8 in this line names(which(TableImpALL > 8) with a more appropriate number. The output from this script will tell you the index of those genes that are predicted as imprinted by the number of models you have specified (e.g., 8).

2. To determine which gene the index corresponds to, transfer the index number of those genes that are predicted as imprinted to a new Excel file, with one index number on each line of column one. Save this file in tab-delimited text form (.txt) and call it something appropriate (e.g., List.txt). Also, make sure that you have saved this file in the same directory where you have installed Perl (see Note 5). Then, paste or type the following program into Notepad++ (see Note 6):

```
#!/usr/local/bin/perl -w
# tabulate number of occurrences of one file within another file
my $usage = 'GeneNames.pl
Extracts information from one file using search criteria form another file.
USAGE:
./GeneNames.pl list.txt genes.txt output.txt
';
# the next 3 lines tell the program that the user will enter the input and output files to use
my $list = shift @ARGV || die "$usage\n";
my $genes = shift @ARGV || die "$usage\n";
my $output = shift @ARGV || die "$usage\n";
open (LIST, "< $list")
or die "can't open file to search within"; #opens the first input file or dies trying
open (OUT, "> $output")
or die "can't open OUTPUT file"; #opens the output file or dies trying
print "Running...\n";
while (<LIST>) {#tells the program what to do while the first file is open
chomp; #removes any new line symbols from the end of each line
(@LIST) = split/\t/; # splits each line on tabs
$line = $LIST [0]; #defines a variable
open (GENES, "< $genes")
or die "can't open file to search with"; #opens the second input file or dies trying
while (<GENES>) {#loops through 2nd file
```

```
chomp; #removes any new line symbols from the end of each line
if ($. == $line) {
print (OUT $_, "\n"); #prints matches if found
}
}
close (GENES);
}
print "Done!\n";
close (LIST); #closes first input file
close (GENES); #closes second input file
close (OUT); #closes output file
exit 0; #closes program
```

Save the file as GeneNames. To save as a Perl file, select "Perl source file (*.pl, *.pm, *.plx)" from the drop-down menu next to "Save as type." Make sure that the file is saved in the same directory as Perl. You will also need to check that the file called Genes.txt, which you created in Subheading 3.1, step 3 above, is saved in this directory as a .txt file. Open Command Prompt, or similar command line program. In Windows, Command Prompt can be found by following Start ->All Programs ->Accessories ->Command Prompt. Change directory from the current directory to the folder where you have installed Perl (see Note 5). To change a directory, type cd followed by the new directory (e.g., cd C:\Perl). Your Command Prompt should now read C:\Perl>, or something very similar. Type perl GeneNames.pl List.txt Genes.txt GNOUT.txt and press enter. This will create a file called "GNOUT.txt." Depending on what you decided to do in Subheading 3.1, step 2 above, either your first or your last column should contain gene names.

3. Once you have your candidate list of imprinted genes, it is essential to determine whether any are actually imprinted using traditional methods of analysis.

4. Notes

1. The choice of which track to use depends on your particular situation and the difference is that UCSC Genes contain gene entries from several sources, including RefSeq, UniProt, GenBank, CCDS, and comparative genomics. Other data sets, for example the RefSeq genes data set, contain genes from a single source, such as the NCBI RNA reference sequences collection, only.

2. To save a file as a text (.txt) file, choose the option of Text (Tab Delimited) from the drop-down menu next to "Save as type."
3. Depending on your version of Excel, "sort ascending" may be different. In some versions it may be represented as an A on top of a Z, both of which are positioned next to an arrow pointing down, and in others it may be called "sort A to Z."
4. For your paste to be effective, you will have to paste values only. To do this, use the "paste special" function and select values.
5. To determine where Perl is installed, follow Start ->Search and search for "Perl" among "All files and Folders." Once the search has finished running, look for a folder named "Perl." If there is more than one folder named "Perl," look for the one that says C:\ next to it. Now, pull up your Command Prompt window next to your search window so that you can see both at the same time. Type cd into Command Prompt, but do not press Enter. Click on the "Perl C:\" and drag and drop the folder into your Command Prompt window. Command Prompt should now read C:\Perl>. Press Enter. You have successfully changed your directory.
6. The # symbol in Perl or R scripts indicates an author's comment and therefore is not run by the program. In the scripts provided, I have tried to give a brief summary of what each line in the script does, and these explanations can be found following the # symbol.
7. Often, equivalent releases of the same genome assembly are identified by different names. For a list of equivalent genome assemblies, see http://genome.ucsc.edu/FAQ/FAQreleases.html.
8. To obtain genomic coordinates for predicted G-quartet sites, download the QUADPARSER program by directing your Web browser here: http://www.quadruplex.org/?view=quadparser. Choose the correct operating system for your computer and follow the prompts to install the program. Download your genome of interest from the UCSC Genome Browser Web site at http://hgdownload.cse.ucsc.edu/downloads.html. Click on the link for your organism of interest. On the next page, find the assembly you are interested in and click on the link for "Full data set." Download "chromFa.tar.gz," and unzip the files to the same directory you have installed QUADPARSER. Open CommandPrompt. Change directory to the location where QUADPARSER is installed. To run the program, type quadparser –n, followed by your input and output file names (e.g., quadparser –n Chr1.txt Chr1GQ.txt) and press enter. If you run into problems, QUADPARSER has a help page available at http://www.quadruplex.org/?view=quadparser_instructions. A help function is also available by entering

quadparser –h into CommandPrompt and pressing Enter. Once you have output files (e.g., Chr1GQ), for each chromosome, upload them all to Galaxy, one at a time by clicking on "Get Data" and selecting "Upload File" from the left-hand menu. Click "Browse" and select the file you wish to upload. At the top of the window will be a drop-down menu under "File Format." Select "bed" from this menu. Above the "Execute" button, there should be a drop-down menu which allows you to select the genome. Make sure to select both the correct species and genome assembly. Once all of your chromosome files have been uploaded, select "Text Manipulation" from the left-hand menu and click on "Concatenate queries." Select your first file from the drop-down menu under "Concatenate query." Then, click on the "Add new querys" button above the "execute" button and select your second file from the drop-down menu. Repeat this process until you have added all of your chromosome files and press "Execute." Once your files have concatenated, cut all columns except the first three, which contain information regarding genomic location. In the left-hand "Tools" menu, click on "Text Manipulation," and then "Cut columns from a table." Type the columns you wish to keep into the box next to "Cut columns" (e.g., c1, c2, c3). Press execute and wait for your job to finish. When it has finished running, click on the name of the job. This should expand the file window and allow you to save the file by clicking on the disk icon. Locate the file and open it with Excel. Delete the header row, by right-clicking on the "1" to the left of the header row and selecting delete. Then, save the file as a .txt file, making sure to give it an appropriate name (e.g., GQs.txt), and close the file.

9. When using this program, the input file name is the file Perl is going to read DNA sequences from in order to calculate GC% of each region you are interested in. You downloaded these files in Subheading 3.2, step 8, and saved each file (the example file name given was CGintron) from Galaxy. The file format is .bed, so the full name of you input file will have .bed at the end (e.g., CGintron.bed). Your output file is the file to which Perl will write the calculated GC%. Name this something appropriate (e.g., CGintronOUT) and use .txt as the file type, as these can be easily opened in Excel (e.g., CGintronOUT.txt).

10. When using this program, the first input file name is the file Perl is going to loop through line by line. This first input file will contain your list of all known genes (e.g., Genes.txt). It is very important that this is your first input file, because you want your output to be in the same order as in your file containing the list of all know genes (e.g., Genes.txt). Your second input file name is the file containing the information Perl will

try to match within your first file. This second file will be your list of all known imprinted genes (e.g., KnownImprinted.txt). Your output file is the file to which Perl will write the information regarding whether each gene is imprinted. Name this something appropriate (e.g., Imprinted) and use .txt as the file type, as these can be easily opened in Excel (e.g., Imprinted.txt). As the information regarding whether each gene is imprinted will not vary between genomic regions, this program needs to be run only once.

11. When using this program, the first input file name is the file Perl is going to loop through line by line. This first input file will contain one of these lists: your list of introns, exons, 5′ UTRs, or 3′ UTRs. It is very important that one of these four files is your first input file for two reasons. The first is because you want your output to be in the same order as it is in these files. The second reason is because this program is meant to be run before the program called count (Subheading 3.4, step 2). This program identifies those introns, exons, 5′ UTRs, or 3′ UTRs that contain each of your features of interest, and reports these data in a format that the count program can use. The count program, in turn, identifies which genes the introns, exons, 5′ UTRs, or 3′ UTRs belong to and tallies the number of occurrences of your features of interest. Your second input file name is the file containing the information Perl will try to match within your first file. This second file will contain data on your features of interest (e.g., GC%, miRNA clusters, CTCF binding sites, histone modification data, etc.). Your output file is the file to which Perl will write the information regarding whether each gene is imprinted. Name this something appropriate (e.g., miRNAexonsOUT_1) and use .txt as the file type, as these can be easily opened in Excel (e.g., miRNAexonsOUT_1.txt). The _1 at the end of the file name is added because the files we create here will be sent through the count program and we will want to distinguish the output files from the two programs (e.g., miRNAexonsOUT_1.txt versus miRNAexonsOUT.txt).

12. The files containing the histone ChipSeq data are very large. In fact, some may contain millions of data points. Therefore, running programs that utilize these files may take quite a long time on standard desktop computers. To speed up the process, you may want to look into using a cluster, if one is available for you to use.

13. When using this program, the first input file name is the file Perl is going to loop through line by line. This first input file will contain one of these lists: your list of all known genes, or any upstream or downstream genomic regions you are interested in. It is very important that one of these types of files is your first input file, because you want your output to be in the

same order as it is in these files. It is also important that you do not use your list of introns, exons, 5′ UTRs, or 3′ UTRs as your first input file. This will be explained further in a few sentences. Your second input file name is the file containing the information Perl will try to match within your first file. For your list of all known genes, and any upstream or downstream genomic regions, the second file will contain data on your features of interest (e.g., GC%, miRNA clusters, CTCF binding sites, histone modification data, etc.). However, for introns, exons, 5′ UTRs, and 3′ UTRs, this formula differs. In these cases only, you will be using the output file from Subheading 3.4, step 1 (e.g., miRNAexonsOUT_1.txt). These files contain the data in a form that the count program can use effectively. Your output file is the file to which Perl will write the information regarding whether each gene is imprinted. Name this something appropriate (e.g., miRNAgenesOUT) and use .txt as the file type, as these can be easily opened in Excel (e.g., miRNAgenesOUT.txt).

14. To save a file as a comma-delimited file (.csv), choose the option of CSV (Comma Delimited) from the drop-down menu next to "Save as type."
15. Depending on where you have installed Perl, the line where you set your working directory may have to be altered. To determine where Perl is installed, see Note 5 above.

References

1. Nikaido I, Saito C, Mizuno Y et al (2003) Discovery of imprinted transcripts in the mouse transcriptome using large-scale expression profiling. Genome Res 13:1402–1409
2. Luedi PP, Hartemink AJ, Jirtle RL (2005) Genome-wide prediction of imprinted murine genes. Genome Res 15:875–884
3. Wang X, Sun Q, McGrath SD et al (2008) Transcriptome-wide identification of novel imprinted genes in neonatal mouse brain. PLoS One 3:e3839
4. Babak T, Deveale B, Armour C et al (2008) Global survey of genomic imprinting by transcriptome sequencing. Curr Biol 18:1735–1741
5. Gregg C, Zhang J, Butler JE et al (2010) Sex-specific parent-of-origin allelic expression in the mouse brain. Science 329:682–685
6. Gregg C, Zhang J, Weissbourd B et al (2010) High-resolution analysis of parent-of-origin allelic expression in the mouse brain. Science 329:643–648
7. Brideau CM, Eilertson KE, Hagarman JA et al (2010) Successful computational prediction of novel imprinted genes from epigenomic features. Mol Cell Biol 30:3357–3370
8. Crespi B (2008) Genomic imprinting in the development and evolution of psychotic spectrum conditions. Biol Rev Camb Philos Soc 83:441–493
9. Mackay DJ, Callaway JL, Marks SM et al (2008) Hypomethylation of multiple imprinted loci in individuals with transient neonatal diabetes is associated with mutations in ZFP57. Nat Genet 40:949–951
10. Shao WJ, Tao LY, Gao C et al (2008) Alterations in methylation and expression levels of imprinted genes H19 and Igf2 in the fetuses of diabetic mice. Comp Med 58:341–346
11. Xie T, Chen M, Gavrilova O, Lai EW, Liu J, Weinstein LS (2008) Severe obesity and insulin resistance due to deletion of the maternal $G_{s\alpha}$ allele is reversed by paternal deletion of the $G_{s\alpha}$ imprint control region. Endocrinology 149:2443–2450
12. Hatada I, Sugama T, Mukai T (1993) A new imprinted gene cloned by a methylation-sensitive genome scanning method. Nucleic Acids Res 21:5577–5582
13. Hayashizaki Y, Shibata H, Hirotsune S et al (1994) Identification of an imprinted U2af

binding protein related sequence on mouse chromosome 11 using the RLGS method. Nat Genet 6:33–40

14. Kaneko-Ishino T, Kuroiwa Y, Miyoshi N et al (1995) Pegl/Mest imprinted gene on chromosome 6 identified by cDNA subtraction hybridization. Nat Genet 11:52–59
15. Kuroiwa Y, Kaneko-Ishino T, Kagitani F et al (1996) Peg3 imprinted gene on proximal chromosome 7 encodes for a zinc finger protein. Nat Genet 12:186–190
16. Luedi PP, Dietrich FS, Weidman JR et al (2007) Computational and experimental identification of novel human imprinted genes. Genome Res 17:1723–1730
17. Maeda N, Hayashizaki Y (2006) Genome-wide survey of imprinted genes. Cytogenet Genome Res 113:144–152
18. Plass C, Shibata H, Kalcheva I et al (1996) Identification of Grfl on mouse chromosome 9 as an imprinted gene by RLGS-M. Nat Genet 14:106–109
19. Pollard KS, Serre D, Wang X et al (2008) A genome-wide approach to identifying novel-imprinted genes. Hum Genet 122:625–634
20. Ruf N, Bähring S, Galetzka D et al (2007) Sequence-based bioinformatic prediction and QUASEP identify genomic imprinting of the KCNK9 potassium channel gene in mouse and human. Hum Mol Genet 16:2591–2599
21. Schulz R, Menheniott TR, Woodfine K et al (2006) Chromosome-wide identification of novel imprinted genes using microarrays and uniparental disomies. Nucleic Acids Res 34:E88
22. Smith RJ, Dean W, Konfortova G et al (2003) Identification of novel imprinted genes in a genome-wide screen for maternal methylation. Genome Res 13:558–569
23. Wolf JB, Cheverud JM, Roseman C et al (2008) Genome-wide analysis reveals a complex pattern of genomic imprinting in mice. PLoS Genet 4:e1000091
24. Wood AJ, Roberts RG, Monk D et al (2007) A screen for retrotransposed imprinted genes reveals an association between X chromosome homology and maternal germ-line methylation. PLoS Genet 3:e20
25. Shibata H, Hirotsune S, Okazaki Y et al (1994) Genetic mapping and systematic screening of mouse endogenously imprinted loci detected with restriction landmark genome scanning method (RLGS). Mamm Genome 5:797–800
26. Shibata H, Yoshino K, Muramatsu M et al (1995) The use of restriction landmark genomic scanning to scan the mouse genome for endogenous loci with imprinted patterns of methylation. Electrophoresis 16:210–217
27. Hayward BE, Kamiya M, Strain L et al (1998) The human GNAS1 gene is imprinted and encodes distinct paternally and biallelically expressed G proteins. Proc Natl Acad Sci U S A 95:10038–10043
28. Kamiya M, Judson H, Okazaki Y et al (2000) The cell cycle control gene ZAC/PLAGL1 is imprinted–a strong candidate gene for transient neonatal diabetes. Hum Mol Genet 9:453–460
29. Meissner A, Gnirke A, Bell GW et al (2005) Reduced representation bisulfite sequencing for comparative high-resolution DNA methylation analysis. Nucleic Acids Res 33:5868–5877
30. Lister R, O'Malley RC, Tonti-Filippini J et al (2008) Highly integrated single-base resolution maps of the epigenome in Arabidopsis. Cell 133:523–536
31. Guenther MG, Levine SS, Boyer LA et al (2007) A chromatin landmark and transcription initiation at most promoters in human cells. Cell 130:77–88
32. Heintzman ND, Stuart RK, Hon G et al (2007) Distinct and predictive chromatin signatures of transcriptional promoters and enhancers in the human genome. Nat Genet 39:311–318
33. Wen B, Wu H, Bjornsson H et al (2008) Overlapping euchromatin/heterochromatin-associated marks are enriched in imprinted gene regions and predict allele-specific modification. Genome Res 18:1806–1813
34. Li E, Beard C, Jaenisch R (1993) Role for DNA methylation in genomic imprinting. Nature 66:362–365
35. Wu MY, Tsai TF, Beaudet AL (2006) Deficiency of Rbbpl/Arid4a and Rbbpl1l/Arid4b alters epigenetic modifications and suppresses an imprinting defect in the PWS/AS domain. Genes Dev 20:2859–2870
36. Delaval K, Govin J, Cerqueira F et al (2007) Differential histone modifications mark mouse imprinting control regions during spermatogenesis. EMBO J 26:720–729
37. Lindroth AM, Park YJ, McLean CM et al (2008) Antagonism between DNA and H3K27 Methylation at the Imprinted Rasgrfl Locus. PLoS Genet 4:e1000145
38. Mikkelsen TS, Hanna J, Zhang X et al (2008) Dissecting direct reprogramming through integrative genomic analysis. Nature 454:49–55
39. Nagano T, Mitchell JA, Sanz LA et al (2008) The Air noncoding RNA epigenetically silences transcription by targeting G9a to chromatin. Science 322:1717–1720
40. Bell AC, Felsenfeld G (2000) Methylation of a CTCF-dependent boundary controls imprinted expression of the Igf2 gene. Nature 405: 482–485

41. Hark AT, Schoenherr CJ, Katz DJ et al (2000) CTCF mediates methylation-sensitive enhancer-blocking activity at the H19/Igf2 locus. Nature 405:486–489
42. Hikichi T, Kohda T, Kaneko-Ishino T et al (2003) Imprinting regulation of the murine Megl/Grb10 and human GRB10 genes; roles of brain-specific promoters and mouse-specific CTCF-binding sites. Nucleic Acids Res 31: 1398–1406
43. Kanduri C, Pant V, Loukinov D et al (2000) Functional association of CTCF with the insulator upstream of the H19 gene is parent of origin-specific and methylation-sensitive. Curr Biol 10:853–856
44. Takada S, Paulsen M, Tevendale M et al (2002) Epigenetic analysis of the Dlk1-Gtl2 imprinted domain on mouse chromosome 12: implications for imprinting control from comparison with Igf2-H19. Hum Mol Genet 11:77–86
45. Yoon BJ, Herman H, Hu B et al (2005) Rasgrf1 Imprinting is regulated by a CTCF-dependent methylation-sensitive enhancer blocker. Mol Cell Biol 25:11184–11190
46. LaSalle JM, Lalande M (1996) Homologous association of oppositely imprinted chromosomal domains. Science 272:725–728
47. Murrell A, Heeson S, Reik W (2004) Interaction between differentially methylated regions partitions the imprinted genes Igf2 and H19 into parent-specific chromatin loops. Nat Genet 36:889–893
48. Kato Y, Sasaki H (2005) Imprinting and looping: epigenetic marks control interactions between regulatory elements. Bioessays 27:1–4
49. Ling JQ, Li T, Hu JF et al (2006) CTCF mediates interchromosomal colocalization between Igf2/H19 and Wsb1/Nf1. Science 312:269–272
50. Zhao Z, Tavoosidana G, Sjolinder M et al (2006) Circular chromosome conformation capture (4 C) uncovers extensive networks of epigenetically regulated intra- and interchromosomal interactions. Nat Genet 38:1341–1347
51. Qiu X, Vu TH, Lu Q et al (2008) A complex deoxyribonucleic acid looping configuration associated with the silencing of the maternal Igf2 allele. Mol Endocrinol 22:1476–1488
52. Barski A, Cuddapah S, Cui K et al (2007) High-resolution profiling of histone methylations in the human genome. Cell 129:823–837
53. Lieberman-Aiden E, van Berkum NL, Williams L et al (2009) Comprehensive mapping of long-range interactions reveals folding principles of the human genome. Science 326:289–293
54. Rhead B, Karolchik D, Kuhn RM et al (2010) The UCSC Genome Browser database: update 2010. Nucleic Acids Res 38:D613–D619
55. Bao L, Zhou M, Cui Y (2008) CTCFBSDB: a CTCF-binding site database for characterization of vertebrate genomic insulators. Nucleic Acids Res 36:D83–D87
56. Blankenberg D, Von Kuster G, Coraor N, et al (2010) Galaxy: a web-based genome analysis tool for experimentalists. Curr Prot in Mol Biol, Chapter 19, Unit 19.10.1–21
57. Goecks J, Nekrutenko A, Taylor J et al (2010) Galaxy: a comprehensive approach for supporting accessible, reproducible, and transparent computational research in the life sciences. Genome Biol 25:R86
58. Huppert JL, Balasubramanian S (2005) Prevalence of quadruplexes in the human genome. Nucleic Acids Res 33:2908–2916
59. R Development Core Team (2008) R: A language and environment for statistical computing. R Foundation for Statistical Computing, Vienna, Austria. URL http://www.R-project.org

Part III

Identifying the Regulatory Features of Imprinted Domains

Chapter 8

Engineering of Large Deletions and Duplications In Vivo

Louis Lefebvre

Abstract

Gene targeting in embryonic stem (ES) cells coupled with the site-specific Cre/*lox*P recombination system offers unique opportunities to identify and analyze the roles of *cis*-acting sequences in the regulation of imprinted gene expression. Although several different approaches have been described to engineer large chromosomal rearrangements in ES cells, these strategies can be labor-intensive and often require several subcloning of the original stem cells, therefore limiting the chances of obtaining germ line transmission of the mutation introduced. Here we describe an alternative approach which is based on in vivo recombination, therefore limiting the number of steps performed in ES cells and allowing to take advantage of the growing number of *lox*P insertional mutations already available in transgenic mice.

Key words: Gene targeting, Cre/*lox*P recombination, Embryonic stem cells, Targeted meiotic recombination, TAMERE

1. Introduction

The Cre recombinase of bacteriophage P1 can catalyze site-specific recombinations between *lox*P site in mammalian cells (1). Each *lox*P site is a short 34-bp sequence consisting of a unique core of 8 bp, flanked by 13-bp inverted repeats: ATAACTTCGTATAat gtatgcTATACGAAGTTAT (2). It is the relative orientation of the core element which will determine the outcome of the recombination event: when present on the same chromosome *in cis*, *lox*P sites inserted in the same orientation will give rise to a deletion of the intervening sequences, whereas *lox*P sites in opposite orientations will generate inversion of the *lox*P-flanked region (3). These basic properties of the Cre/*lox*P system have been exploited in a number of different targeted and random approaches to generate large sets of deletions in ES cells (4–6). The main disadvantages of these approaches is that they typically require three subsequent subcloning of ES cells: targeting of the two *lox*P sites defining the breakpoints of the rearrangement, followed by Cre-mediated

Nora Engel (ed.), *Genomic Imprinting: Methods and Protocols*, Methods in Molecular Biology, vol. 925,
DOI 10.1007/978-1-62703-011-3_8, © Springer Science+Business Media, LLC 2012

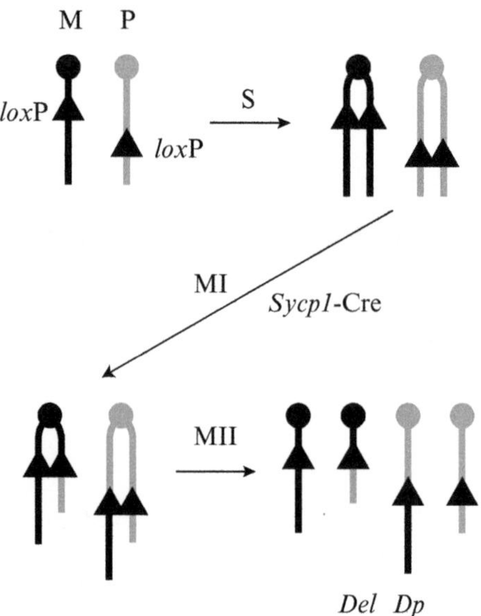

Fig. 1. Targeted meiotic recombination (TAMERE). Targeted *loxP* site insertions are generated at nonallelic positions on the same chromosome, and in the same orientation relative to the centromere (*circle* at *top* of acrocentric chromosome). Triple transgenic males carrying the loxP site insertions on the maternal (M, *black*) and paternal (P, *gray*) homologues of the targeted chromosome and the *Sycp1*-Cre transgene will undergo a Cre-mediated recombination during chromosome pairing at meiosis I (MI). After meiosis II (MII) mature gametes carrying the deletion of the intervening sequence (Del) or its duplication (Dp) will be recovered in mature sperm and these new mutations can be established in a mouse line by direct breeding of these "*trans*-loxing" males.

recombination via transient production of Cre from an expression vector. Depending on the frequency at which the recombinants are recovered, such approaches often rely on strong genetic selection conferred by the reconstitution of a functional selectable marker at the recombination breakpoint.

An alternative approach overcoming some of these limitations has been developed to obtain recombination between *loxP* sites targeted at nonallelic positions on the two parental homologues in vivo (7). Termed TAMERE, for targeted meiotic recombination, this strategy takes advantage of transgenic lines expressing Cre in germ cells. In its original form, TAMERE was based on the *Sycp1*-Cre transgene, producing Cre recombinase in primary spermatocytes, at a stage when *lox*P sites on homologous chromosomes are brought in close proximity upon pairing at meiosis I. In males carrying the Cre transgene as well as *lox*P sites on the two parental homologues (termed *trans*-loxing males), Cre can catalyze the recombination between the two *lox*P site *in trans* and produce the expected deletion and duplication in germ cells (Fig. 1). These new mutations are independently recovered in the progeny of *trans*-loxing males without selection and frequencies of 0.1–20% have been reported (7–10).

This chapter describes the steps required to introduce *loxP* site insertions in the mouse genome by homologous recombination in ES cells and the generation of novel duplication and deletion alleles in vivo.

2. Materials

All ES cell work should be performed in a dedicated tissue culture facility equipped with humidified incubators (37 °C, 5% CO_2), a laminar flow cabinet, tabletop centrifuge to spin down cells, an inverted microscope for cell observation, and a stereomicroscope with light base for colony picking. An electroporator for mammalian cells (e.g., Bio-Rad GenePulser Xcell with CE Module 165-2661) with electroporation cuvettes (0.4 cm electrode gap; e.g., VWR 89047-210, Bio-Rad 165-2088) is also required.

2.1. MEF and ES Cell Culture

1. Mouse embryonic fibroblast (MEFs) resistant to neomycin, hygromycin, and puromycin can be prepared from transgenic mouse lines (e.g., DR4; JAX #003208 (11)) following established protocols (12) or purchased (e.g., Stemcell Technologies, http://www.stemcell.com/).
2. MEF medium (MEFM): Dulbecco's modified Eagle's medium (DMEM) supplemented with 10% fetal bovine serum (FBS) (see Note 1). Some authors also add 2 mM L-glutamine (Invitrogen 25030-081) and antibiotics (100 units penicillin/100 μg streptomycin; Invitrogen 15140-122).
3. ES medium (ESM) 500-ml bottle: 400 ml DMEM high glucose (4.5 g/l; e.g., Millipore SLM-120-B), 2 mM L-glutamine (5 ml of 200 mM stock; Invitrogen 25030-081), 0.1 mM nonessential amino acids (5 ml of 10 mM stock; Invitrogen 11140-050), 0.1 mM 2-mercaptoethanol (5 ml of 10 mM stock; Sigma M7522; 70 μl of 14.3 M stock in 100 ml PBS), 1 mM sodium pyruvate (5 ml of 100 mM stock; Invitrogen 11360-070), 100 units penicillin/100 μg streptomycin (5 ml of 10,000 U/10,000 μg stock; Invitrogen 15140-122), 15% of FBS (75 ml; ES qualified, e.g., Hyclone SH30070, StemCell Technologies 06952) (see Note 2), and leukemia inhibitory factor (LIF) at 10 μg/l (e.g., StemCell Technologies 02740, Santa Cruz Biotech. sc-4989) (see Note 3).
4. 0.25% Trypsin: Add 10 ml of 2.5% trypsin (Invitrogen 15090-046) to 90 ml of autoclaved EDTA in Hank's buffered saline: 0.35 g $NaHCO_3$, 0.4 g KCl, 0.06 g KH_2PO_4, 0.01 g phenol red, 1.0 g glucose, 8.0 g NaCl, 0.09 g $Na_2HPO_4{\cdot}7H_2O$, and 0.2 g EDTA to 900 ml with distilled water.

5. PBS (Ca^{2+} and Mg^{2+} free): 137 mM NaCl (8 g/l), 2.7 mM KCl (0.2 g/l), 8.1 mM $Na_2HPO_4 \cdot 2H_2O$ (1.44 g/l), and 1.76 mM KH_2PO_4 (0.2 g/l) in distilled water. Adjust pH to 7.2 with HCl and bring to 1 l with distilled water.
6. Mitomycin C: Prepare 100× (1 mg/ml) stock by dissolving 2 mg of Mitomycin C (Sigma M0503) in 2 ml PBS. Store at –20 °C in 100 μl aliquots.
7. Gelatin: 0.1% in water (0.5 g in 500 ml), autoclaved. Store at 4 °C. To prepare gelatinized plates, add enough 0.1% gelatin to cover the surface of the plate (4 ml per 10 mm plates, 100 μl per well of 96-well plate), let sit briefly, aspirate gelatin solution, and let dry for 5 min.
8. Puromycin: Prepare 1 mg/ml stock (500×) by dissolving 10 mg of puromycin (Sigma P8833) in 10 ml of sterile distilled water. Store at –20 °C in 1-ml aliquots. For R1 ES cells, we use puromycin at a final concentration of 2 μg/ml (1 ml of 500× stock per 500 ml bottle of ESM). Puromycin is also available as a 10 mg/ml (5,000×) solution (Invitrogen A11138-02).
9. Geneticin (G418): Prepare 100 mg/ml stock (500×) by dissolving 5 g of G418 (e.g., Gibco 11811, Sigma G5013) in 50 ml of sterile distilled water. Store at –20 °C in 1-ml aliquots. For R1 ES cells, we use G418 at a final concentration of 200 μg/ml (1 ml of 500× stock per 500 ml bottle of ESM).

2.2. TAMERE

1. PCR primers to genotype the Cre mice. Forward: ATGTCC AATTTACTGACCGTAC. Reverse: GTTTCACTGGTTATG CGGCG. These primers amplify the first 356 bp of the Cre coding region.

3. Methods

3.1. Gene Targeting: Electroporation

1. Using recombinant DNA or recombineering (13) techniques, build a targeting vector to generate each *lox*P site insertions. The general features of such vectors are shown in Fig. 2. The arms of homology can be subcloned from an available genomic clone, amplified from genomic DNA isogenic to the ES cell used (see Note 4), or recovered from genomic BAC clones by gap repair (14).
2. Linearize 50 μg of the targeting vector using a restriction enzyme with a single cut site, positioned outside of the arms of homology. Heat inactivate the enzyme, phenol extract the digest and recover the linear DNA by ethanol precipitation. Wash the pellet with a large volume of 70% ethanol, dry the pellet, and resuspend in 25 μl of sterile distilled water (see Note 5).

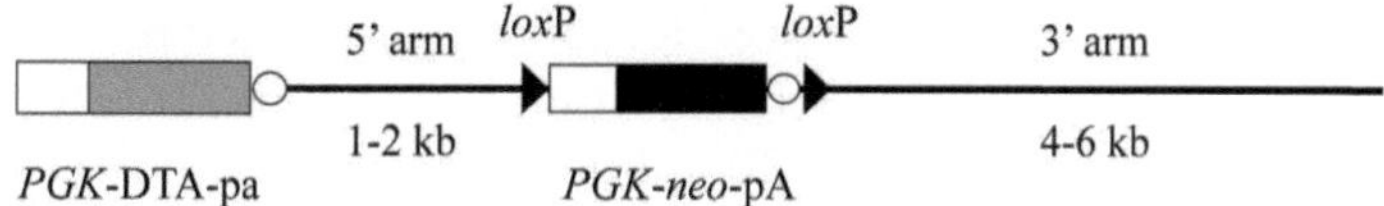

Fig. 2. Targeting vector for *loxP* site insertion. Each of the chromosomal breakpoints of the TAMERE rearrangements (deletion and duplication) is defined by a *loxP* site insertion on the desired chromosome. Any existing *loxP* site insertion already available in transgenic mouse can be used as one of the breakpoint. Additional *loxP* site insertions can be generated by targeting in ES cells. A typical targeting vector will contain a selectable marker cassette (e.g., *PGK-neo*-pA) inserted in a continuous fragment of genomic DNA of 6–8 kb. This cassette can be flanked by loxP sites, as shown here. This will allow to monitor excision of the cassette in vivo and will eliminate residual foreign genetic elements at the rearrangement breakpoint. Enrichment for targeted events can also be provided by the addition of a negative selection cassette (e.g., diphtheria toxin-A chain, DTA) outside the smallest arm of homology.

Confirm the completion of digestion and estimate the final concentration by agarose gel electrophoresis.

3. Thaw a vial of MEFs by heating rapidly in your hand or in your 37 °C incubator (see Note 6). Using a 1-ml pipette gently collect MEFs from the vial and add drop by drop to a 14-ml tube containing 10 ml of MEFM. Gently invert the tube and pellet MEFs by centrifugation at 300 × *g* for 5 min. Aspirate supernatant and resuspend the MEF pellet in MEFM to reach a cell density of 2×10^5 cells/ml. Plate 10 ml on a 100 mm dish (see Note 7).
4. Grow MEFs until they reach confluence (2–3 days) and split cells onto two 150 mm dishes (~1:5 passage) as follows. Aspirate MEFM and rinse 100 mm dish twice with PBS. Add trypsin to cover the surface of the dish (~2 ml) and incubate 2–3 min at 37 °C. Recover MEFs in a 14-ml tube by adding 6 ml of MEFM to the dish to inactivate the trypsin and gently pipetting up and down to generate a single cell suspension. Pellet MEFs, aspirate supernatant, and resuspend cells in 4 ml MEFM. Add 2 ml of MEFs to each of two 150 mm dishes containing 23 ml of MEFM. Rock plates back and forth, right and left and incubate at 37 °C, 5% CO_2.
5. Grow MEFs until they reach confluence (2–3 days). Feeders are prepared by mitotically inactivating MEFs with mitomycin C. Aspirate MEFM and add 10 ml MEFM per plate. Add mitomycin C to 10 μg/ml (100 μl aliquot of 1 mg/ml stock), swirl plates, and incubate for 2 h at 37 °C (see Note 8).
6. Aspirate MEFM and rinse plates twice with PBS. Trypsinize as described in step 4, and resuspend combined feeder pellet in MEFM to bring to 2×10^5 cells/ml (~100 ml from two 150 mm plates).

7. Seed feeders on gelatinized plates as follows: One 60 mm plate (5 ml), four 100 mm plates (10 ml each), two 96-well plates (200 μl per well), using a multichannel pipette (see Note 9).
8. Thaw a vial of germ line-competent ES cells as described in step 3, using ESM. Resuspend the ES cell pellet in ESM to reach a cell density of 10^6 cells/ml. Seed 1 ml on the 60 mm feeder plate containing 4 ml of ESM. Gently rock the plate and incubate at 37 °C, 5% CO_2.
9. Grow ES cells for 48 h, changing medium every day. Trypsinize as described in step 4, using ESM (see Note 10). Resuspend pellet in 4 ml of ESM and seed 2 ml of ES cell suspension per 100 mm feeder plate containing 8 ml of ESM. This is a ~1:5 passage. Gently rock the plate and incubate at 37 °C, 5% CO_2.
10. Grow ES cells for 48 h, changing medium every day.
11. The day of the electroporation, change medium in the morning. In the afternoon, trypsinize ES cells as above and resuspend the combined ES cell pellets in 2 ml of cold PBS. Count cells from a 1:10 dilution in PBS using a hemocytometer. For each electroporation, pellet 5.6×10^6 cells and resuspend in 0.8 ml of cold PBS (7×10^6 cells/ml) (see Note 11).
12. Put electroporation cuvettes (0.4 cm gap size) on ice and set the electroporator for one pulse of 250 V at 500 μF.
13. Using a 1-ml pipette transfer 0.8 ml of ES cells to each cuvette and add the DNA (20–40 μg of linear targeting vector). Gently mix.
14. Deliver the electroporation pulse (time constant should be of 1.7–1.9 ms) and let the cuvette sit on ice for 30 min.
15. Transfer the cells to a 14-ml tube containing 5 ml of ESM and seed two 100 mm feeder plates containing 8 ml of ESM with 2.8 ml of the ES cell suspension. Incubate at 37 °C, 5% CO_2.
16. The next day, change the medium and start drug selection. Use puromycin at 2.0 μg/ml or G418 at 200 μg/ml (see Note 12).
17. Change the medium every day. Colonies big enough for picking should appear 8–10 days after electroporation (see Note 13).

3.2. Gene Targeting: Growth and Analysis of ES Clones in 96-Well Plates

Although the description of colony picking, analysis and expansion is beyond the scope of this chapter, as well as the formation of germ line chimeras, the reader can refer to detailed description of these steps in the literature (12, 15).

3.3. TAMERE

The generation of new deletion and duplication alleles in vivo by TAMERE requires three simple breeding steps involving mice carrying the *Sycp1*-Cre transgene (see Note 14) as well as the two

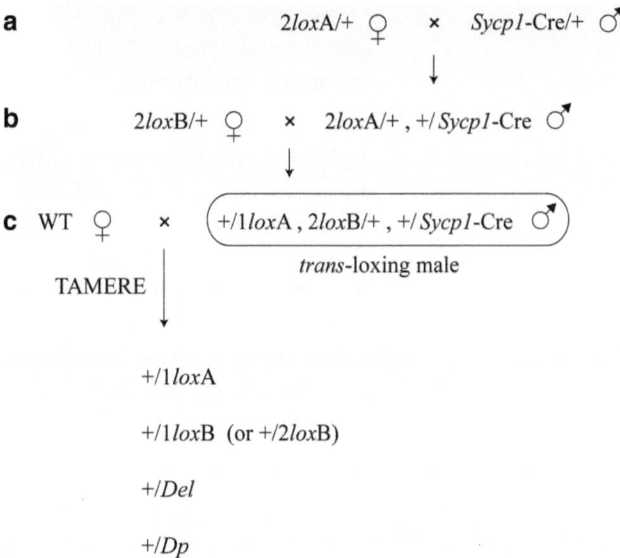

Fig. 3. Generation of new deletion (*Del*) and duplication (*Dp*) alleles in vivo by TAMERE. In the breeding scheme shown here the two nonallelic loxP site insertions are denoted *lox*A and *lox*B. The number of *lox*P sites present at each allele is referred to as 1*lox* and 2*lox* for single *lox*P site insertions and floxed alleles, respectively. If one or both alleles already carry a single *lox*P site, the same steps are required. (**a**) First, mice carrying the *lox*A allele are crossed with *Sycp1*-Cre transgenics. The reciprocal cross can also be done. The goal is to obtain a male carrying the *lox*A allele and the Cre transgene. (**b**) In this male, Cre is active in germ cells so transmission of 2*lox*A will delete the selectable marker at high frequency. Consequently, close to half of the progeny from this hemizygous male should contain the 1*lox*A allele and simple PCR genotyping distinguishing 1*lox*A from 2*lox*A can be used to confirm efficient germ line Cre activity. By breeding this male with females carrying the 2*lox*B allele (hemizygotes, or homozygotes), *trans*-loxing males carrying the two *lox*P site insertions (compound hemizygotes) and the Cre transgene can be obtained. (**c**) When bred to wild-type females (WT), *trans*-allelic recombination will occur in the germ cells of *trans*-loxing males. PCR genotyping of the progeny can be used to identify hemizygotes for each of the possible alleles. Segregation of the Cre transgene is not shown here. Note that for all genotypes, the maternal allele is shown first.

nonallelic *lox*P site insertions on the same chromosome (Fig. 3). For simplicity, the two *lox*P site insertion alleles are referred to as *lox*A and *lox*B, and the number of *lox*P site present at each allele (1 or 2, for flowed alleles), as 1*lox* and 2*lox*. All animals are genotyped by genomic PCR on ear punch lysates. Segregation of the *Sycp1*-Cre transgene is monitored using the Cre primers given in Subheading 2.

1. Cross 2*lox*A carrying mice with *Sycp1*-Cre transgenics to obtain double transgenic males (Fig. 3a; see Note 15). Since Cre is only active in germ cells in the *Sycp1*-Cre line (16), these animals are mosaic and only germ cells should contain the 1*lox*A allele (see Notes 16 and 17).
2. Cross *lox*A *Sycp1*-Cre double transgenic males to females carrying the second *lox*P site insertion, *lox*B, to obtain triple transgenic,

trans-loxing males (Fig. 3b). In these male progeny the 2*lox*A allele should have been efficiently converted to 1*lox*A during germ line transmission.

3. Cross the *trans*-loxing males carrying both *lox*P site insertions and the Cre transgene to wild type inbred or outbred females and genotype progeny at weaning. Most progeny should be hemizygous for either of the *lox*P insertion (1*lox*A and 1*lox*B). PCR reactions should also be established to specifically identify the two recombination products, the deletion and duplication alleles. Unless they are associated with lethal phenotypes in heterozygotes, they should be recovered at approximately the same frequency (see Note 18).

4. Notes

1. The quality of FBS is less critical for MEFs and we use regular (cheaper) tissue culture grade FBS for all MEF and feeder growth.
2. ES cells are particularly affected by the quality of the serum used. For small scale experiments, it is preferable to spend the money on ES qualified serum (pretested). Otherwise, different batches of serum can be tested for maintenance of good ES cell morphology; order the optimum serum in large quantities (15).
3. Recombinant LIF can also be produced in bacteria as a GST fusion.
4. In the case of ES cell lines established from F1 embryos, pure genomic DNA from one of the two parental strains should be used for PCR amplification.
5. Presence of excess salt in the DNA solution can cause arcing during electroporation. Consequently the 70% ethanol washes are critical.
6. To limit cell death, it is important to thaw the frozen vial rapidly and immediately add its content in culture media to dilute the freezing medium which contains DMSO.
7. Although this should not be required, using gelatinized plates can increase the adherence of MEFs and feeders.
8. Alternatively, MEFs can be mitotically inactivated by exposure to 6,000–10,000 rads in a gamma cell irradiator after trypsinization.
9. The 60 mm feeder plate can be used within 2 h, but ideally the feeders are allowed to settle overnight.
10. Since ES cells form tight colonies, increase the trypsin incubation to 4–5 min at 37 °C. Pipette the ES cell suspension several

times once ESM has been added to the plate, to disaggregate cell clumps.

11. Alternatively, commercial electroporation buffers can be used (e.g., Millipore/Chemicon ES-003-D).
12. Each drug selection has its own kinetics and cell death is not expected on the same days. Puromycin is fat-acting and sensitive cells with be killed within 2 days of drug selection. For G418, this will take 4–5 days.
13. Colonies large enough for picking can clearly be seen with the naked eye when the plate is observed from below.
14. Although all the published reports of TAMERE have relied on the Sycp1-Cre transgenic line, there is no reason to believe that other lines expressing Cre in the germ line could not also be suitable for this approach. In fact, *trans* recombination can also be obtained using a ubiquitous CMV-Cre line (L.L. unpublished observation).
15. If only one of the two *lox*P insertions is present in mice as a floxed allele (2*lox*), it is advantageous to introduce it first on the Cre background, as conversion of this allele from *2lox* to 1*lox* can be used to monitor the efficiency of the germ line Cre in the second generation.
16. We have previously demonstrated this conditional germ cell-specific deletion by Southern blot analysis of genomic DNA isolated from liver (all 2*lox*A), or testis and sperm (mixture of 2*lox*A and 1*lox*A). This analysis also revealed that in some *Sycp1*-Cre male, ubiquitous excision is observed in all tissue analyzed, suggesting leakiness of the Cre transgene and recombination early in development (9).
17. A previous study has reported the inactivation of *lox*P site by DNA methylation following *Sycp1*-Cre mediated recombination in the male germ line (17). However, these observations have not been confirmed by others, and we have not observed such loss of excision in second and even third generation recombination events in *Sycp1*-Cre males.
18. The frequency of TAMERE appears quite variable from one locus to another and might also depend on the size of the rearrangements expected. In four published studies using TAMERE, each recombinant allele was recovered in 0.1–19% of the progeny genotyped (7–10).

Acknowledgment

This work was supported by CIHR grant MOP-82863 and a Canada Research Chair to LL.

References

1. Sauer B, Henderson N (1988) Site-specific DNA recombination in mammalian cells by the Cre recombinase of bacteriophage P1. Proc Natl Acad Sci U S A 85:5166–5170
2. Nagy A (2000) Cre recombinase: the universal reagent for genome tailoring. Genesis 26:99–109
3. Branda CS, Dymecki SM (2004) Talking about a revolution: the impact of site-specific recombinases on genetic analyses in mice. Dev Cell 6: 7–28
4. Bilodeau M, Girard S, Hébert J, Sauvageau G (2007) A retroviral strategy that efficiently creates chromosomal deletions in mammalian cells. Nat Methods 4:263–268
5. Ramírez-Solis R, Liu P, Bradley A (1995) Chromosome engineering in mice. Nature 378: 720–724
6. Zheng B, Sage M, Sheppeard EA, Jurecic V, Bradley A (2000) Engineering mouse chromosomes with Cre-loxP: range, efficiency, and somatic applications. Mol Cell Biol 20: 648–655
7. Hérault Y, Rassoulzadegan M, Cuzin F, Duboule D (1998) Engineering chromosomes in mice through targeted meiotic recombination (TAMERE). Nat Genet 20:381–384
8. Kmita M, Fraudeau N, Hérault Y, Duboule D (2002) Serial deletions and duplications suggest a mechanism for the collinearity of Hoxd genes in limbs. Nature 420:145–150
9. Lefebvre L, Mar L, Bogutz A, Oh-McGinnis R, Mandegar MA, Paderova J, Gertsenstein M, Squire JA, Nagy A (2009) The interval between Ins2 and Ascl2 is dispensable for imprinting centre function in the murine Beckwith-Wiedemann region. Hum Mol Genet 18:4255–4267
10. Tarchini B, Huynh TH, Cox GA, Duboule D (2005) HoxD cluster scanning deletions identify multiple defects leading to paralysis in the mouse mutant Ironside. Genes Dev 19: 2862–2876
11. Tucker KL, Wang Y, Dausman J, Jaenisch R (1997) A transgenic mouse strain expressing four drug-selectable marker genes. Nucleic Acids Res 25:3745–3746
12. Nagy A, Gertsenstein M, Vintersten K, Behringer R (2003) Manipulating the mouse embryo: a laboratory manual. Cold Spring Harbor Laboratory Press, Cold Spring Harbor, NY
13. Sharan SK, Thomason LC, Kuznetsov SG, Court DL (2009) Recombineering: a homologous recombination-based method of genetic engineering. Nat Protoc 4:206–223
14. Copeland NG, Jenkins NA, Court DL (2001) Recombineering: a powerful new tool for mouse functional genomics. Nat Rev Genet 2:769–779
15. Wurst W, Joyner A (1993) Production of targeted embryonic stem cell clones. In: Gene targeting: a practical approach
16. Vidal F, Sage J, Cuzin F, Rassoulzadegan M (1998) Cre expression in primary spermatocytes: a tool for genetic engineering of the germ line. Mol Reprod Dev 51:274–280
17. Rassoulzadegan M, Magliano M, Cuzin F (2002) Transvection effects involving DNA methylation during meiosis in the mouse. EMBO J 21:440–450

Part IV

Epigenetics of Imprinted Regions

Chapter 9

Methylated DNA Immunoprecipitation (MeDIP) from Low Amounts of Cells

Julie Borgel, Sylvain Guibert, and Michael Weber

Abstract

Methylated DNA immunoprecipitation (MeDIP) is an immunocapturing approach for unbiased enrichment of DNA that is methylated on cytosines. The principle is that genomic DNA is randomly sheared by sonication and immunoprecipitated with an antibody that specifically recognizes 5-methylcytidine (5mC), which can be combined with PCR or high-throughput analysis (microarrays, deep sequencing). The MeDIP technique has been originally used to generate DNA methylation profiles on a genome scale in mammals and plants. Here we provide an optimized version of the MeDIP protocol suitable for low amounts of DNA, which can be used to study DNA methylation in cellular populations available in small quantities.

Key words: DNA methylation, MeDIP, Cytosine, CpG, Profiling, Epigenomics, Microarrays, Deep sequencing

1. Introduction

DNA methylation occurs on the carbon 5 of cytosines and plays essential roles in genome regulation in a variety of organisms and in both normal and disease contexts (1). To better understand the role of this epigenetic mark, several strategies have been developed to assess the distribution of cytosine methylation at a genome-wide scale (2). Some of these technologies use methylation-sensitive (e.g., *Hpa*II) or methylation-specific (e.g., *Mcr*BC) restriction enzymes, with the caveat that they are biased towards specific restriction motifs. Other methods combine sodium bisulfite conversion and deep sequencing, which offers a powerful readout at a single-nucleotide resolution but requires large sequencing efforts when applied genome-wide (3). Alternative strategies use affinity purification of methylated DNA that can be coupled to microarray hybridization or deep sequencing. These are based on the use of

Nora Engel (ed.), *Genomic Imprinting: Methods and Protocols*, Methods in Molecular Biology, vol. 925,
DOI 10.1007/978-1-62703-011-3_9, © Springer Science+Business Media, LLC 2012

methyl-binding protein domains (MBD) that recognize methylated DNA, or in the case of MeDIP on the use of antibodies that specifically recognizes 5-methylcytidine (5mC). These affinity methods provide valuable tools for a rapid and unbiased profiling of DNA methylation at more limited costs.

The principle of MeDIP is that genomic DNA is randomly sonicated and immunoprecipitated with a monoclonal antibody directed against 5mC (4). The methylated fraction of the genome can be analyzed at a single-gene resolution by conventional PCR and real-time PCR, or on a genome-wide scale by microarray hybridization or deep sequencing. It is however important to keep in mind that enrichment-based methods also have certain limitations. First, they offer an incomplete resolution (defined by the size of sonicated fragments), and for this reason bisulfite sequencing still remains the method of choice when detailed methylation information at single nucleotide resolution is required. Second, there is a confounding effect of the DNA sequence because the methylation enrichment also depends on the local CpG concentration. Indeed, low MeDIP enrichments can indicate either an unmethylated state or the absence of sufficient CpG targets in very CpG-poor regions of the genome. This effect can be corrected by applying bioinformatics normalization to obtain absolute cytosine methylation levels with a relatively good accuracy (5–7). As a consequence, it also appears that the accuracy of MeDIP measurements decreases in regions that are very CpG-poor. The classical MeDIP protocol was originally designed to work with relatively large amounts of DNA (at least 2 μg) (8). Here we describe an optimized protocol that can be used to immunoprecipitate methylated DNA from as low as 20,000 cells (9) (Fig. 1).

2. Materials

2.1. Isolation of Genomic DNA

1. Eppendorf LoBind® and standard 1.5 ml microtubes.
2. Lysis buffer: 20 mM Tris pH 8.0, 4 mM EDTA, 20 mM NaCl, 2% SDS.
3. Proteinase K, 10 mg/ml stock. Store at −20 °C.
4. Dry heating block for 1.5 ml microtubes.
5. PCI (phenol–chloroform–isoamyl alcohol 25:24:1).
6. Linear polyacrylamide (LPA), 5 mg/ml stock, used as a coprecipitant.
7. Refrigerated microcentrifuge.
8. Qubit® Fluorometer (Invitrogen) for quantification of low amounts of nucleic acids.

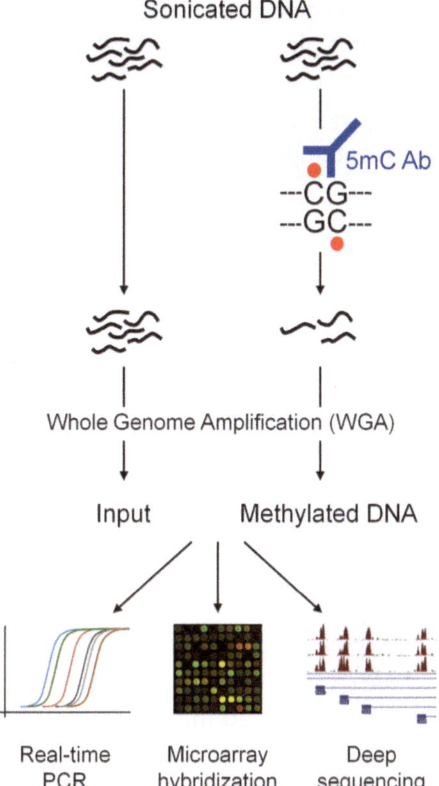

Fig. 1. Principle of MeDIP (methylated DNA immunoprecipitation). Genomic DNA is randomly sheared by sonication and immunoprecipitated with an antibody that recognizes 5-methylcytidine (5mC Ab). A portion of the sonicated DNA is left untreated and serves as input control. When MeDIP is performed on low amounts of starting DNA, a whole genome amplification (WGA) step is performed on the input and methylated DNA. Enrichments in the methylated fraction can be measured at a single gene resolution by real-time PCR, or on a global scale by microarray hybridization and deep sequencing. The deep sequencing image capture is reprinted by permission from Macmillan Publishers Ltd: *Nature Biotechnology* 28:1097-105, © 2010 -12).

2.2. Sonication

1. Eppendorf LoBind® 1.5 ml microtubes.
2. Diagenode Bioruptor® sonicator (standard model), with an automated cooling system that allows for continuous cooling of the water bath.
3. Equipment for small size agarose gel electrophoresis.

2.3. Immuno-precipitation of Methylated DNA

1. Eppendorf LoBind® 2 and 1.5 ml microtubes, and standard 1.5 ml microtubes.
2. Magnetic rack for microtubes, used for recovering the magnetic beads.
3. Dry heating block for 2 ml microtubes, with shaking.
4. IP buffer 10×: 100 mM Na-phosphate pH 7.0, 1.4 M NaCl, 0.5% Triton X-100. Store at room temperature.

5. 1 M Na-phosphate pH 7.0 buffer: Mix 39 ml 2 M monobasic sodium phosphate (NaH_2PO_4) (276 g/l), 61 ml 2 M dibasic sodium phosphate (Na_2HPO_4) (284 g/l), and 100 ml H_2O.
6. IP buffer 1×: Dilute 1 ml IP buffer 10× in 9 ml H_2O. Store at room temperature.
7. Mouse anti 5-methylcytidine monoclonal antibody, clone 33D3, available at a 1 mg/ml concentration from various suppliers such as Eurogentec or AbD Serotec. Other mouse monoclonal antibodies such as the ones developed by Diagenode work equally well. Store the antibody as 5 μl aliquots at −20 °C.
8. Vortex Genie 2 shaker with a platform for microtubes, placed at room temperature.
9. Overhead rotator for microtubes, placed in a 4 °C cold room.
10. Magnetic beads: Dynabeads M-280 Sheep anti-mouse IgG (Invitrogen).
11. PBS-BSA 0.05%: Mix 9.5 ml PBS with 0.5 ml BSA at 10 mg/ml concentration.
12. Proteinase K digestion buffer: 50 mM Tris pH 8.0, 10 mM EDTA, 0.5% SDS.
13. Proteinase K, 10 mg/ml stock. Store at −20 °C.
14. PCI (phenol–chloroform–isoamyl alcohol 25:24:1).
15. Linear polyacrylamide (LPA), 5 mg/ml stock, used as a coprecipitant.
16. Refrigerated microcentrifuge.

2.4. Amplification and Analysis

1. Genomeplex® complete whole genome amplification kit WGA2 (Sigma-Aldrich).
2. Real-time PCR reagents and apparatus.

3. Methods

3.1. Isolation of Genomic DNA

This protocol is suitable for isolating genomic DNA from 20,000 to 200,000 mammalian cells. If extracting DNA from higher number of cells, please refer to the standard MeDIP protocol (8). The use of LoBind microtubes in the initial step allows to minimize the loss of DNA during the procedure.

1. Resuspend the cells in a LoBind 1.5 ml microtube in 300 μl lysis buffer containing 20 μl proteinase K (10 mg/ml stock) (see Note 1).
2. Incubate at 55 °C in the dry heating block for 3 h.

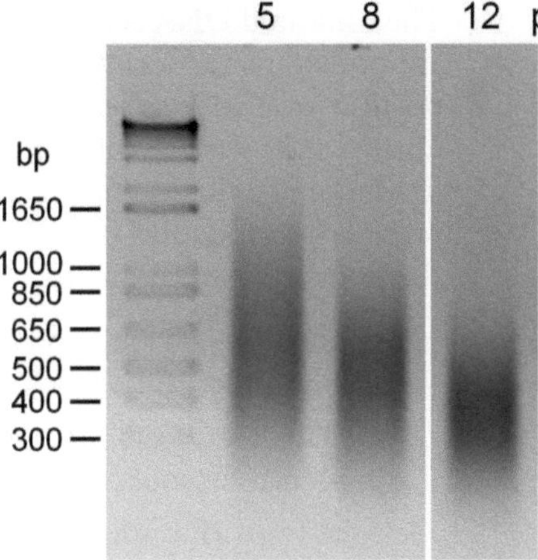

Fig. 2. Example of sonicated DNA migrating on a 1% agarose gel and stained with ethidium bromide. In this experiment, we sonicated 1 μg mouse genomic DNA in a volume of 150 μl H_2O and loaded 100 ng on the agarose gel. The numbers above the gel indicate the number of 30 s sonication pulses, which shows that 12 pulses leads to an optimal sonication under these conditions. Ideally, sheared DNA fragment should have an average size of 400 bp and be no longer than 1,000 bp.

3. Extract with 1 volume PCI. Transfer the upper phase in a new standard microtube (see Note 2).
4. Precipitate the DNA with 3 volumes (900 μl) ethanol containing 300 mM NaCl. Add 1 μl LPA if the amount of cells is <100,000. Store at −20 °C overnight.
5. Centrifuge for 40 min at full speed at 4 °C.
6. Wash the pellet with 500 μl ethanol 70% and centrifuge for 20 min at full speed at 4 °C.
7. Resuspend the pellet in 30 μl H_2O.
8. Quantify the amount of DNA with the Qubit fluorometer (see Note 3).

3.2. Sonication

Genomic DNA is randomly sheared by sonication to generate fragments between 200 and 1,000 bp. This step is crucial because the size of the DNA fragments will determine the resolution of the MeDIP assay.

1. Dilute 100–1,000 ng of genomic DNA in 150 μl H_2O in a 1.5 ml LoBind microtube.
2. Sonicate 12 times for 30 s (with 30 s intervals) with the Bioruptor (see Note 4).
3. If possible, verify the efficiency of the sonication by loading at least 50 ng on a 1% agarose gel. Ideally, the sheared fragments should have an average size of 400 bp and be no larger than 1,000 bp (Fig. 2) (see Note 5).

3.3. Immunoprecipitation of Methylated DNA

The sonicated DNA is then immunoprecipitated with a monoclonal antibody directed against 5-methylcytidine (5mC). Importantly, a portion of the sonicated DNA (at least 10 ng) should be left untreated to serve as input control. We describe here on optimized immunoprecipitation protocol for 75–100 ng sonicated DNA. For immunoprecipitation of larger amounts of DNA, you can refer to the standard MeDIP protocol (8).

1. Dilute 75–100 ng sonicated DNA in 135 µl H_2O in a 2 ml LoBind microtube (see Note 6).
2. Denature at 95 °C for 10 min in the dry heating block, and immediately cool on ice.
3. Add 15 µl IP buffer 10×.
4. Add 1/5 µl 5mC antibody (see Note 7).
5. Incubate for 2 h at 4 °C on the overhead rotator (see Note 8).
6. Prewash 2 µl magnetic beads with 500 µl PBS-BSA 0.05% in a standard 1.5 ml microtube. Incubate 5 min at room temperature with vortexing on the vortex Genie 2 (see Note 9).
7. Collect the magnetic beads on the magnetic rack and repeat the washing step with 500 µl PBS-BSA 0.05%.
8. Collect the magnetic beads on the magnetic rack and wash briefly with 500 µl IP buffer 1× to eliminate the traces of PBS-BSA.
9. Collect the magnetic beads on the magnetic rack and resuspend in 2 µl IP buffer 1× (see Note 9).
10. Transfer the magnetic beads to the sample.
11. Incubate for 2 h at 4 °C on the overhead rotator (see Note 10).
12. Collect the magnetic beads on the magnetic rack and wash with 700 µl IP buffer 1× by incubating 10 min at room temperature with vortexing on the vortex Genie 2.
13. Repeat the washing step with 700 µl IP buffer 1× twice.
14. Collect the magnetic beads on the magnetic rack and resuspend in 250 µl proteinase K digestion buffer.
15. Add 5 µl proteinase K and incubate 30 min at 50 °C in the dry heating block with shaking (see Note 11). Transfer the sample in a 1.5 ml LoBind microtube.
16. Extract with one volume (250 µl) PCI. Transfer the upper phase in a new standard 1.5 ml microtube (see Note 2).
17. Precipitate the DNA with 3 volumes (750 µl) ethanol containing 300 mM NaCl and 1 µl LPA. Store at −20 °C overnight.
18. Centrifuge for 40 min at full speed at 4 °C.
19. Wash the pellet with 500 µl ethanol 70% and centrifuge for 20 min at full speed at 4 °C.
20. Resuspend the pellet in 10 µl H_2O and store at −20 °C.

3.4. Amplification and Analysis

Enrichments in the MeDIP fraction can be measured by real-time PCR or microarray hybridization. If the MeDIP is performed with small amounts of starting material, it is necessary to perform a nonspecific amplification to increase the amount of DNA for downstream analyses, with the caveat that it might introduce amplification biases. We routinely use the Genomeplex® complete whole genome amplification kit WGA2 (Sigma-Aldrich), following the manufacturer's protocol on 10 ng of input DNA and the entire MeDIP product. This amplification step does not alter the MeDIP profiles at most targets, however we and others experienced that it can introduce amplification biases in the MeDIP enrichments at certain targets, especially the ones that are CpG-rich (9, 10). The MeDIP procedure can be validated after the whole genome amplification by performing real-time PCR in the input and MeDIP fraction on endogenous methylated and unmethylated controls. Real-time PCR can be performed on 10 ng of input and MeDIP DNA. Methylated controls are typically imprinting control regions or promoters of germline-specific genes that appear methylated in most somatic cells, whereas unmethylated negative controls are CpG island promoters that remain constitutively unmethylated in most cells or regions that contain very few CpGs. Table 1 gives a number of primers for real-time PCR in methylated and unmethylated controls that can be used to validate the MeDIP in human and mouse. Enrichments in the MeDIP fraction are calculated relative to one unmethylated negative control with the following formula: enrichment $= (IP_{target}/IN_{target})/(IP_{nc}/IN_{nc})$, with IP_{target} and IN_{target} representing the amount of the target sequence in the MeDIP and input fraction, and IP_{nc} and IN_{nc} representing the amount of the unmethylated negative control sequence in the MeDIP and input fraction. An example of typical enrichment profile obtained by MeDIP on small amounts of human or mouse genomic DNA is given in Fig. 3. Keep in mind to interpret the MeDIP results with caution because apparent low MeDIP enrichments can reflect in some cases an absence of sufficient CpGs in the target or a bias introduced during the whole genome amplification. For these reasons, we highly recommend to complement the MeDIP results with bisulfite sequencing at selected targets whenever possible. For genome-wide analyses, WGA-amplified input and MeDIP fractions can be differentially labeled with Cy3 and Cy5 and cohybridized to high-density oligonucleotides microarrays. It is also possible to couple the MeDIP procedure with deep sequencing; however, it is important to keep in mind that the MeDIP generates single-stranded DNA that cannot directly be used to generate libraries for deep sequencing. This can be circumvented by ligating the library adapter oligos between the sonication and the immunoprecipitation step (5, 7, 11, 12).

Table 1
Primer sequences for real-time PCR validation of MeDIP in somatic cells

Organism	Gene name	Comment	Sequence
Mouse	*Dpep3*	Methylated in somatic cells	Forward: GCAGGTTACCCACAGAGACG Reverse: GTGACCAAGACTGAGCACCA
Mouse	*Prss21*	Methylated in somatic cells	Forward: CAAGACGTTGGTGCCACTG Reverse: CACTGCCCCCAGTCTCAC
Mouse	*H19* ICR	Partially methylated in somatic cells	Forward: GCATGGTCCTCAAATTCTGCA Reverse: GCATCTGAACGCCCCAATTA
Mouse	IGd	Control sequence with very few CpGs	Forward: CCCTCTGGCCCTGAATTTAT Reverse: CACCCAGCAATGCTTCAGT
Mouse	*Tbx15*	Unmethylated CpG island	Forward: TCCCCCTTCTCTTGTGTCAG Reverse: CGGAAGCAAGTCTCAGATCC
Human	*TSH2B*	Methylated in somatic cells	Forward: CAGACATCTCCTCGCATCAA Reverse: GGAGGATGAAAGATGCGGTA
Human	*BRDT*	Methylated in somatic cells	Forward: CCCTTTGGCCTTACCAACTT Reverse: GCCCTCCCTTGAAGAAAAAC
Human	IG5	Control sequence with very few CpGs	Forward: GACCATGTCCAGGCAAAAGT Reverse: AGGCTCCTACAGACGTGGAA
Human	*UBE2B*	Unmethylated CpG island	Forward: CTCAGGGGTGGATTGTTGAC Reverse: TGTGGATTCAAAGACCACGA

4. Notes

1. Due to the viscosity of the solution after cell lysis, we recommend to add the proteinase K to the lysis buffer before mixing it with the cells.
2. We recommend not using a LoBind microtube for the ethanol precipitation step because we experienced that it hinders the recovery of the pellet.
3. Typically, up to 100 ng genomic DNA can be recovered from 20,000 diploid mammalian cells.
4. Because the sonication efficiency varies with DNA quality, quantity and sonicator settings, it is highly recommended to first verify the efficiency of the sonication on nonprecious DNA. For this, sonicate nonprecious DNA in the same conditions and verify the size of the sheared DNA by loading at least 50 ng sonicated DNA on a 1% agarose gel. If the sonication needs to be optimized for <50 ng, you can sonicate several tubes in parallel and then check the size of the pooled DNA.

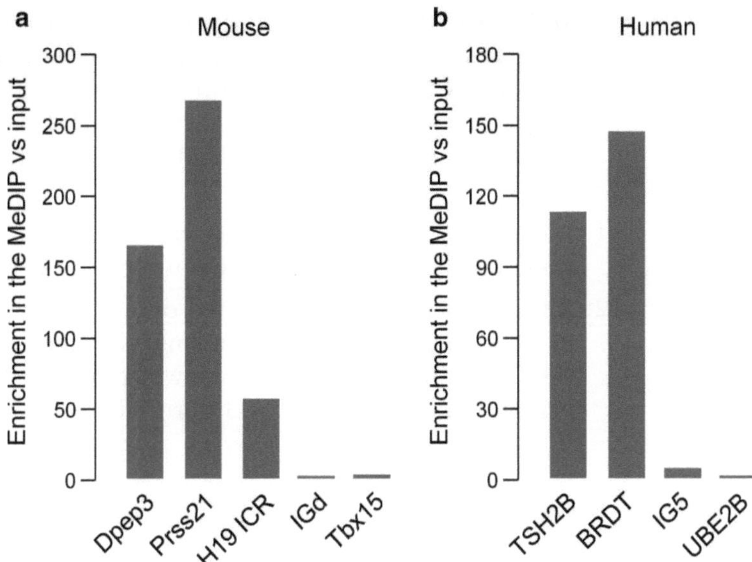

Fig. 3. Examples of MeDIP enrichment profiles measured by real-time PCR. MeDIP was performed with 200 ng sonicated DNA from mouse E9.5 embryos (**a**) or human primary fibroblasts (**b**), followed by whole genome amplification. The graphs show the enrichment in the MeDIP versus input fraction of methylated sequences over unmethylated negative controls. Values are normalized with the formula $(IP_{target}/IN_{target})/(IP_{nc}/IN_{nc})$ to the unmethylated negative controls (nc) IGd (mouse) or *UBE2B* (human), whose ratios are set to 1 (see Note 12).

To ensure a better consistency in the sonication, we also recommend always filling up the sonicator tube holder with empty microtubes containing 150 μl H_2O.

5. If sonicating small amounts of DNA, it is possible to monitor the efficiency of the sonication by sonicating in parallel an equal amount of nonprecious DNA and loading it on a 1% agarose gel.
6. We use 2 ml instead of 1.5 ml microtubes because it allows for a better mixing of small volumes.
7. For a better reproducibility, we suggest to first dilute 1 μl 5mC antibody in 4 μl IP buffer 1×, and then add 1 μl of the dilution to the sample. This amount of antibody has been optimized for somatic cells with a standard methylation level. It can be adjusted for cells with an unusual hyper- or hypomethylation state.
8. Verify that the sample is properly mixed in the microtube. Alternatively, you can also perform gentle horizontal shaking on a vortex Genie 2 with a platform for microtubes placed in a 4 °C cold room.
9. If performing several MeDIPs in parallel, you can wash all the beads together. For a better consistency, we also recommend to wash more beads than necessary for the experiment. For instance in the case of 5 MeDIPs, wash 15 μl of beads, resuspend in 15 μL IP buffer 1×, and use 2 μl per MeDIP reaction.

10. Verify that the sample is properly mixed in the microtube. Alternatively, you can also perform gentle horizontal shaking on a vortex Genie 2 with a platform for microtubes placed in a 4 °C cold room. In that case, the horizontal shaking should be strong enough to prevent the sedimentation of the magnetic beads.
11. The shaking speed must be sufficient to prevent the sedimentation of the magnetic beads. We routinely use 900 rpm.
12. Enrichment values of methylated over unmethylated controls in the MeDIP fraction can sometimes vary drastically from 100 to 10,000 between experiments, which reflects the stochastic nature of the MeDIP and WGA procedures.

Acknowledgments

This work was supported by the CEFIC Long Research Initiative (LRI-EMSG49-CNRS-08).

References

1. Meissner A (2010) Epigenetic modifications in pluripotent and differentiated cells. Nat Biotechnol 28:1079–1088
2. Laird PW (2010) Principles and challenges of genomewide DNA methylation analysis. Nat Rev Genet 11:191–203
3. Lister R, Pelizzola M, Dowen RH et al (2009) Human DNA methylomes at base resolution show widespread epigenomic differences. Nature 462:315–322
4. Weber M, Davies JJ, Wittig D et al (2005) Chromosome-wide and promoter-specific analyses identify sites of differential DNA methylation in normal and transformed human cells. Nat Genet 37:853–862
5. Down TA, Rakyan VK, Turner DJ et al (2008) A Bayesian deconvolution strategy for immunoprecipitation-based DNA methylome analysis. Nat Biotechnol 26: 779–785
6. Pelizzola M, Koga Y, Urban AE et al (2008) MEDME: an experimental and analytical methodology for the estimation of DNA methylation levels based on microarray derived MeDIP-enrichment. Genome Res 18: 1652–1659
7. Chavez L, Jozefczuk J, Grimm C et al (2010) Computational analysis of genome-wide DNA methylation during the differentiation of human embryonic stem cells along the endodermal lineage. Genome Res 20:1441–1450
8. Mohn F, Weber M, Schubeler D et al (2009) Methylated DNA immunoprecipitation (MeDIP). Methods Mol Biol 507:55–64
9. Borgel J, Guibert S, Li Y et al (2010) Targets and dynamics of promoter DNA methylation during early mouse development. Nat Genet 42:1093–1100
10. Jia J, Pekowska A, Jaeger S et al (2010) Assessing the efficiency and significance of methylated DNA immunoprecipitation (MeDIP) assays in using in vitro methylated genomic DNA. BMC Res Notes 3:240
11. Li N, Ye M, Li Y et al (2010) Whole genome DNA methylation analysis based on high throughput sequencing technology. Methods 52:203–212
12. Harris RA, Wang T, Coarfa C et al (2010) Comparison of sequencing-based methods to profile DNA methylation and identification of monoallelic epigenetic modifications. Nat Biotechnol 28:1097–1105

Chapter 10

Chromatin Immunoprecipitation to Characterize the Epigenetic Profiles of Imprinted Domains

Purnima Singh and Piroska E. Szabó

Abstract

Imprinted genes are marked by parental allele specific DNA methylation and histone modifications which regulate their monoallelic expression. Chromatin immunoprecipitation (ChIP) is the technique of choice to characterize the histones associated with either maternal or paternal chromosomes. To study allele-specific chromatin composition at imprinted regions, the method has to be efficient to work on limiting amount of starting material, and specific enough to recognize one of the parental alleles. We optimized the commonly used ChIP technique for efficient recovery of one parental allele from small number of cells. We provide examples to show that this ChIP protocol can specifically distinguish between parental alleles in mouse embryo fibroblasts carrying maternal and paternal duplication of mouse distal Chr7 and also in normal mouse embryo fibroblasts carrying single nucleotide polymorphism at imprinted regions.

Key words: Genomic imprinting, Monoallelic, Differentially methylated regions, Chromatin immunoprecipitation, Histone modifications

1. Introduction

Genomic imprinting is a form of epigenetic regulation whereby the allele inherited from either the mother or father is functional (1, 2). Imprinted genes often occur in clusters and are often associated with differentially methylated regions (DMRs). DMRs are classified as imprinting control regions (ICRs) when they determine the parental allele-specific expression of the imprinted genes in their respective domain. The promoters of imprinted genes and also the DMRs and ICRs exhibit parental allele specific DNA methylation and post translational histone modifications. Chromatin immunoprecipitation (ChIP) (3) is performed to determine allele specific histone modifications at the maternal or paternal allele

Nora Engel (ed.), *Genomic Imprinting: Methods and Protocols*, Methods in Molecular Biology, vol. 925, DOI 10.1007/978-1-62703-011-3_10,

at a DNA sequence in vivo. Histones are cross-linked to DNA with formaldehyde followed by shearing the DNA into smaller fragments by sonication. Antibodies against specific histones are used to precipitate the complex. The cross-linking is reversed, and the bound proteins are digested using proteinase K to release the DNA fragments which can be analyzed for enrichment using locus specific PCR primers (4, 5).

One challenge in analyzing allele-specific epigenetic features is to differentiate between the parental alleles. One possible approach is to use cells with uniparental duplications, where two copies of a chromosome segment are inherited from one parent and no copies from the other parent. Chromatin from primary mouse embryo fibroblast (MEFs) carrying maternal and paternal duplication of distal Chr7, MatDup.dist7 and PatDup.dist7, can be used to characterize maternal and paternal allele specific histone modifications, respectively at the *H19/Igf2* and *Cdkn1c/Kcnq1* domains (1, 2), because these are both located along the duplicated region of Chr7 (6, 7). The *H19-Igf2* region is controlled by an ICR, which is methylated in the paternal chromosome but is associated with the insulator protein CTCF in the maternal chromosome. The *Cdkn1c/Kcnq1* domain is controlled by the KvDMR1 that carries DNA methylation in the maternal allele and functions as the promoter for the noncoding regulatory RNA, *Kcnq1ot1* in the paternal allele. Another possible approach for allele-specific chromatin analysis utilizes DNA sequence differences, such as single nucleotide polymorphisms (SNPs) between the mother and father to distinguish the parental alleles in normal cells (8, 9).

The other challenge in analyzing imprinted genes is, that the starting material can be limiting. General ChIP protocols suggest using 25 μg of chromatin or 10^7 cells per reaction, which is hard to obtain when only small embryos or small populations of purified embryonic cells are available. The following ChIP protocol is routinely performed in our laboratory using 4 μg of chromatin or 400,000 cells. Using this optimized technique we demonstrate here that an active histone modification mark, H3K4me2, is associated with the unmethylated maternal and paternal allele of *H19/Igf2* ICR and KvDMR1, respectively and a repressive mark, H3K9me3, is associated with the reciprocal, methylated alleles at these DMRs. This ChIP protocol has been used in combination with tiling microarrays to reveal chromosome-wide allele-specific features along distal Chr7 and distal Chr15 (7). It was also used in combination with multiplex single nucleotide primer extension (SNuPE) assays in the sequenom-allelotyping platform to generate histone modification profiles across multiple DMRs in the mouse genome (8, 9). We also provide an even smaller scale version (*), which gives reproducibly allele-specific results using only 100,000 purified mouse embryonic germ cells (10).

2. Materials

All buffers are prepared in tissue culture grade water, filter-sterilized, and stored at appropriate temperature.

1. DMEM (Dulbecco's modified eagle medium).
2. FBS (Fetal bovine serum).
3. Formaldehyde 37%.
4. PBS (phosphate buffered saline without calcium and magnesium).
5. Complete protease inhibitor cocktail tablets, (Roche applied science) 50×, one tablet dissolved in 1 ml water.
6. ChIP Dilution Buffer: 0.01% SDS, 1.1% Triton X-100, 1.2 mM EDTA, 16.7 mM Tris–HCl pH 8.1, and 167 mM NaCl. Store at RT. Add 4 μl/ml of 50× complete protease inhibitor cocktail before use.
7. ChIP Lysis Buffer: 1% SDS, 10 mM EDTA, and 50 mM Tris–HCl, pH 8.1. Store at RT. Add complete protease inhibitor cocktail to 1× concentration before use.
8. Preblocked A/G Beads: 2 ml A/G Beads (Santa Cruz# sc-2003), 20 μl herring sperm DNA (Invitrogen# 15634-017), 56 μl 50× complete protease inhibitor cocktail, 280 μl 10 mg/ml BSA (PENTEX), and 2 ml ChIP dilution Buffer. Preblock the beads by gentle rotation for 2 h in the cold room in 15 ml falcon tube. Spin at 250 × *g* for 5 min. Discard supernatant and resuspend the beads in 2 ml of ChIP dilution buffer supplemented with 40 μl of 50× complete protease inhibitor cocktail. Store at 4 °C for up to a month (see Notes 1–3).
9. Buffer A (Low-salt Buffer): 0.1% SDS, 1.0% Triton X-100, 2.0 mM EDTA, 20 mM Tris–HCl, pH 8.1, 150 mM NaCl. Store at 4 °C.
10. Buffer B (High-salt Buffer): 0.1% SDS, 1.0% Triton X-100, 2.0 mM EDTA, 20 mM Tris–HCl, pH 8.1, 500 mM NaCl. Store at 4 °C.
11. Buffer C (LiCl Buffer): 0.25 M LiCl, 1.0% Igepal-CA630 (also known as NP-40), 1.0% Sodium deoxycholate, 1.0 mM EDTA, 10 mM Tris–HCl, pH 8.1. Store at 4 °C.
12. ChIP Elution Buffer: 1% SDS, 0.1 M $NaHCO_3$.
13. Proteinase K.
14. TE Buffer: 1.0 mM EDTA, 10 mM Tris–HCl, pH 8.0.
15. 3 M Sodium Acetate (pH 5.5).
16. Phenol (equilibrated to pH 8.0).
17. Chloroform.
18. Linear polyacrylamide (11).
19. Ethanol.

3. Methods

3.1. Cross-linking and Cell Harvesting

1. Grow primary mouse embryo fibroblasts (MEFs) close to confluency in a 150 mm dish containing 25 ml of culture medium (DMEM supplemented with 6% fetal bovine serum) to obtain about 2.5×10^7 cells per plate. Prepare chromatin in two ways, using N-ChIP or X-ChIP conditions, to test out new antibodies. *Trypsinized and flow-sorted cells, such as primordial germ cells, are cross-linked in suspension (see Notes 4 and 5).

3.1.1. Preparation of N-Link Chromatin

1. Add formaldehyde directly into the media to a final concentration of 1% (675 μl of 37% formaldehyde per 25 ml medium) inside a fume hood. It is best to tilt the plate and quickly add formaldehyde by touching the side of the plate where medium level is highest and without delay mixing up the medium by swirling. *For formaldehyde cross-linking of 100K cells, first pellet cells at $728 \times g$ for 10 min in an eppendorf tube, then remove medium and add 500 μl of PBS with 27 μl of 37% formaldehyde. Suspend cells and mix by pipetting.
2. Incubate the plate at room temperature with gentle agitation for exactly 2 min on a horizontal shaking platform. *Tubes are placed on a rotating platform for 2 min.
3. Stop the cross-linking reaction by adding 2.6 ml of 1.25 M glycine solution, mix well immediately. *For 100K cells, add 25 μl of 1.25 M glycine and spin at $728 \times g$ for 10 min.
4. Remove media inside a fume hood by pouring it from the plate into a glass beaker and blotting the last drop off on a paper towel. Wash cells twice with 20 ml ice-cold PBS. *Remove liquid from 100K pellet by pipetting. To wash 100K cells, suspend the cell pellet in 700 μl of cold PBS and spin.
5. Aspirate PBS completely after the second wash. Place cell culture dish or *eppendorf tube on ice.
6. Scrape cells off the plate in 5 ml ice-cold PBS with a plastic scraper and collect into a 15 ml falcon tube by pipetting.
7. Pellet the cells by centrifugation for 5 min at $250 \times g$ at 4 °C. Aspirate supernatant without touching the cell pellet, keep cell pellet on ice. *For 100K cells, aspirate the last drop using a pulled pasteur pipette attached to a mouthpiece under a dissecting microscope so as not to dislodge the cell pellet.

3.1.2. Preparation of X-Link Chromatin

1. X-ChIP is the same as N-ChIP, except for step 2.
2. Incubate at room temperature for exactly 10 min.

3.2. Sonication

1. Resuspend pelleted cells in 750 μl ChIP Lysis buffer containing complete protease inhibitors. *Add 100 μl of lysis buffer to 100K cells, mix, snap-freeze in liquid N_2, and store at −80 °C. These small samples are sonicated on the day of use.
2. Sonication is carried out in 1.7 ml eppendorf tubes using Branson Sonifier cell Disruptor-350 with a micro tip. Pulse three to four times for 10 s at 40% duty cycle and 4 output control. Keep samples on ice at all times. In our experience, N-ChIP and X-ChIP samples need to be sonicated three and four times, respectively.
3. If using Diagenode Bioruptor UCD-200, sonicate at high "H" setting with 15 s "ON" 1 min "OFF" for 5 min. Repeat six times with 1 min on ice between each cycle. Diagenode Bioruptor is preferred for sonicating 100K cells in 100 μl volume to prevent frothing and loss of chromatin on the sonicator tip.
4. Centrifuge the sheared chromatin at 15,710×*g* for 5 min at 4 °C. Transfer the supernatant containing the chromatin to a fresh tube. Store chromatin on ice in cold room during quantification/before freezing at −80 °C or use directly for ChIP.

3.3. Determination of Chromatin DNA Concentration and Fragment Size

1. To determine the concentration of DNA in the sonicated chromatin, take a small aliquot and reverse-cross-link it. For a 25 μl chromatin preparation, add 25 μl of ChIP lysis buffer, add NaCl to a final concentration of 0.3 M (3 μl of 5 M NaCl) and incubate at 65 °C for 4 h.
2. Add 1 μl each of 0.5 M EDTA, 1 M Tris–HCl, pH 6.5, and 20 mg/ml proteinase K. Incubate at 55 °C for 1 h.
3. Bring the volume to 100 μl by adding 50 μl of TE (pH 8.0). Extract with phenol–chloroform. Add 50 μl of phenol and mix vigorously. Add 50 μl of chloroform and mix well. Centrifuge at 15,710×*g* for 10 min. Collect the top, aqueous phase and also the interphase at this point. Add 100 μl of chloroform and mix well. Spin at 15,710×*g*. Collect only the top aqueous phase this time and transfer it to a fresh tube.
4. Precipitate DNA: add 3 μl of linear polyacrylamide and 1/10 volume of 3 M sodium acetate, pH 5.5 and mix. Add 2.5 volumes of chilled 100% ethanol, mix, and freeze sample at −80 °C for 30 min. To collect DNA pellet, centrifuge sample at 15,710×*g* for 15 min. Wash DNA pellet with chilled 70% ethanol. Resuspend DNA in 25 μl Tris-EDTA buffer (TE), pH 8.0.
5. Check DNA concentration using Nanodrop (see Note 6).

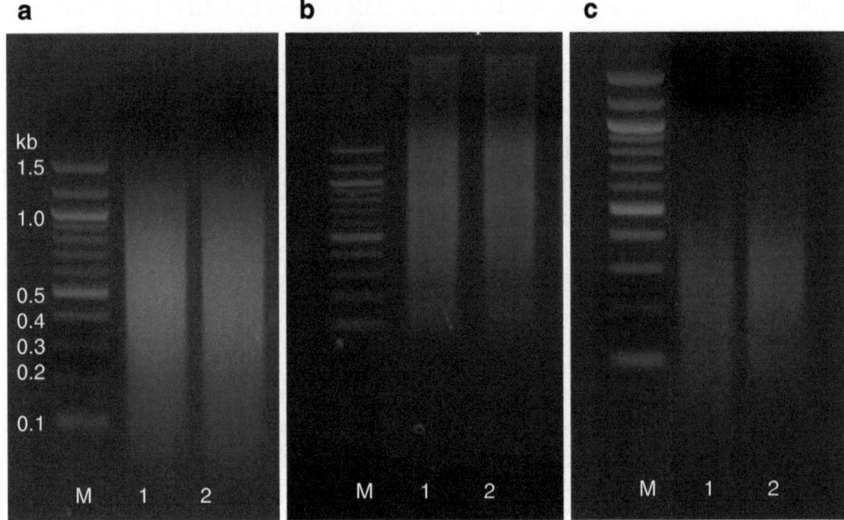

Fig. 1. Testing the efficiency of chromatin sonication. Agarose gel electrophoresis of sonicated DNA to determine the fragment size of the reverse cross-linked chromatin (**a**) Ideal DNA sonication is shown with fragment size ranging up to 1,200 bp with median fragment length of ~500 bp. The small fraction <200 bp on this gel is RNA and can be eliminated by RNAse treatment. (**b**) Undersonication. High molecular weight unfragmented DNA can be seen on top of the gel. (**c**) Oversonication. The majority of fragments are below 400 bp. (*lane 1*, M: 1 kb-ladder, *lane 2*, MatDup.dist7, and *lane 3*, PatDup.dist7 DNA).

6. Load 1 μg of DNA on a 1.5% agarose gel and run it in 1×TAE (Tris-Acetate-EDTA) buffer to check the size of chromatin fragments. They should range from 0.2 to 1.5 kb. (Fig. 1) (see Note 7).
7. Dilute chromatin preparations (which were stored on ice to this point) to 0.4 μg/μl concentration with ChIP lysis buffer containing 1× complete protease inhibitors. You will use 10 μl (4 μg nucleic acid equivalent chromatin) per ChIP reaction. At this point, chromatin preparations can be used right away or 100 μl aliquots can be snap-frozen in liquid nitrogen for long-term storage at −80 °C.

3.4. Immunoprecipitation

1. Quickly defrost frozen sheared chromatin sample on the day of ChIP. *100K cell aliquots can be defrosted, combined and sonicated in a larger batch on the day of ChIP. Sonication efficiency of small aliquots is similar to larger samples. This can be optimized and tested beforehand.
2. Take out 10 μl sheared chromatin for "input."
3. Use 10 μl of chromatin for each IP. Add 90 μl ChIP Lysis Buffer with 1× complete protease inhibitors to the 10 μl chromatin fraction in an eppendorf tube. Dilute tenfold with ChIP Dilution Buffer containing 4 μl/ml 50× complete. *100K cells had been already suspended in 100 μl ChIP Dilution Buffer containing 4 μl/ml of 50× complete. Dilute these tenfold with ChIP dilution buffer containing 4 μl/ml 50× complete.

4. Preclear chromatin with preblocked Protein A/G agarose beads. Add 50 μl Protein A/G agarose into each reaction. Rotate at 4 °C for 1 h. Pellet beads by centrifugation for 5 min at 250 × *g*. Transfer supernatant chromatin to a new tube (see Note 8).
5. Add 4 μg of the antibody to the supernatant and incubate overnight at 4 °C on a rotating platform. Set up one sample with normal rabbit IgG antibody as negative control (see Notes 9–11).
6. To collect the antibody–chromatin complexes, the next morning add 50 μl Protein A/G agarose beads to each IP. Rotate for 2 h at 4 °C.
7. To collect beads centrifuge for 1 min at 92 × *g* and discard supernatant.
8. Wash agarose beads on a rotating platform for 5 min with 1 ml of the ice-cold buffers A, B and C. Collect beads by centrifugation after each wash at 1,000 rpm for 1 min at 4 °C.
   (a) Buffer A with 2 μl/ml 50× complete in cold room.
   (b) Buffer B with 2 μl/ml 50× complete in cold room.
   (c) Buffer C with 2 μl/ml 50× complete in cold room.
   (d) TE Buffer with 2 μl/ml 50× complete at room temperature (repeat this wash two more times)
9. Remove as much buffer as possible without disturbing the beads (see Notes 12–14).

3.5. Elution, Reverse Cross-linking, and Proteinase K Treatment

1. Add 150 μl freshly made ChIP Elution buffer to the washed beads (see Note 15). Incubate for 15 min at room temperature with gentle agitation.
2. Pellet beads by centrifugation at 1,487 × *g* for 2 min. Transfer the ChIP eluate supernatant to a new tube.
3. Repeat elution. Combine the second eluate with the first one.
4. Reverse-cross-link the eluted DNA by adding 18 μl 5 M NaCl per 300 μl eluate followed by incubation for 4 h at 65 °C (see Note 16).

   (Remember to reverse-cross-link the reserved "input" DNA samples also at this point)
5. Add 6 μl each of 0.5 M EDTA, 1 M Tris–HCl, pH 6.5, and 20 mg/ml proteinase K. Incubate for 1 h at 55 °C (see Note 17).

3.6. DNA Isolation and Purification

1. Add 1.5 ml of Binding buffer QG (QIAquick Kit, QIAGEN).
2. Purify DNA according to the manufacturer's instruction. Wash with PE four times(see Note 18).
3. Recover DNA with 100 μl warm elution buffer. Store at −20 °C. Samples are ready for subsequent qPCR assays or sequenom allelotyping (Figs. 2 and 3) (see Notes 19–23).

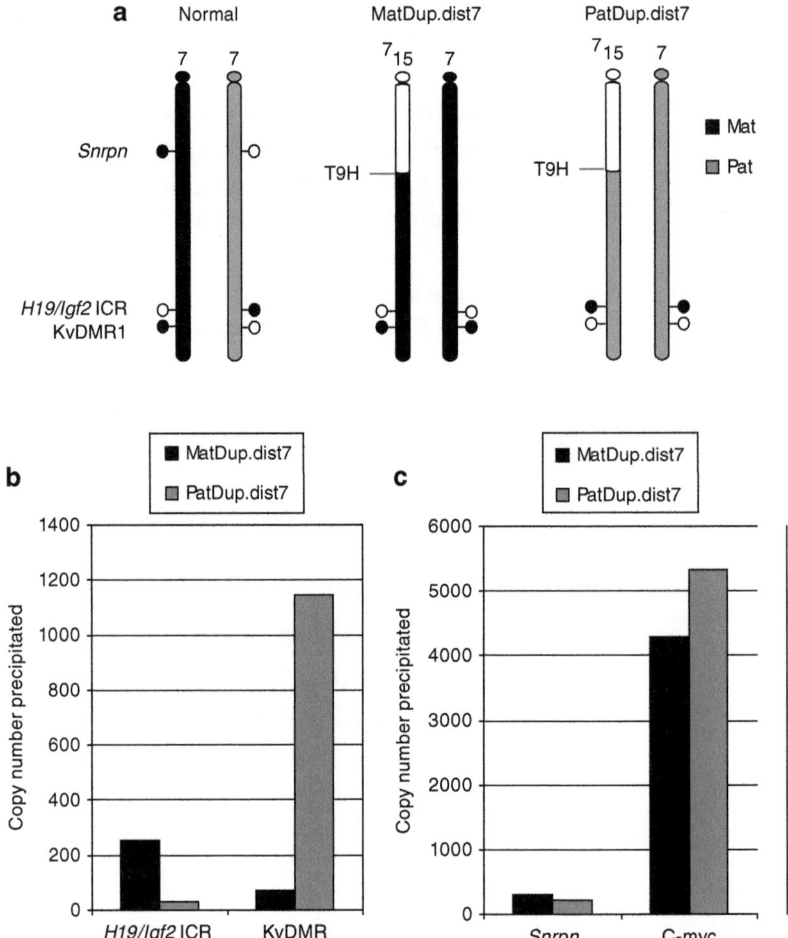

Fig. 2. Detecting allele-specific H3K4me2 enrichment at imprinted regions using cells with uniparental duplications of distal Chr7. (**a**) MatDup.dist7 and PatDup.dist7 MEFs carry two maternal (*black*) or two paternal (*grey*) copies of chromosome 7 regions, located distally to the T9H translocation breakpoint. These cells allow the analysis of allele-specific marks at the *H19/Igf2* ICR and at the KvDMR1 along the maternally or paternally duplicated distal chromosome 7 region. Parental allele-specific methylation and hypomethylation of the DMRs is shown by *closed* and *open lollipops*, respectively. (**b**) Real-time PCR was used to quantify an active chromatin mark, H3K4me2, levels at specific loci in MatDup.dist7 and PatDup.dist7 MEFs at two imprinted regions. The paternally methylated *H19/Igf2* ICR shows H3K4me2 in the MatDup.dist7 MEFs, whereas the maternally methylated KvDMR1 exhibits H3K4me2 enrichment in PatDup.dist7 MEFs. (**c**) Control regions. The control *Snrpn* DMR is located outside of the duplicated chromosome region, therefore the allele-specificity of positive H3K4me2 enrichment cannot be discerned at this locus. The control c-myc promoter is constitutively active and enriched in H3K4me2.

3.7 Multiplex Quantitative Real-Time PCR for Allele Specific Histone Modifications

1. Dissolve all the primers and probes (Table 1) in TE (pH 8.0) to a final concentration of 100 μM. The oligo tubes from IDT are spun briefly before dissolving, as the pellet may dislodge during shipment. Mix equal volumes of the upper and lower (U + L) primers. These can be stored at −20 °C (see Note 24).
2. Prepare MIQ 5× Buffer by mixing equal volumes of 10× iTaq Buffer and 5 mM $MgCl_2$ from the iTaq DNA Polymerase kit (Bio-Rad #170-8875).

Table 1
Primers for multiplex real-time PCR

DMR	Probe sequence and dye	PCR primers
H19-Igf2 ICR-FAM	ACATTCACACGAGCATCCA GGAGGC FAM	CACTTACACCCAGGACTCAAAGG GCGTATAAACCCCACAACTGATTC
KvDMR1	CCGCAGTGGCTCCGTATTCGTT TEX	CGGCTGGGCTCCATCTTC CGACCTCGGGGCTCAAAG
Snrpn promoter	CATGCGTCCCAGGCAATGGCTGC TAMRA	TCCTTTTGGTAGCTGCCTTTTGG CCGCAATGGCTCAGGTTTGTC
c-myc promoter	CTGCCTCGCTCCACACAA TACGCCA Cy5	AGATAACTCATTCGTTCGTCCTTCC TGTGTTCTTGCCCTGCGTATATC

3. For amplification standards use a sonicated mouse genomic DNA dilution series: 10 ng/μl, 1 ng/μl, 0.1 ng/μl, 0.01 ng/μl, 0.001 ng/μl, and a no-template control. In the iQ5 real-time plate setup menu, define units as copy numbers (10,000; 1,000; 100; 10; 1; and 0 copies, respectively). This calculation considers that one diploid mouse genome equals 6 pg of genomic DNA. Set up the multiplex PCR reaction as follows:

	X1 (μl)
MIQ 5× Buffer	5.0
dNTP mix (25 mM each)	0.45
iTaq (Biorad) 5 U/μl	0.325
Primer (U+L) (4)	0.2 Each pair (up to 5 total pairs)
Probes (4)	0.075 Each probe (up to 5 total probes)
DNA	3 μl of ChIP eluate or 30 ng input or standard genomic DNA
H_2O	Up to 25 μl

The following PCR parameters are used to run the reactions in Bio-Rad iQ5-Thermal Cycler:

95 °C	3 min
40 cycles	
95 °C	30 s
55 °C	45 s

To illustrate the applicability of this ChIP protocol to allele-specific analysis, we precipitated chromatin from MatDup.dist7 and PatDup.dist7 MEFs using the H3K4me2 antibody and amplified the *H19/Igf2* ICR, the KvDMR, as well-as the control *Snrpn* promoter and c-myc promoter regions. Four sets of primers and probes (Table 1) were used in a multiplex real-time PCR reaction (Fig. 2). We also performed ChIP with H3K4me2 and H3K9me3 antibodies using chromatin from normal 129XJF1 and JF1X129 MEFs and submitted aliquots of the ChIP elution for multiplex sequenom allelotyping as we have done earlier (8, 9). The percent H3K4me2 and H3K9me3 enrichment in the total immunoprecipitation was measured utilizing SNPs between the parental alleles (Fig. 3a, b). The

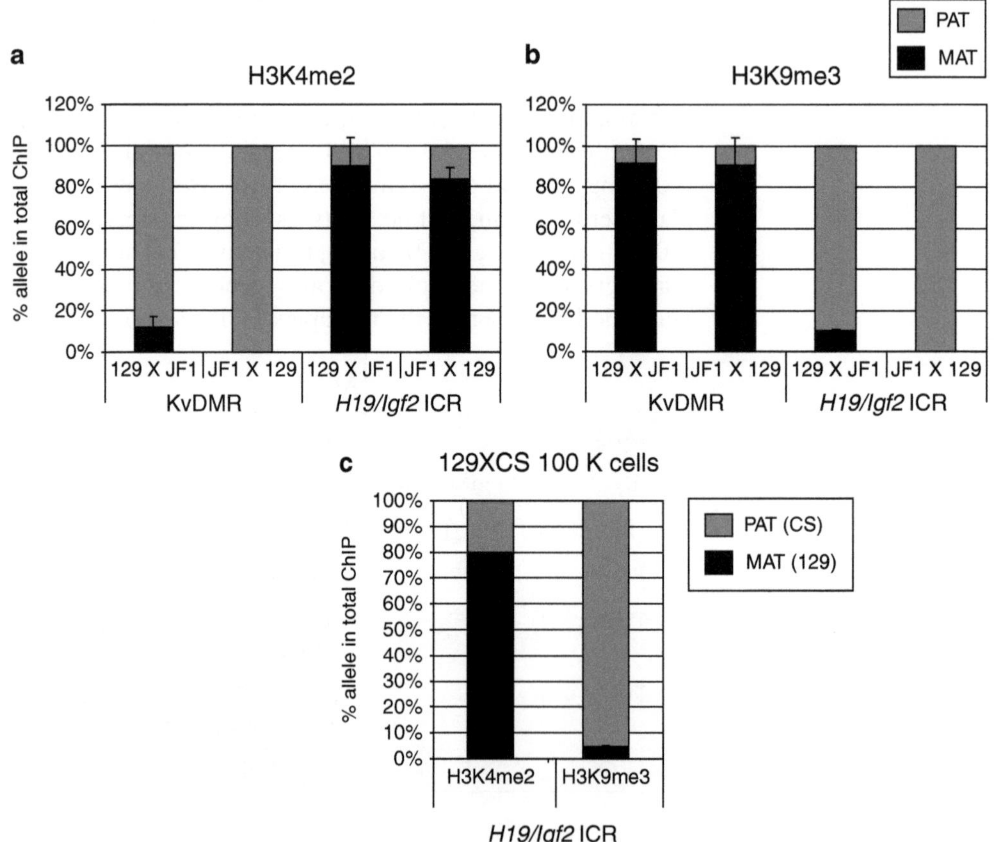

Fig. 3. Detecting allele-specific chromatin in normal cells. Chromatin was prepared from 129XJF1 (129 mother and JF1 father) and JF1X129 (JF1 mother and 129 father) MEFs and was subjected to ChIP using (**a**) H3K4me2 and (**b**) H3K9me3 antibodies. Sequenom allelotyping assays were used to measure the percent maternal and paternal component in the total ChIP DNA at the KvDMR1 and at the *H19/Igf2* ICR. H3K4me2 is enriched in the unmethylated paternal allele (PAT) at the KvDMR1 and maternal allele (MAT) at the *H19/Igf2* ICR. H3K9me3 shows enrichment at the reciprocal, methylated, alleles. (**c**) ChIP using 100,000 MEFs. These MEFs were obtained by mating a 129 mother to a CAST/Ei (CS) father. Sequenom allelotyping shows correct allele specific enrichment in 129XCS MEFs for H3K4me2 and H3K9me3 at the *H19/Igf2* ICR similar to ChIP from 4 μg chromatin.

results for the H3K4me2 antibody were in agreement with the results of the MatDup.dist7-PatDup.dist7 experiment, showing active chromatin-specific H3K4me2 enrichment in the unmethylated DMR alleles. The repressing mark, H3K9me3, however, was present in the CpG-methylated alleles at both DMRs. The ChIP protocol using 100K 129XCS MEFs also revealed the correct allele specific bias (Fig. 3c). These experiments demonstrate that our optimized ChIP protocol efficiently and quantitatively recovers the correct parental alleles of imprinted regions from chromatin even when the cell numbers are limited.

4. Notes

1. Protein A/G agarose beads are preferred over either Protein A or G because they bind nearly all isotypes.
2. The protein A/G beads are preblocked to reduce nonspecific binding.
3. Gently swirl the beads to uniformly resuspend them before taking an aliquot. Vigorous shaking should be avoided, because the beads may stick to the side of tube and dry. Also, vortexing or high speed centrifugation may break the beads.
4. For most histone modifications N-linked chromatin works well. For certain nonhistone proteins like CTCF insulator and transcription factors X-linking is preferred. The optimal cross-linking strength should be experimentally determined for each antibody by ChIP, real-time PCR and allele-specific protocols.
5. Mouse embryos/organs can be used for chromatin preparation. The embryo/organ is first suspended in PBS and cross-linked in suspension (9).
6. Keep in mind that RNAse step is not necessary in this DNA preparation. Therefore, the DNA will have some contaminating RNA (4 μg nucleic acid equivalent chromatin has less than 4 μg DNA; 100K fetal germ cells contain 600 ng DNA and ~1 μg total nucleic acids).
7. A <1.2 kb smear on the gel is considered optimal shearing with ~500 bp as median fragment size. It is critical to obtain optimal size distribution of sonicated chromatin as fragment size >1.2 kb may result in high background from neighboring chromosomal regions. Fragment size <200 bp may result in inadequate labeling yields for downstream ChIP-chip experiments. Fragmentation needs to be optimized with each sonication device and cell type being used for the experiment.

Table 2
Volumes for preclearing the chromatin

	Chromatin	Lysis buffer	Dilution buffer	Preblocked A/G beads
Sonicated chromatin from plate	10 μl (~4 μg)	90 μl	900 μl	50 μl
Chromatin from 100K cells	100 μl (~1 μg)	0 μl	900 μl	50 μl

8. The preclearing step inhibits nonspecific DNA and protein binding to Protein A/G beads and thereby reduces the background. Preclearing can be done in large amount for multiple ChIPs by multiplying the volumes in Table 2.
9. Set up your first ever ChIP using 4 μg DNA equivalent chromatin and well-characterized antibodies, such as those in Fig. 3, which give reliable allele-specific enrichment at DMRs.
10. It is a good practice to test new batches of commercial antibodies for specificity in a dot blot before using them in ChIP (9).
11. A1:1 (4 μg antibody and 4 μg chromatin) ratio usually works for histone modifications. However, it is best to determine optimal concentrations for each antibody. Polyclonal antibodies give better results than monoclonal ones in ChIP.
12. Washing buffers can be prepared beforehand and stored at 4 °C. Add 50× complete protease inhibitor just before use.
13. 1 ml pipette tips can be used for washes. Gently dispense washing buffers from the side of the tubes. Avoid touching the beads when taking out supernatant after washes.
14. Remove buffer leaving ~50 μl behind, spin again, and use a 100 μl pipette tip to take out as much buffer as possible.
15. Prepare elution buffer not more than 10 min before use.
16. Reverse cross-linking removes the protein and allows purifying genomic DNA. Proteinase K treatment digests proteins in the DNA-protein mix.
17. 65 °C and 55 °C temperatures should be maintained during reverse cross-linking and proteinase K treatments for best results. Rotating of the samples is not required.
18. ChIP protocols suggest stopping/freezing after reverse cross-linking or proteinase K treatment and continuing the process the next day. In our hands finishing all the steps on the second day gives the best results.

19. If washes with PE buffer are done less than four times, SDS from the elution buffer may remain and you may see SDS crystals in the eluate.
20. Repeat freeze-thawing or long-term storage of ChIP-precipitated DNA should be avoided.
21. The ChIP-precipitated DNA can be extracted using phenol–chloroform and precipitated using linear polyacrylamide or glycogen (small scale ChIP). However, if ChIP DNA is to be amplified for ChIP-on chip, it is best to use Qiagen columns for purification, which removes excess salts.
22. The copy number of immunprecipitated DNA should be higher for specific antibodies than for nonspecific IgG. Using real-time PCR we generally measure less than 10 copies for nonspecific IgG and above 20 copies, up to 10,000 copies for specific antibodies from a 3 μl ChIP eluate (Fig. 2) (9). The downstream allele-specific measurement is more accurate with higher copy numbers for the specific region in the ChIP elution.
23. ChIPs and downstream allele-specific allelotyping assays should be done in duplicates or triplicates. Small standard deviation in these tests (Fig. 3) indicates that the precipitation worked and there is real enrichment with a specific antibody at a specific site. High standard deviation is a warning signal suggesting that the PCR randomly amplified background precipitation.
24. Reciprocal mouse crosses substantiate the allele-specific findings (Fig. 3).
25. Use the Beacon Designer software for four-color real-time qPCR probe and primer design for other regions or consult our paper for more DMR sets (9). Multiplexing in the real-time PCR provides internal controls, saves time and importantly, saves the majority of the ChIP elution for downstream processes, such as sequenom-allelotyping or amplification for microarray hybridization.

Acknowledgments

We thank Jeff Mann for the MatDup.dist7 and PatDup.dist7 MEFs, Li Han for her work on the initial phases of this project, and Diana Tran for her comments on the manuscript. This work was supported by a Public Health Service grant (GM064378) from the National Institute of General Medicine to P.E.S.

References

1. Ideraabdullah FY, Vigneau S, Bartolomei MS (2008) Genomic imprinting mechanisms in mammals. Mutat Res 647:77–85
2. Koerner MV, Barlow DP (2010) Genomic imprinting—an epigenetic gene-regulatory model. Curr Opin Genet Dev 20:164–170
3. Solomon MJ, Larsen PL, Varshavsky A (1988) Mapping protein-DNA interactions in vivo with formaldehyde: evidence that histone H4 is retained on a highly transcribed gene. Cell 53:937–947
4. Orlando V, Strutt H, Paro R (1997) Analysis of chromatin structure by in vivo formaldehyde cross-linking. Methods 11:205–214
5. Hebbes TR, Thorne AW, Crane-Robinson C (1988) A direct link between core histone acetylation and transcriptionally active chromatin. EMBO J 7:1395–1402
6. McLaughlin KJ, Szabo P, Haegel H, Mann JR (1996) Mouse embryos with paternal duplication of an imprinted chromosome 7 region die at midgestation and lack placental spongiotrophoblast. Development 122:265–270
7. Singh P, Wu X, Lee DH et al (2011) Chromosome-wide analysis of parental allele-specific chromatin and DNA methylation. Mol Cell Biol 31:1757–1770
8. Singh P, Han L, Rivas GE et al (2010) Allele-specific H3K79 Di- versus trimethylation distinguishes opposite parental alleles at imprinted regions. Mol Cell Biol 30: 2693–2707
9. Singh P, Cho J, Tsai SY, Rivas GE, Larson GP, Szabo PE (2010) Coordinated allele-specific histone acetylation at the differentially methylated regions of imprinted genes. Nucleic Acids Res 38:7974–7990
10. Lee DH, Singh P, Tsai SY et al (2010) CTCF-dependent chromatin bias constitutes transient epigenetic memory of the mother at the H19-Igf2 imprinting control region in prospermatogonia. PLoS Genet 6:e1001224
11. Gaillard C, Strauss F (1990) Ethanol precipitation of DNA with linear polyacrylamide as carrier. Nucleic Acids Res 18:378

Chapter 11

Quantitative Chromosome Conformation Capture

Raffaella Nativio, Yoko Ito, and Adele Murrell

Abstract

It is becoming increasingly apparent that chromatin is not randomly folded into the nucleus, but instead is highly organized into specific conformations within the nucleus. One consequence of such higher order structure is that chromatin looping can bring together genomic elements which are separated by several hundreds of kilobases, such as enhancers and promoters, and functionally facilitate their interaction. The Chromosome Conformation Capture (3C) assay is a powerful technique to detect looping structures and assess the probability of interaction between distant genomic elements (1–3). Here we describe the 3C methodology, its power, and limitations, together with the controls and normalization steps required for an accurate analysis.

Key words: 3C: Chromosome Conformation Capture, Cross-link, Digestion, Intramolecular ligation, 3C product, 3C template, PCR standard template, Association frequency

1. Introduction

The 3C technique detects the proximal association frequencies between distant genomic elements. If these elements represent regulatory regions, it is possible that they functionally interact. The 3C procedure consists of five experimental steps (2) (Fig. 1). The first step involves cross-linking of DNA and protein complexes. Formaldehyde cross-linking is commonly used, and such fixation enables a snapshot of interactions between DNA and proteins. The second step involves digestion of the cross-linked chromatin complexes with a restriction enzyme that cuts DNA in the regions under 3C analysis. This step is followed by ligation of the restricted chromatin at low DNA concentration so as to favor intramolecular ligation. The chromatin complexes are then reverse cross-linked and the DNA is purified. The purified DNA (3C template) contains religated sites in the genome that are enriched because of their proximity through looping interactions as well as religated

Nora Engel (ed.), *Genomic Imprinting: Methods and Protocols*, Methods in Molecular Biology, vol. 925,
DOI 10.1007/978-1-62703-011-3_11, © Springer Science+Business Media, LLC 2012

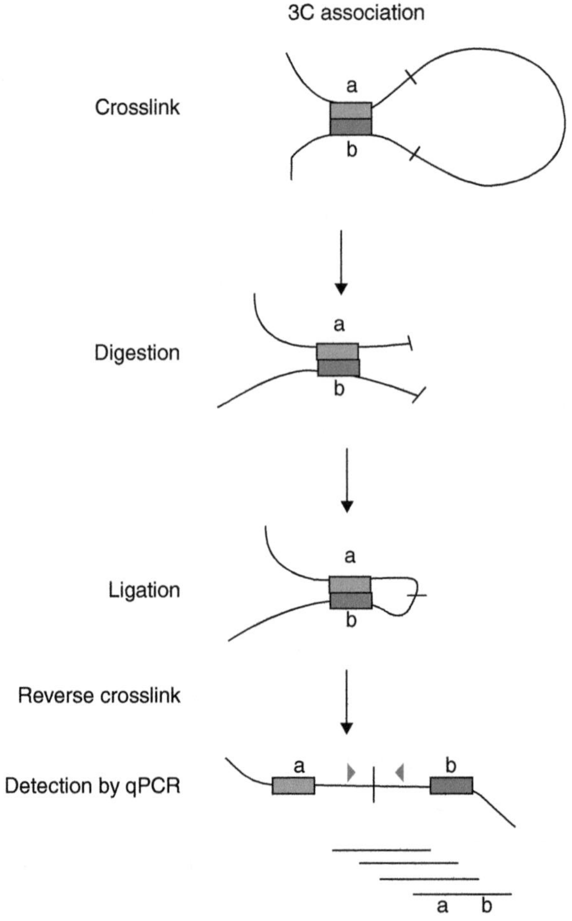

Fig. 1. 3C procedure. Schematic representation of a 3C assay. *Light grey* and *dark grey boxes* represent two interacting genomic elements, a and b, that are separated by a long intervening region (*black curved line*). The 3C assay starts with a cross-linking step using formaldehyde to capture protein–protein and protein–DNA interactions. A second step consists of enzymatic digestion with a restriction enzyme known to recognize sites in the investigated regions. In the third step, the cross-linked complex is religated under conditions that favor intramolecular ligation. Lastly, the cross-links are reversed by heat treatment; the DNA is then purified and the resulting 3C product is detected by qPCR.

sites at lower abundance that represent regions associated at lower frequency. A final PCR step with two primers that amplify across the ligated site is used to quantitatively detect the 3C product. Generally, when examining a specific locus for looping interactions, a specified region is chosen as an anchor and the anchor PCR primer is tested for associations between other locations where bait primers (in the same orientation as the anchor primer) are placed.

A crucial interpretation of the 3C assay depends on how the association frequency is calculated and the ability to distinguish between a specific interaction between two genomic sequences and the background of random ligations due to the flexibility of the chromatin fiber. Thus the detection of an association between two elements does not necessarily mean that they are associated in vivo, since the flexibility of the chromatin fiber enables random collisions between sites on the same chromosome. However, such random ligation events are inversely proportional to the distance between two sites and can be plotted as a random ligation curve, while specific associations can be detected as "spikes" above the random ligation curve (2). In a simpler definition, two sites are considered associated when their association frequency value is higher than the one measured in the intervening sites, given that these are not engaged in similar associations. Therefore, it is important when assessing the association between two elements to first establish the baseline of random interactions.

Furthermore, it is important to bear in mind that associations detected by 3C need to be genetically or biochemically verified prior to any conclusions regarding a functional association. Depending on the restriction enzymes used, the 3C assay still has low resolution and often multiple adjacent restriction sites within a 5 kb stretch of DNA will associate with similar frequencies to a distant restriction site used as an anchor. Therefore the 3C signals indicate proximities to associations, rather than pinpointing the exact sequences involved in the association.

Since 3C captures all associations taking place in a cell population at a given time, it is possible that the associations detected between different elements do not all take place at the same time and in the same cell. For example, it has been shown that the associations of a particular site with four other sites take place preferentially in a pair-wise manner when analyzed by single cell imaging (4). The heterogeneity of cell populations can further decrease the sensitivity in the detection of an association frequency, especially when the associations are specific to a particular cell condition. In order to detect the association, cells need to be selected and enriched for the appropriate state. This means that low-frequency associations can still be specific and biologically significant although they are not detectable as spikes over other associations.

Once all the 3C controls have been considered, the 3C analysis is able to provide important insights into the higher order organization of the analyzed locus.

2. Materials

2.1. Reagents and Buffers

1. Phosphate-Buffered Saline (PBS).
2. 37% formaldehyde (Sigma).
3. Lysis buffer: 50 mM Tris–HCl pH 8, 1% SDS, 10 mM EDTA.

4. Protease Inhibitor Cocktail (PI) (Sigma).
5. Triton-X100 (Sigma).
6. Digestion buffer (New England Biolabs, NEB).
7. Restriction enzyme (NEB).
8. T4 ligase buffer (NEB).
9. T4 ligase (NEB).
10. Proteinase K (Sigma).
11. Phenol.
12. Chloroform.
13. 100% Ethanol.
14. Glycogen (Invitrogen).
15. 3 M Sodium Acetate.
16. qPCR primers (Sigma).
17. Picogreen (Invitrogen).
18. 1× Sybr-green master mix (Power SYBR, ABI).
19. 384 well real-time machine (7900HT Fast Real time PCR system, ABI).
20. Commercially obtained genomic DNA (Becton Dickinson).

2.2. Equipment

1. Optical microscope for qualitative and quantitative assessment of nuclei.
2. Infinite M200 Tecan System for Picogreen-based quantification of the 3C products.
3. DNA purification columns (DNA Clean α Concentrator™ 25 Kit, Zymo Research) for purification of the PCR standard amplicons.
4. 2 ml Eppendorf tubes.
5. 15 ml falcon tubes.
6. Table microcentrifuge.
7. Macrocentrifuge.

3. Methods

3.1. Cell Cross-Linking and Nuclei Preparation

3C is a multistep process, the main steps of which are cross-linking of cells and nuclei preparation; chromatin digestion and religation; DNA template preparation; and identification and analysis of chromatin interactions.

1. Grow adherent cells in a 15 cm dish at semi-confluence.
2. Remove the culture medium and wash the cells with 10 ml of ice-cold PBS.

3. Cross-link the cells with 20 ml of 1% formaldehyde (diluted in PBS) at 37 °C for 10 min (see Notes 1 and 2).
4. Remove the formaldehyde and wash the cells three times with 10 ml of ice-cold PBS. Leave 2 ml of PBS from the last wash into the dish to scrape the cells off the dish.
5. Scrape the cells off with a spatula and transfer into a 2 ml Eppendorf tube.
6. Pellet the cells by centrifugation into a table microcentrifuge at 100×*g* for 5 min.
7. Remove the PBS from the tube and add 1 ml of ice-cold lysis buffer (to release the nuclei from the cytoplasm) supplemented with fresh Proteinase Inhibitor Cocktail and allow the cells to lyse for 10 min on ice (see Note 3).
8. Confirm that lysis is complete by examining 10 μl aliquot of the lysate microscopically. Lysis is complete when nuclei can be seen under the microscope.
9. After lysis is complete pellet the nuclei by centrifugation at 900×*g* for 5 min.
10. Remove carefully the lysis buffer and resuspend the nuclei in 200 μl of 1× digestion buffer (see Note 4).
11. Count the nuclei under the microscope and resuspend 1.5×10^6 nuclei in 300 μl of 1× digestion buffer. Nuclei can be stored at −80 °C at this stage.

3.2. Chromatin Digestion and Religation

1. Add 1.8% Triton-X to sequester the remaining traces of SDS and put the nuclei to shake at 37 °C for 1 h on a shaker.
2. Put aside 30 μl of the nuclei suspension to use as the "undigested control" later (freeze sample at −20°C and purify together with the 3C sample) (see Note 7).
3. Digest the remainder of the nuclei with 1,000 U of the chosen restriction enzyme and incubate at 37 °C on a shaker for 8 h or overnight (see Notes 5 and 6).
4. After digestion put aside a 30 μl aliquot of the reaction to use as the "digestion control" to measure the percent of digestion (freeze sample at −20 °C and purify together with the 3C sample) (see Note 7). Only templates with digestion >70% can be used to proceed to the next steps (see Note 8).
5. Dilute the digested chromatin to 2.5 ng/μl by using 1× of T4 ligase buffer and add 3,200 U of T4 ligase (high concentration) in a total volume of 1.5 ml in order to favor intramolecular ligation. Leave ligation to proceed for 8 h or overnight (see Notes 9 and 10).
6. Digest the chromatin with 1,000 U of a second enzyme (high concentration) which recognizes sites different from the first

enzyme and that digests outside of the regions that are assessed for interaction (see Notes 11 and 12).

7. Reverse cross-link the DNA–protein complexes by treatment with 100 μg/ml of Proteinase K for 4 h or overnight at 65 °C.

3.3. DNA Template Purification

1. Extract the relegated DNA with phenol and chloroform treatment. Add 1 volume of phenol, mix, and centrifuge samples at 14,000 rpm for 10 min (see Note 13).
2. Collect the aqueous phase in a new tube and repeat the phenol treatment.
3. Transfer the aqueous phase to a new tube, add 1 volume of chloroform, mix, and centrifuge samples at 14,000 rpm for 10 min.
4. Place the aqueous phase in a new tube, add 1 volume of H_2O, 2 volumes of 100% ethanol, 1/10 volume of 3 M Sodium Acetate, and 1/200 volume of 20 μg/μl of glycogen.
5. Allow the DNA to precipitate for 2 h at −80 °C or overnight at −20 °C.
6. Wash the DNA with 1 ml of 70% of ethanol and centrifuge at 14,000 rpm for 5 min.
7. Repeat the 70% ethanol wash.
8. Dry the pellet at 37 °C and resuspend the DNA into 250 μl of H_2O (see Note 14).

3.4. Identification and Analysis of Chromatin Interactions

Chromatin interactions are measured by determining the ligation frequencies between nonadjacent restriction sites. The quantification is done by qPCR followed by normalization for differences in genomic copy number and digestion–ligation efficiency in the 3C template. Considerations of 3C primer design, production of standard curves for 3C targets, and quantitative 3C analysis are described.

1. Use a primer design tool (e.g., Primer Express Software V3.0, ABI) to design a series of primers with similar melting temperatures and PCR efficiencies (see Note 15) that bind next to the restriction sites analyzed by 3C. For locus-wide analysis of the association of a particular site, a common primer (anchor primer) adjacent to this site can be used in combination with specific primers for each of the sites tested for an association. Since two sites are scored as being associated when their association frequency is higher than those of intervening sites, it is important to design primers also for testing the intervening region. It is also important to test genome-wide associations in a reciprocal manner by setting the anchor primer in one of the elements previously tested for association. This is a control to ensure the validity of the detected associations.

The PCR efficiency of the 3C primers is assessed by real-time PCR on a PCR standard template.

2. Prepare a PCR standard template that contains equimolar amounts of all possible religation products from the region of interest. A PCR standard template can be easily made by digestion and religation of small genomes. However, in the case of humans and mice, where genomic size is large, direct use of genomic DNA would produce too many combinations of religated products, making the detection by PCR of a specific 3C product difficult. In this case a PCR standard template can be prepared by PCR amplification of the genomic regions spanning each restriction site analyzed by 3C. The resulting amplicons are purified by column and quantified using a Picogreen assay (follow manual instructions). Equimolar amounts of amplicons are mixed together, digested with the same enzyme used in the 3C experiment, religated, phenol/chloroform extracted, ethanol precipitated, and dissolved in H_2O.
3. Quantification of the 3C product is done by qPCR and is based on standard curves for each primer set. Make standard curves using ¼ serial dilutions of the PCR standard template. The 3C values are calculated with the parameters of the standard curves for each 3C primer set as follows: 3C value = $10^{(Ct-i)/s}$, where Ct represents the cycle threshold of the PCR reaction and i and s represent the intercept and the slope of the standard curve, respectively.
4. The qPCR conditions are as follows: use 1.5 μl of the 3C or PCR standard template in a qPCR reaction. The PCR mixture contains 200 nM of FW and REV primers and 1× Sybr-green master mix made up to a final volume of 12.5 μl with H_2O. Use a 96- or 384-well real-time machine for the PCR amplification. Carry out the qPCR as follows: after an initial 2-min preincubation step at 50 °C and 10 min at 95 °C, run 40 amplification cycles, each consisting of 95 °C for 15 s and 60 °C for 1 min. Obtain the melting curve by a 15-min incubation step at 95 °C followed by a 15-min step at 60 °C.
5. When comparing different 3C biological replicates or biological conditions it is important to be able to normalize the 3C products for differences in genomic copy number and digestion–ligation efficiency of the 3C template (Fig. 3). Detect the genomic copy number of the 3C template by qPCR amplification of a region that does not contain any of the restriction sites recognized by the enzyme. The genomic copy number is also used to ensure that the amount of the 3C template is within the range of the standard curve for any given 3C product. The genomic copy number of the 3C template used in each qPCR reaction is 10,000 copies that correspond to approximately 70 ng of DNA.

6. Detect the digestion–ligation efficiency of the 3C template by measuring the religation frequency of two adjacent sites. The religation between two adjacent sites produces a circular molecule of DNA; therefore the circularization frequency of this DNA fragment is representative of the efficiency of both digestion and religation. The circularization frequency is measured by qPCR amplification across the religated site. As example, the circularization of the i–j fragment is shown in Fig. 2c.

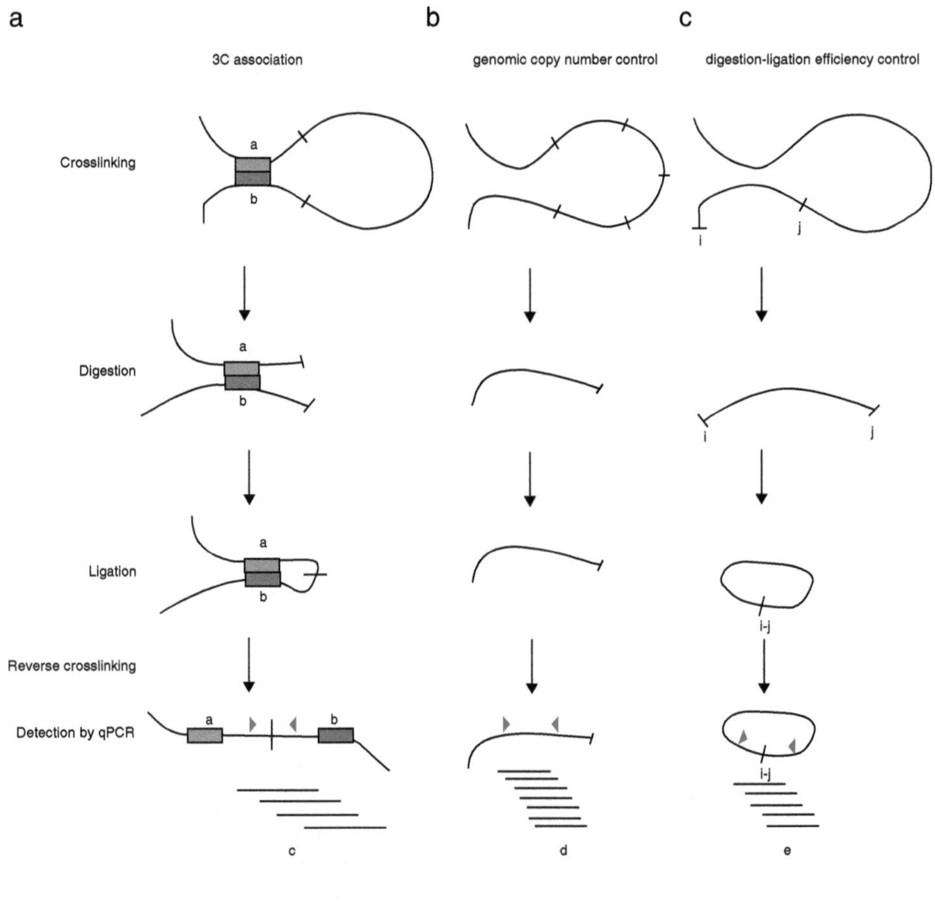

Fig. 2. Normalization for genomic copy number and digestion–ligation efficiency. Schematic on how to normalize a 3C product for genomic copy number and digestion–ligation efficiency. (**a**) The first column represents the 3C procedure to detect the interaction between two genomic elements, a and b (a is indicated as *purple box*; b is indicated by a *pink box*), which are distant along the primary genomic sequence (*black curved line* indicated genomic separation). (**b**) The second column shows the qPCR product that is used to normalize for genomic copy number. This product derives from amplification of a genomic region that lacks *Bam* HI sites and therefore is not digested when treated with the enzyme. (**c**) The third column shows the qPCR product that is used to normalize for digestion–ligation efficiency. Normalization for digestion–ligation efficiency is done using the circularization frequency of two adjacent sites (i–j). (**d**) The formula to calculate the Association Frequency between two elements AF(a–b) is the 3C association value divided by the circularization frequency value and genomic copies of the 3C template: $AF(a–b) = c/(d \times e)$. The application of this normalization allows the analysis of the same interactions among biological replicates or different biological conditions.

7. The Association Frequency between two elements a and b (AF_{a-b}) normalized for genomic copy number and digestion–ligation efficiency is calculated as follows:

$$AF_{a-b} = c / (d \times e),$$

where c represents the 3C product derived by amplification of the religated sites a and b, d represents the genomic copy number of the 3C template, and e the circularization frequency of two adjacent sites (i and j) (Fig. 2a–c).

Once the association frequency values of a particular element with other sites are normalized for genomic copy number and digestion–ligation efficiency, they can be plotted on a graph and a chromatin looping model can be drawn to get insights into the function of higher order chromatin organization at the locus (Fig. 3).

4. Notes

1. Formaldehyde cross-linking is used to capture associations between distant genomic elements by cross-linking protein–protein and protein–DNA molecules by their amino and imino groups. Cross-linking is preferably done on cells in order to capture the interactions in the natural state of a cell. If cross-linking is done on nuclei it is possible that some interactions may have been lost during the procedure of nuclei preparation. The amount, time, and temperature at which cross-linking is carried out can affect the efficiency of cross-linking. Also, cell size is important since it affects the formaldehyde concentration in the nucleus. The detection of stable associations may require less cross-linking than weak or less frequent associations. If cross-linking is not efficient, the chromatin interactions are lost during the following 3C steps and therefore they cannot be detected.
2. If cells grow in suspension, transfer them in a 50 ml falcon tube; wash once with 20 ml ice-cold PBS and cross-link with 1% formaldehyde at 37 °C for 10 min on a rotor. Pellet cells in a centrifuge at 100 × *g* for 5 min and remove the formaldehyde. Wash cells three times with 20 ml of ice-cold PBS. Resuspend cells in 2 ml of PBS left from the last wash and transfer them into a 2 ml falcon tube. Proceed with the lysis as for adherent cells.
3. Cell size is one of the factors to consider when lysing cells. Larger cells are more difficult to lyse, especially when cross-linked. The epithelial breast cell line HB2 and the fibroblast cell line HS27 cells have high cytoplasmic content and therefore

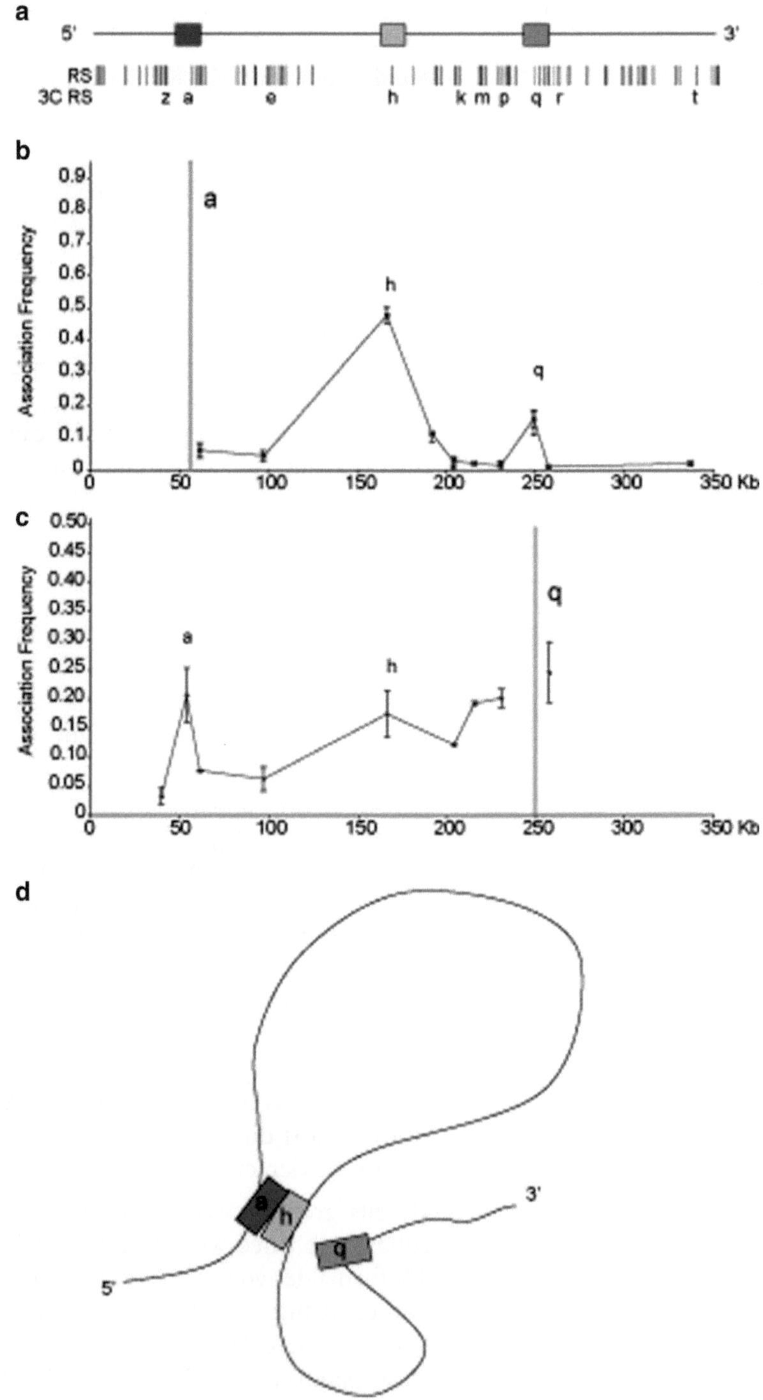

Fig. 3. 3C data and chromatin looping model. Example of a 3C experiment results: the association frequency values of two different anchor primers within a specific locus are shown in a graph. (**a**) Schematic representation of the 350 kb genomic region including three different genomic elements (*different shaded boxes*). *Vertical lines* indicate the position of the Restriction Sites (RSs) recognized by the selected restriction

they require strong lysis with 1% of SDS. Cells with high cytoplasm content such as fibroblasts require to be passed a few times through a 27 G syringe needle. However, cells of smaller size such as the breast cancer cell line SUM 159 and CAL51 can be lysed by using only 0.5% of SDS. The process of lyses and nuclei preparation can be checked by inspecting the lysate under the microscope. In most 3C procedures, the lysis step is followed by a nuclear permeabilization step but in this case it is not necessary since the SDS also permeabilizes the nuclear membrane.

4. Nuclei can easily clump. Avoid nuclei clumping by tapping the pellet nuclei before addition of digestion buffer and add the digestion buffer in aliquots of 100 μl per time and gently pipette up and down each time you add the digestion buffer.
5. The choice of the restriction enzyme is important in the design of the 3C assay. This is because association between two elements is measured by the association frequency of the restriction sites within or close to the elements under study. If the enzyme does not cut within or in the proximity of the two elements that are to be tested for association, then no association can be studied. Restriction sites have to be present also in the region between the two elements since an association between two elements is significant only when it is detected at a higher frequency than that measured with sites in the intervening region. Enzymes that digest very frequently (four base pair cutters) increase the resolution of 3C and are therefore used to analyze associations between elements in a small region while enzymes that digest less frequently (six base pair cutters) can be used to analyze elements separated by long distances. In this case, a limitation of the 3C assay is that, independently of how many restriction sites there are in between two associated elements, it is not possible to assess the association between two elements closer than 30 kb. This is because the association between the two sites would be masked by random collisions of the chromatin fiber due to its flexibility within the nuclear space. The assessment of associations with different restriction

Fig. 3. (continued) enzyme and letters point out the RSs analyzed by 3C (3C RS) in the graphs and histograms below. *Thick vertical lines* in *panels* **b** and **c** link the positions of the restriction sites used as anchors in each *panel* with the overview of the locus in (**a**). The *X*-axis is labeled according to genomic position and position 0 is arbitrarily fixed 42 kb upstream of the element a. (**b**) The anchor restriction site is a. A peak of association is detected with the h and q elements. (**c**) Associations detected with the anchor site in q. An association is present with the a and h elements. The specificity of the interaction is confirmed by the detection of the a–h interaction with both the anchor primers. (**d**) A model of the chromatin looping interactions at the locus is based on the 3C analysis reported in **b** and **c**. The three elements a, h, and q contact each other at the base of the chromatin loops.

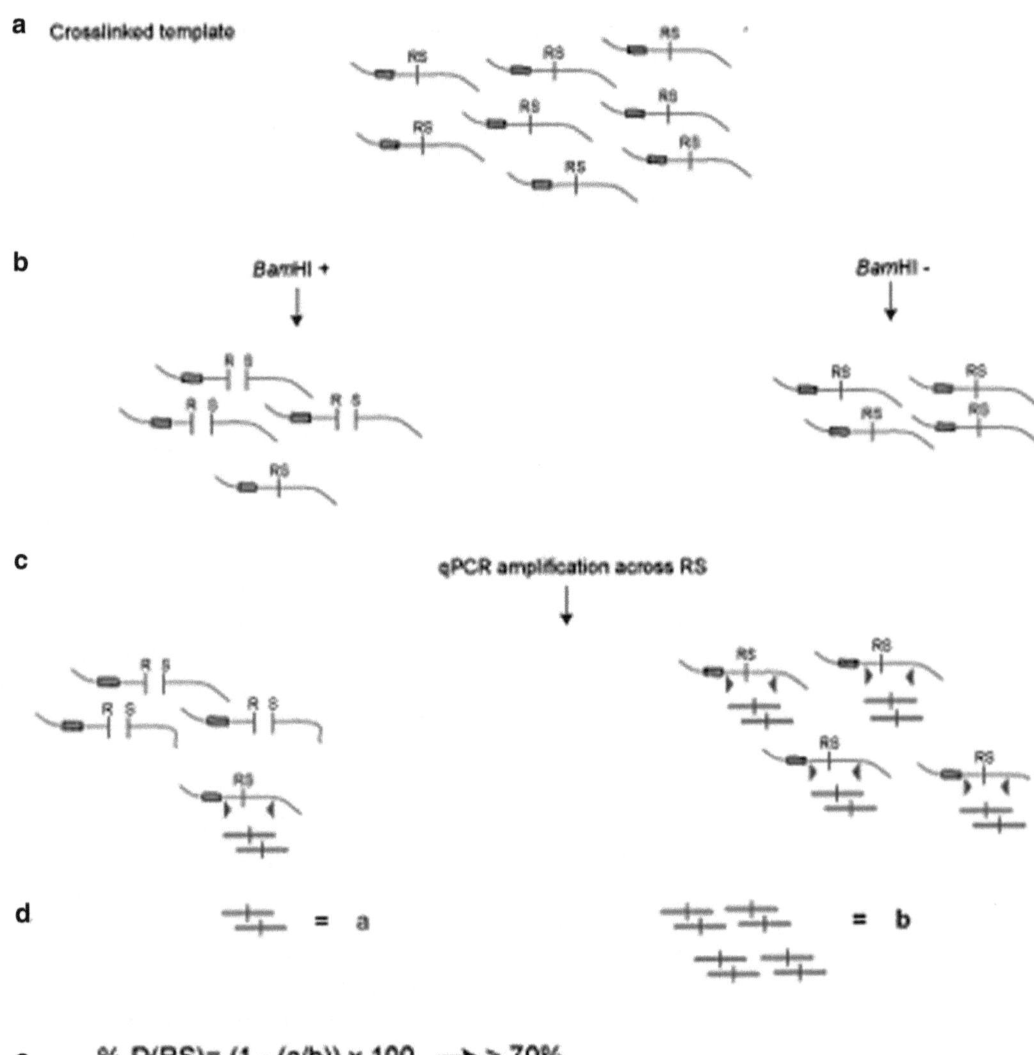

Fig. 4. Digestion efficiency of the 3C template. The assessment of the digestion efficiency of the 3C template at a specific restriction site (RS) is schematically represented. (**a**) The cross-linked chromatin is split into two equal aliquots, (**b**) one is digested with the selected restriction enzyme, *Bam* HI (*Bam* HI +) in this example, and the other is put aside and used as the undigested control (*Bam* HI –). (**c**) Primers binding adjacent to the RS are used to amplify by qPCR the undigested sites both in the *Bam* HI-treated and *Bam* HI-untreated samples. The products resulting from these amplifications (**d**) are used to calculate the percent of digestion at the selected site. (**e**) The formula to calculate the percent of Digestion of the RS is $\%D(RS) = (1-(a/b))\times 100$, where a is the qPCR product from amplification across the RS in the *Bam* HI-treated sample and b derives from amplification in the *Bam* HI-untreated sample.

enzymes gives more confidence in the interpretation of the 3C results.

6. Add the enzyme in aliquots of three and let the sample digest for 2 h before adding the following aliquot. After the addition of the last aliquot of enzyme, the digestion can be left to proceed overnight.

7. The digested and undigested fractions are purified similarly to the 3C template: phenol and chloroform extraction, ethanol precipitation, and resuspension in H_2O (80 μl).
8. An important qualitative control is the assessment of digestion efficiency at the restriction sites analyzed by 3C. Poor enzymatic digestion results in low religation and subsequently low amplification of the 3C product representing the association between two sites. Moreover the digestion efficiency has to be similar between different sites analyzed by 3C; otherwise the associations between these different sites cannot be compared. The digestion efficiency of the 3C template is assessed by qPCR across each restriction site as shown in Fig. 4. The percent digestion is determined by comparing qPCR amplification between digested and undigested fractions present in same genomic copy number.
9. High dilution of chromatin favors intramolecular ligation. Intramolecular ligation is required to religate the digested chromatin fragments that have been captured in the same cross-linked complex. The starting chromatin concentration is estimated by counting nuclei before digestion and by knowing that 1×10^6 nuclei contain 7.2 μg of DNA.
10. More ligations can be carried out at this point and samples can be pooled for DNA extraction.
11. The secondary digestion step is required to avoid the formation of long DNA templates formed during the ligation steps as they can negatively affect the efficiency of the PCR reaction.
12. Add the enzyme in aliquots of three, similarly to the treatment with the first enzyme.
13. Before phenol treatment, the religated samples from the same condition can be pooled in a 15 ml falcon tube.
14. If the DNA has been precipitated in several Eppendorf tubes, aliquot the 250 μl of H_2O among these tubes and recollect the dissolved DNA into a single tube.
15. It is possible to compare interaction between different sites only when the corresponding primer sets have similar efficiency.

References

1. Splinter E, Grosveld F, de Laat W (2004) 3C technology: analyzing the spatial organization of genomic loci in vivo. Methods Enzymol 375:493–507
2. Dekker J (2006) The three 'C's of chromosome conformation capture: controls, controls, controls. Nat Methods 3:17–21
3. Miele A, Gheldof N, Tabuchi TM, Dostie J, Dekker J (2006) Mapping chromatin interactions by chromosome conformation capture. Curr Protoc Mol Biol Chapter 21, Unit 21.11
4. Chang HY, Cuvier O, Dekker J (2009) Gene dates, parties and galas. Symposium on Chromatin Dynamics and Higher Order Organization. EMBO Rep 10:689–693

Chapter 12

Genome-Wide Analysis of DNA Methylation in Low Cell Numbers by Reduced Representation Bisulfite Sequencing

Sébastien A. Smallwood and Gavin Kelsey

Abstract

Development of high-throughput sequencing technologies now enables genome-wide analysis of DNA methylation of mammalian cells and tissues. Here, we present a protocol for Reduced Representation Bisulfite Sequencing (RRBS) applicable to low amounts of starting material (from 200 to 5,000 cells). RRBS is a cost-effective and powerful technique offering the advantages of absolute DNA methylation quantification and single nucleotide resolution while covering mainly CpG islands. Typically one sequencing experiment using the Illumina Genome Analyser IIx platform provides information on the DNA methylation status of more than half of the CpG islands of the mouse genome.

Key words: DNA methylation, CpG islands, Bisulfite sequencing, Reduced representation bisulfite sequencing, High-throughput sequencing, Genomic imprinting, Illumina genome analyser, Bismark

1. Introduction

DNA methylation is an epigenetic modification of the mammalian genome that plays a critical role in regulating gene expression and thereby the definition of specific cellular identity. Notably DNA methylation at CpG islands (CGIs), which are associated with the majority of gene promoters, plays a crucial role during embryonic development, cellular differentiation, and genomic imprinting (1–3). The development of deep-sequencing technologies is having a major impact on epigenetic studies in general and on DNA methylation more specifically. It is now possible to comprehend DNA methylation dynamics in a genome-wide manner, for all mammalian organisms with a reference genome. Several techniques to study DNA methylation now exist, each with their strengths and weaknesses depending on the biological question asked and the resources available in each laboratory (4, 5).

Nora Engel (ed.), *Genomic Imprinting: Methods and Protocols*, Methods in Molecular Biology, vol. 925,
DOI 10.1007/978-1-62703-011-3_12, © Springer Science+Business Media, LLC 2012

DNA methylation can be profiled genome-wide by sequencing after methylated cytosine (5mC) enrichment either using a specific anti-5mC antibody (MeDIP-Seq) or protein domain (methyl-binding domain; MBD-Seq) (6, 7). This type of approach is cost-effective, as one sequencing run provides a good coverage of the entire mouse or human genome. It also provides information on both single-copy loci and repetitive elements and MeDIP-Seq allows as well the distinction between 5mC and 5-hydroxymethyl-cytosine (5hmC) (6). However, these 5mC enrichment approaches do not offer single-nucleotide resolution as they are based on enrichment of fragmented DNA (typically 300 bp length), and more importantly, 5mC methylation quantification is relative. In addition, precipitation-based methods entail a nonspecific background contribution that sets a minimum limit of input material below which signal-to-noise ratios become unacceptable. Alternatively, genome-wide DNA methylation profiles can be studied by bisulfite sequencing (BS-Seq or shotgun BS-Seq) approaches (8). In this case, treatment of DNA with sodium bisulfite converts unmethylated Cytosines into Uracils (Thymines after PCR amplification) while 5mCs are protected. Therefore BS-Seq provides single-nucleotide resolution and absolute quantification of DNA methylation. On the downside, this technique does not allow the distinction between 5mC and 5hmC, yet. In addition, to obtain coverage of the entire genome a large number of sequencing runs is required and therefore this approach is expensive with current technology.

Alternative techniques based on BS-Seq exist, and here we describe in detail a protocol for Reduced Representation Bisulfite Sequencing (RRBS). RRBS combines the advantages of BS-Seq in providing accurate DNA methylation level of individual cytosines, and enrichment in CGIs (4, 9–11). While this technique was originally performed using relatively large amounts of genomic DNA (1 μg), thus limiting its application, it is now possible to perform RRBS with 10–300 ng genomic DNA (12) or even lower (11) (this protocol). One sequencing run on the Illumina Genome Analyser IIx platform typically gives reliable information on ~1 million individual CpGs, ~65 % of them overlapping the CGIs identified by CAP-Seq (13). Since the main biological targets of 5mC are gene promoters, the majority of which reside in CGIs, RRBS is a powerful, informative, flexible, and yet relatively inexpensive technique.

We present here a RRBS protocol suitable for 200–5,000 somatic cells (6 pg DNA per cell). We detail DNA purification, sequencing library generation, quality control, and basic bioinformatic analysis. Using this protocol, we successfully generated and sequenced libraries from as low as 500 pg of genomic DNA (11). How does RRBS work? First, genomic DNA is digested with a restriction endonuclease, in this case MspI (C^CGG), which cuts more frequently within CGIs, thereby providing the basis for CGI enrichment based on appropriate size selection (in frequency, smaller fragments correspond more to CGIs). After MspI digestion, DNA fragments are repaired and 5mC Illumina sequencing adapters are ligated, followed by bisulfite conversion. Then after a first PCR amplification, size selection on agarose gel is performed to recover the smaller fragment fraction. This is followed by a second PCR amplification and purification. The protocol presented here can be easily performed, in 2–3 days, and is particularly efficient as MspI digestion, End-Repair/A-tailing, adapter ligation, bisulfite conversion can be performed within the same PCR tube, with no intermediate purification steps, thus limiting DNA loss.

2. Materials

1. QIAamp DNA Micro Kit (QIAGEN).
2. EB buffer (10 mM Tris–HCl, pH 8.5).
3. Quant-iT PicoGreen dsDNA Assay Kit (Invitrogen).
4. Fluorometer.
5. PCR tubes with individual lids.
6. Water (UltraPure DNase/RNase-Free; Invitrogen).
7. PCR cycler.
8. MspI (Fermentas).
9. Klenow Fragment exo- (Fermentas).
10. PCR-Grade dNTPs: dATP, dCTP, dGTP, dTTP (Roche Applied Science).
11. Nucleotide End-Repair Mix (1 mM dATP; 0.1 mM dGTP; 0.1 mM dCTP).
12. T4 DNA Ligase HC (30 u/μl) (Fermentas).
13. ATP 50 mM (Fermentas).
14. Illumina 5mC Adapters. We are using PE adapters, which provide the flexibility for single-run or paired-end run. Adapters must have 5mC to conserve their sequence upon bisulfite treatment. 5mC adapters can be bought directly from Illumina (Early Access Methylation Adapter Oligos, Illumina). Alternatively, 5mC adapters oligos can be synthesized by another company such as Sigma-Aldrich and annealing done separately.
15. Imprint DNA Modification Kit (Sigma-Aldrich).
16. Ethanol 100 %.
17. Pfu Turbo Cx Hotstart DNA polymerase (Agilent).
18. Illumina Library PCR primers PE1.0 and PE2.0.
19. Agarose (molecular biology grade).

20. TBE buffer (Tris/Borate/EDTA buffer).
21. Ethidium bromide.
22. 100 and 50 bp DNA ladders (no ready-mix).
23. 6× Loading dye.
24. UV Transilluminator.
25. Scalpel blades.
26. QIAquick Gel Extraction kit (QIAGEN).
27. Platinum Pfx DNA Polymerase (Invitrogen).
28. Agencourt AMPure XP (Beckman Coulter Genomics).
29. Bioanalyser and Illumina Sequencing platform (Genome Analyser GIIx).

3. Methods

This protocol describes how to generate RRBS libraries starting from 200 to 5,000 cells corresponding to ~500 pg to 20 ng genomic DNA after purification. Here, we refer to starting material >5 ng (L.A. for Low Amount) or <5 ng (V.L.A. for Very Low Amount). It is recommended to perform a negative control (no DNA) in parallel of your samples. For V.L.A. a positive control (20 ng of genomic DNA) can also be performed. Bench, pipettes and racks used should be thoroughly cleaned with DNA decontamination solution. It is recommended to use low DNA binding plastic ware.

3.1. DNA Purification

1. Use the QIAamp DNA microkit according to the manufacturer's protocol for tissue. Do not use carrier RNA. Elute the DNA using 22 μl of warm EB buffer (see Note 1).
2. Quantify DNA using Quant-iT PicoGreen kit according to the manufacturer's instructions. Use 2 μl of eluted DNA per replicate and perform duplicates. In our hands, this technique is accurate for as little as 100 pg/μl (see Note 2).

3.2. MspI Digestion

In a PCR tube add the following:

- Genomic DNA
- 0.9 μl of MspI
- 1.8 μl of Tango 10× buffer
- H_2O to 18 μl final volume

Incubate for 3 h at 37 °C, followed by enzyme heat inactivation at 80 °C for 20 min.

Proceed to next step or pause (see Note 3).

3.3. End Repair/A-Tailing

In the same PCR tube (Subheading 3.2), directly add:

- 1 μl of Klenow Fragment exo-
- 0.8 μl of nucleotide end-repair mix
- 0.2 μl of Tango 10× buffer

Incubate for 40 min at 37 °C, followed by enzyme heat inactivation at 75 °C for 15 min.

Proceed to next step or pause.

3.4. Adapter Ligation

In the same PCR tube (Subheading 3.3), directly add:

- 1 μl of HC T4 DNA ligase
- 1 μl of 5mC sequencing adapters (For L.A.:1.5 μM; for V.L.A.: 0.5 μM) (see Note 4)
- 0.5 μl of Tango 10× buffer
- 0.5 μl of ATP (50 mM)
- 2 μl of H_2O

Incubate overnight at 4 °C, followed by enzyme heat inactivation at 65 °C for 20 min (see Note 5).

Proceed to next step or pause.

3.5. Bisulfite Modification

Take 24 μl of the previous reaction mix (Subheading 3.4) and use directly the Imprint DNA Modification Kit two-step modification procedure, with the following modifications (see Note 6):

- For L.A.: once DNA modification mix is added, incubate at 65 °C for 90 min, 99 °C for 2 min, 65 °C for 30 min.
- For V.L.A.: once DNA modification mix is added, incubate at 65 °C for 90 min.

Perform purification steps according to the manufacturer's protocol, and elute DNA in 22 μl (2 × 11 μl) of warm EB buffer.

Proceed to next step or pause.

3.6. First Amplification Step

To the BS converted DNA, set up a 25 μl PCR reaction by adding:

- 20.6 μl of bisulfite treated DNA
- 0.4 μl of Pfu Turbo Cx Hotstart DNA polymerase (see Note 7)
- 2.5 μl of Pfu Turbo Cx Hotstart 10× buffer
- 0.5 μl of dNTP mix
- 0.5 μl of primers PE1.0
- 0.5 μl of primers PE2.0

Perform six PCR cycles (95 °C-2 min; 95 °C-20 s, 65 °C-30 s, 72 °C-30 s) (see Note 8).

Proceed to next step or pause.

3.7. Fragment Size Selection

Prepare a 1.7 % agarose gel with Ethidium Bromide (see Note 9). Prepare 50 and 100 bp ladder in Pfu Turbo Cx Hotstart 10× buffer in a 25 μl volume for each sample (see Note 10). Load directly the product from the first amplification step (Subheading 3.6) on the gel using 6× loading dye. Load 50 bp ladder/sample/100 bp ladder and repeat for each sample, leaving a few wells empty in between to avoid cross-contamination. Run at 80 mV for ~45 min. Place the gel still in its plastic case on a UV Transilluminator, and using the DNA ladders as a guide, cut between 150 and 400 bp.

Proceed to Gel Extraction with Qiagen QIAquick following the manufacturer's protocol (see Note 11).

Elute in 40 or 44 μl of warm EB buffer (see Note 12).

Proceed to next step or pause.

3.8. Second Amplification Step and Purification

Set up a 50 μl PCR reaction by adding:

- 36.2 μl of size selected DNA from the previous step
- 0.8 μl of Platinum Pfx DNA Polymerase (see Note 13)
- 5 μl of Platinum Pfx 10× buffer
- 2 μl of $MgSO_4$ (50 mM)
- 2 μl of dNTPs
- 2 μl of primer PE1.0
- 2 μl of primer PE2.0

Do 12 Cycles (94 °C 2 min; 94 °C-20 s, 65 °C-30 s, 68 °C-30 s) (see Note 14).

The 50 μl library is purified using Agencourt AMPure XP magnetic beads according to the manufacturer's protocol, and elution is done in 20 μl EB buffer (see Note 15).

3.9. QC Analysis and General Troubleshooting

Bioanalyser using DNA High Sensitivity chips is performed with 1 μl of purified library. Bioanalyser is important to assess the overall quality of the library. The following points need to be examined. Control samples (no DNA) should be clear: presence of DNA within 150–400 bp size range is a sign of contamination, and a new library should be prepared.

The size distribution of the MspI fragments is not uniform, and some peaks are visible, mainly corresponding to MspI fragments for interspersed repeats (Fig. 1). While the overall intensity of the peaks might vary slightly between samples, their size is nonetheless constant between different experiments. The absence or shift in size of one or more peaks is an indication of a suboptimal library (Fig. 1). Absence of peaks can be the result of inaccurate size selection on the agarose gel; the library can still be sequenced but CGI coverage will be reduced. Shift in size can be the result of strong PCR bias and clonality, and therefore, it is not recommended to sequence this library.

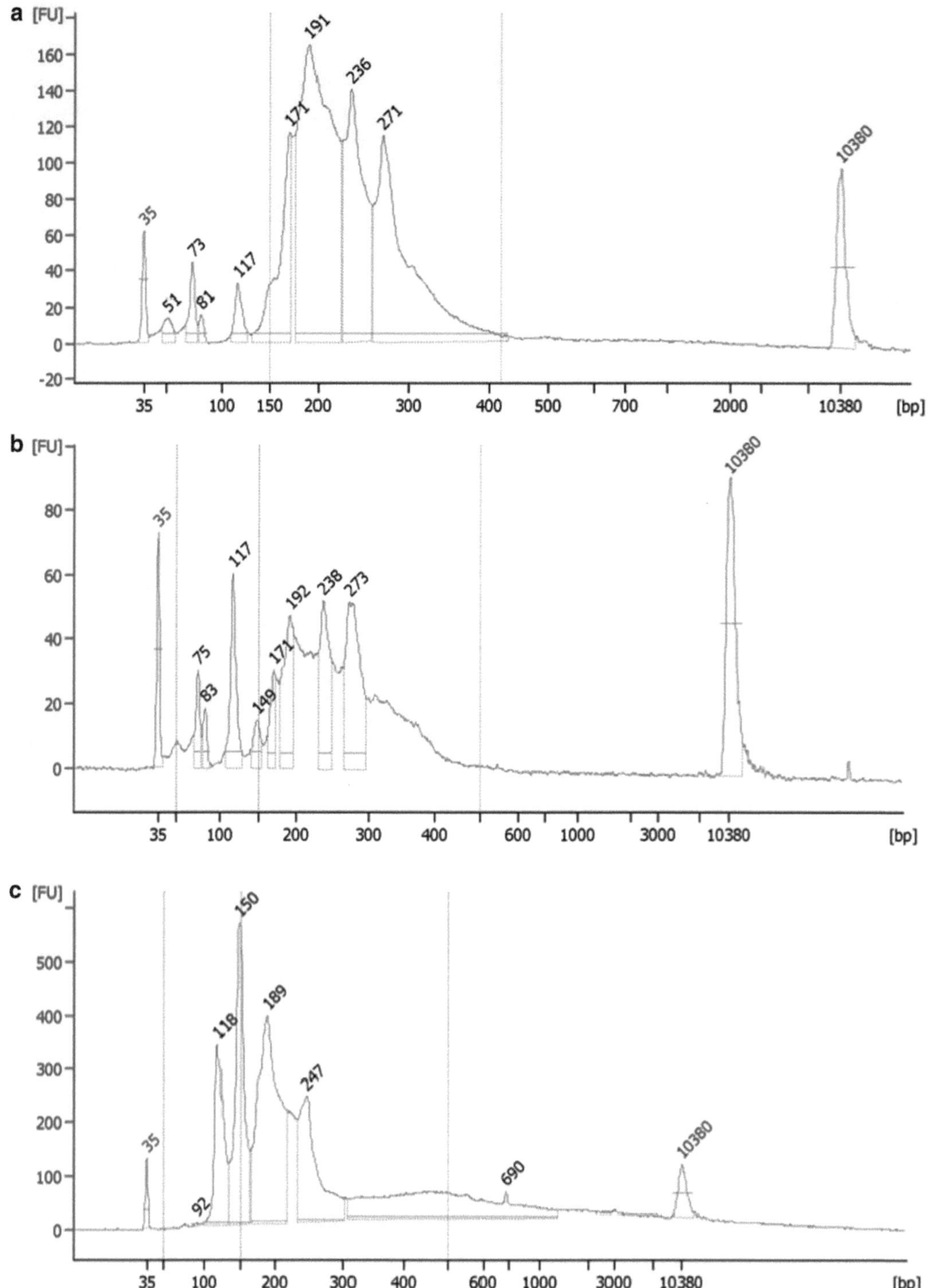

Fig. 1. Quality assessment of RRBS libraries using Bioanalyser. (**a**) Typical Bioanalyser profile of a good quality RRBS library. The library corresponds to size fragments in the 150–400 bp size range, with a characteristic pattern of peaks corresponding to MspI fragments from interspersed repeats, and limited adapter contamination (117 bp). (**b**) Example of Bioanalyser profile of a low quality RRBS library. Peaks are present at the right sizes; however, strong adapter contamination is also present (117 bp). This library can still be sequenced but sequence complexity may be reduced. Alternatively a second gel extraction can be performed to remove adapter dimers (see Subheading 3.9). (**c**) Example of Bioanalyser profile of a poor quality RRBS library. There is obvious bias in peak distribution (strong 150 bp peak and absence of 273 bp peak) and substantial adapter contamination (118 bp). It is not recommended to sequence this library. These libraries were generated from mouse DNA. RRBS libraries from DNA of other species will have a different pattern of peaks from repetitive sequences.

Bioanalyser analysis allows the assessment of adapter dimer contamination in the library, which appears as a ~117 bp peak. Presence of adapter dimers can result from poor size selection, or indication that the amount of genomic DNA used was overestimated. In the presence of a small quantity of dimers, libraries can still be sequenced but sequence complexity may be reduced. If adapter dimers represent the majority of the library in terms of DNA concentration, it is recommended to generate a new library. If this is not feasible, a second gel extraction can be performed to remove adapter dimers, followed by a couple of extra PCR cycles to compensate for the loss of material during this process (see Note 14). Libraries should be purified again and bioanalyser analysis repeated.

3.10. Library Quantification and Sequencing

Bioanalyser analysis also informs on library concentration (3 μl of 10 nM libraries are normally required for Illumina sequencing). If concentration is too low, a few additional PCR cycles can be performed (see Note 14). It is also recommended to determine library concentration by qPCR using as standards Illumina libraries of known concentration (14). Alternatively, Quant-iT PicoGreen dsDNA Assay can also be performed. We usually perform sequencing on the Genome Analyser GAIIx platform with single read analysis and 40 bp read length. RRBS library sequencing can be done with around ~250,000–280,000 clusters per tile. This usually gives ~30 millions sequences after Bareback processing (15) resulting in ~15–18 millions aligned sequences by Bismark (16) (see below).

3.11. Data Analysis

For data analysis, a recent PC or Mac is needed and sufficient; however, 8 Gb or more of RAM memory can be useful (e.g., 64 bit version of Windows). Specific software is required for alignment of RBBS/BS-Seq reads. We are currently using Bismark (16) that can perform both read mapping and DNA methylation calls. In addition Bismark can discriminate between DNA methylation in a CpG context or another context (CHG, CHH). DNA methylation call visualization and analysis can be made in a general program used for high-throughput sequencing analysis. We are using Seqmonk (www.bioinformatics.bbsrc.ac.uk/projects/seqmonk/).

Here is a short summary of the steps required for CGI DNA methylation analysis in Seqmonk. First data need to be imported, in the form of a list of cytosines, their position and their methylation status ("+" strand corresponds to methylated cytosine while "–" strands correspond to unmethylated cytosines). Probes corresponding to individual cytosines are generated using Data/Define probes/Contig probe generator. Then probes are quantified by number of reads to assess the number of time each individual cytosine has been sequenced (Data/Quantification/Read count). Then filter out the probes with less than five reads depth (Filtering/ Filter on value/Individual probes) (this filter gives sufficiently accurate CGI methylation; however, a filter of ten reads can also be used). Once the filtering is done, perform a new probe quantification to give methylation values in percentage: Data/Quantification/Difference quantification/Forward only/As a percentage of/All reads. It is possible to get a CGI methylation value by averaging methylation of individual CpGs within the CGI. To do so, after importing the CGI dataset (13) in Seqmonk as an annotation track (File/Import Annotation), perform a report (Reports/Create feature report/annotate CGIs with any overlapping probes/do not exclude features with no probes). The report (tabulated text file) can be imported in Excel where the table function and filters associated can be used for analysis. CGIs can be filtered by the number of probes associated (which corresponds to the number of informative CpGs per CGIs). We generally consider only CGIs for which at least 10 % of CpGs are informative (i.e., a minimum of five reads), but this number can be increased to 20 or 30 % for more stringent analysis.

4. Notes

1. Obtaining good quality genomic DNA is key to successful RRBS. This can be challenging when starting with very low numbers of cells. Phenol–chloroform purification of DNA is an option and the use of glycogen (20 μg) as carrier in ethanol precipitation does not interfere with the subsequent enzymatic steps. However, we obtained more reproducible results using the QIAamp DNA microkit. Even with fewer than 1,000 cells, recovery is ~50–70 % of the theoretical amount of DNA (6 pg of DNA per somatic cell).
2. For very low amounts of DNA PicoGreen can be optional, and RRBS performed without quantification, based simply on the expected yield of DNA from the initial number of cells. In this case, elute DNA in 20 μl of warm EB buffer.
3. Enzymatic reaction can be paused at 4 °C for up to 16 h. For longer times, freeze the reaction mix at −20 °C.
4. It is important to perform initial tests with different adapter concentrations (using 20 ng genomic DNA), as annealing efficiency of oligos might vary from stock to stock. Using the optional step described in **Note 11**, if after gel analysis a library is not or only weakly visible, increase adapter concentration; if only or large amounts of adapter dimers are observed, reduce adapter concentration.
5. For L.A. conditions, adapter ligation can also be performed at 22 °C for 90 min with good results. Overnight incubation at 4 °C is always recommended for V.L.A. conditions.
6. The main advantage of this bisulfite modification kit versus others is low DNA degradation and therefore better DNA

recovery for low amount of starting material. On the downside the conversion rate of unmethylated cytosine is not optimal. In our experience it is typically ~95 % (for CGI DNA methylation analysis this does not introduce significant/detectable artifacts). It is possible to use carrier DNA at this stage (12), but this has not been tested in our laboratory.

7. The use of Pfu Turbo Cx Hotstart DNA polymerase is critical at this step, in our experience. This enzyme is resistant to uracil stalling and therefore both methylated and unmethylated fragments are amplified with similar efficiencies. The use of another polymerase could result in an overestimation of DNA methylation.
8. This two-step PCR strategy is designed to increase the amount of DNA for gel extraction, which is the major limiting step of our protocol (where most material may be lost). Our tests have shown that this does not introduce major biases in sequence amplification.
9. Prepare a gel using a volume of agarose and combs just sufficient to allow loading of 30 μl volumes. Using combs too big or too much agarose will result in a heavier agarose slice after size selection, resulting in lower recovery during QIAquick gel extraction.
10. It is important to use DNA ladders prepared in Pfu Turbo Cx Hotstart buffer, otherwise differences of migration between ladders and samples might occur, resulting in nonoptimal size selection.
11. Once cut, the gel slices should not weigh more than 300 mg or the DNA recovery might be low. If necessary, adapt comb size and migration time appropriately.
12. It is possible at this stage to perform an optional step to validate all previous steps, without the need to proceed to final library generation and purification with AMPure XP beads. For this, set up a 2×20 μl PCR (with Platinum Pfx DNA Polymerase, and conditions described in Subheading 3.8) with 2 μl of size selected fragments. Amplify for 20 and 25 cycles for L.A. or 25 and 30 cycles for V.L.A. Load directly on a 3 % agarose gel. Assess size selection accuracy and the presence of adapter dimers. If this optional test is not performed the elution volume used in the QIAquick column can be reduced to 40 μl. Alternatively, for initial optimisation of adapter concentration, set up 4×20μl PCRs with 10 μl of size selected fragments: amplify for 15, 20, 25, and 30 cycles.
13. Platinum Pfx DNA Polymerase produces better yields than Pfu Turbo Cx Hotstart polymerase. At this stage uracil stalling due to bisulfite conversion is no longer a problem.
14. Twelve cycles of PCR are generally enough to obtain "sequencable" libraries for both L.A. and V.L.A. conditions. In the case of low library concentration, and if no original sample is

left, extra PCR cycles can be performed without affecting significantly the overall quality (up to four extra cycles). Purification using beads and Bioanalyser analysis should be repeated. More than 22 cycles in total and the overall DNA methylation values might be overestimated, and more than 26 cycles in total and major artifacts will appear.

15. As an alternative to AMPure XP magnetic beads, Qiagen Minielute columns can be used. However, percentage of recovery and overall quality of the library will be lower, especially if adapter dimers are present.

Acknowledgments

This work was supported by the Biotechnology and Biological Sciences Research Council, the Medical Research Council, and the Centre for Trophoblast Research.

References

1. Smallwood SA, Kelsey G (2012) De novo DNA methylation: a germ cell perspective. Trends Genet 28:33–42. doi:10.1016/j.tig.2011.09.004
2. Feng S, Jacobsen SE, Reik W (2010) Epigenetic reprogramming in plant and animal development. Science 330:622–627
3. Ferguson-Smith AC (2011) Genomic imprinting: the emergence of an epigenetic paradigm. Nat Rev Genet 12:565–575
4. Harris RA, Wang T, Coarfa C et al (2010) Comparison of sequencing-based methods to profile DNA methylation and identification of monoallelic epigenetic modifications. Nature Biotechnol 28:1097–1105
5. Bock C, Tomazou EM, Brinkman AB et al (2010) Quantitative comparison of genome-wide DNA methylation mapping technologies. Nat Biotechnol 28:1106–1114
6. Ficz G, Branco MR, Seisenberger S et al (2011) Dynamic regulation of 5-hydroxymethylcytosine in mouse ES cells and during differentiation. Nature 473:398–402
7. Serre D, Lee BH, Ting AH (2010) MBD-isolated Genome Sequencing provides a high-throughput and comprehensive survey of DNA methylation in the human genome. Nucleic Acids Res 38:391–399
8. Lister R, Pelizzola M, Dowen RH et al (2009) Human DNA methylomes at base resolution show widespread epigenomic differences. Nature 462:315–322
9. Meissner A, Mikkelsen TS, Gu H et al (2008) Genome-scale DNA methylation maps of pluripotent and differentiated cells. Nature 454: 766–770
10. Gu H, Bock C, Mikkelsen TS et al (2010) Genome-scale DNA methylation mapping of clinical samples at single-nucleotide resolution. Nat Methods 7:133–136
11. Smallwood SA, Tomizawa S-I, Krueger F et al (2011) Dynamic CpG island methylation landscape in oocytes and preimplantation embryos. Nat Genet 43:811–814
12. Gu H, Smith ZD, Bock C et al (2011) Preparation of reduced representation bisulfite sequencing libraries for genome-scale DNA methylation profiling. Nat Protoc 6:468–481
13. Illingworth RS, Gruenewald-Schneider U, Webb S et al (2010) Orphan CpG Islands identify numerous conserved promoters in the mammalian genome. PLoS Genet 6:e1001134
14. Quail MA, Kozarewa I, Smith F et al (2008) PERSPECTIVE A large genome center's improvements to the Illumina sequencing system. Nat Methods 5:1005–1010
15. Krueger F, Andrews SR, Osborne CS (2011) Large scale loss of data in low-diversity illumina sequencing libraries can be recovered by deferred cluster calling. PLoS One 6:e16607
16. Krueger F, Andrews SR (2011) Bismark: a flexible aligner and methylation caller for Bisulfite-Seq applications. Bioinformatics 27: 1571–1572

Part V

Analysis of Imprinted Expression

Chapter 13

Isolation of RNA and DNA from Single Preimplantation Embryos and a Small Number of Mammalian Oocytes for Imprinting Studies

Sarah Rose Huffman, Md Almamun, and Rocío Melissa Rivera

Abstract

Researchers whose experimental models are mammalian oocytes and preimplantation embryos are often limited by the yield of nucleic acids that can be isolated from such a small sample size. In addition, the limited number of cells from these types of samples makes the simultaneous recovery of RNA and DNA very difficult and often sample pooling is necessary to increase nucleic acid yield. Here we report a simple set of procedures using commercially available kits that results in consistent yield and quality of nucleic acids. After sample lysis, RNA is isolated and converted to a reusable cDNA library. Following RNA isolation, DNA is precipitated, isolated, and bisulfite converted for DNA methylation studies. Our results demonstrate the feasibility of isolating RNA and DNA from a small number of cells with repeatability of results.

Key words: Mammalian embryo, Blastocyst, Oocyte, RNA isolation, DNA isolation, Bisulfite mutagenesis

1. Introduction

One of the challenging tasks of researchers who use models that contain small numbers of cells, such as mammalian preimplantation embryos, is the ability to isolate both DNA and RNA from them. Depending on the model organism these may be hard to obtain. For example, under natural cyclic conditions, a mouse could produce ~6–15 embryos, while a cow would only produce one or two embryos. The difficulties are further exacerbated by the developmental stage (e.g., 2-cell vs. blastocyst) at which embryos are used. Mouse preimplantation-stage embryos range in number of cells from 1 to approximately 80 (1). In addition, a blastocyst-stage mouse embryo will contain ~2 ng of total RNA and ~500 pg

Nora Engel (ed.), *Genomic Imprinting: Methods and Protocols*, Methods in Molecular Biology, vol. 925,
DOI 10.1007/978-1-62703-011-3_13, © Springer Science+Business Media, LLC 2012

of DNA. Therefore, often times several embryos need to be pooled in order to have the necessary yield of nucleic acids to perform experiments. While conclusions made from such an experimental design are valid, they can only pertain to groups of embryos or even groups of female donors.

RNA and DNA isolation from individual samples is essential in the case where allelic determinations of gene expression and DNA methylation are sought after when doing genomic imprinting studies. Genomic imprinting is an epigenetic modification that results in parent-specific gene expression (2). Imprinted genes are found in clusters (3) and are often regulated by a differentially methylated region known as an imprinting control region (ICR). We (4) and others (5) have demonstrated that manipulation of embryos during the preimplantation stage results in loss-of-imprinting during mid-gestation. In those studies a correlation was made between biallelic expression of imprinted genes and loss of methylation at ICRs. Similar studies are not very feasible using single preimplantation embryos due to their paucity of cells. Here we provide the procedures we use to isolate RNA and DNA from single embryos and small groups of oocytes. The nucleic acids can be then used for downstream expression and methylation analyses with good and consistent results.

Caveat: The protocol presented in this chapter uses oligo dT magnetic beads which only allows for the hybridization of polyadenylated transcripts. However, we have been able to consistently isolate one nonpolyadenylated transcript that contains long stretches (>20) of adenosines within the transcriptional unit.

2. Materials

1. TE (Tris EDTA)—1 mM EDTA and 10 mM Tris–HCl in RNAse/DNAse free water. Autoclave.
2. TIT (TE, IGEPAL, Tween 20)—equal parts of TE, 10% IGEPAL, and 10% Tween 20. Autoclave.
3. Commercially available Kits.

2.1. RNA Isolation

1. Dynabeads mRNA DIRECT kit (Invitrogen).
   (a) *Lysis buffer/binding buffer* (100 mM Tris–HCl, pH 7.5, M EDTA, pH 8, 1% LiDS, 5 mM dithiothreitol (DTT)).
   (b) *Washing Buffer A*—(10 mM Tris–HCl, pH 7.5, 0.15 M LiCl, 1 mM EDTA, 0.1% LiDS).
   (c) *Washing Buffer B*—(10 mM Tris–HCl, pH 7.5, 0.15 M LiCl, 1 mM EDTA).

   Alternatively, Oligo $d(T)_{25}$ magnetic beads (NEB)

(a) *Lysis/Binding Buffer* (20 mM Tris–HCl (pH 7.5), 500 mM LiCl, 0.5% LiDS, 1 mM EDTA, 5 mM DTT).

(b) *Wash Buffer I* (20 mM Tris–HCl (pH 7.5), 500 mM LiCl, 0.1% LiDS, 1 mM EDTA, 5 mM DTT).

(c) *Wash Buffer II* (20 mM Tris–HCl (pH 7.5), 500 mM LiCl, 1 mM EDTA).

2. Other materials—1.5 ml microcentrifuge magnet for affinity procedures (available from several vendors).

2.2. DNA Bisulfite Conversion Kit

1. Sigma Imprint DNA Modification Kit.

3. Methods

Foreword: There are a few procedures we utilize which have increased the yield of nucleic acids obtained from small samples. These are:

1. When possible, centrifuge tubes between steps to collect droplets from the lid and sides of tubes.
2. Use the smallest pipet tip possible (e.g., when pipetting <10 μl use a 10 μl tip instead of 20 μl) and pipet slowly.
3. Perform all procedures (from RNA isolation to bisulfite mutagenesis) in one day (see Note 1).

3.1. Sample Collection

1. Prepare the necessary number of 1.5 ml RNAse/DNAse free microcentrifuge tube containing 100 μl of lysis/binding buffer.
2. Single embryos or groups of oocytes are placed in the lysis-buffer containing tubes (see Note 2).
3. Centrifuge the tubes and proceed to RNA isolation or store at -80.0 °C.

3.2. mRNA Isolation by Affinity Purification

1. All procedures are done at room temperature.
2. Remove the bottle containing the oligo dT-bound magnetic beads and gently resuspend the beads by swirling and inverting the bottle.
3. Pipet out 10 μl of beads for every sample plus one for a blank control and transfer to an RNase/DNase free 1.5 microcentrifuge tube. For example, if analyzing five embryos then pipet out 60 μl of bead solution.
4. Centrifuge briefly to collect all beads at the bottom of the tube.
5. Place tube on a microcentrifuge tube holder magnet for at least 1 min or until the fluid is perfectly clear.

6. Remove the supernatant using a 100–200 RNase/DNase free filtered pipet tip while tube is in the magnet (see Note 3).
7. Add 100 μl lysis buffer to the beads and rinse by vortexing continuously for 5 min (Note 4).
8. Place tubes on the magnet until the supernatant looks clear.
9. Pipet out the supernatant and remove the tube from the magnet.
10. Resuspend beads in lysis buffer. The amount of lysis buffer added should equal the original volume of bead solution removed from the bead-stock bottle included with the kit (i.e., 60 μl).
11. Thaw embryo lysate and centrifuge briefly.
12. Add 10 μl of bead suspension to each of the samples to be analyzed.
13. The remaining beads will serve as a blank control. Add 100 μl of lysis buffer to the blank control tube.
14. Place the microcentrifuge tubes containing embryo lysates and beads on a vortexer for 5 min (medium to low speed) to allow RNA to anneal to the beads (see Note 5).
15. Place the tubes on a tube rack and incubate for 5 min.
16. Briefly centrifuge tubes and place on magnet for 3–5 min until the supernatant is completely clear (i.e., beads are all collected on the back of the tube).
17. Remove supernatant and transfer to a new 1.5 ml microcentrifuge tube and save until use (see Note 6; see Subheading 3.3).
18. Wash mRNA/beads complexes by vortexing continuously for 5 min with 100 μl Washing Buffer A.
19. Briefly centrifuge tubes and place on the magnet for 3–5 min.
20. Remove supernatant and repeat wash with Buffer A (steps 15–17).
21. Wash mRNA/beads complexes by vortexing continuously for 5 min with 100 μl of Washing Buffer B.
22. Briefly centrifuge tubes and place on the magnet for 5 min.
23. Repeat wash with Buffer B (steps 18–20; see Note 7)
24. Add Buffer B a third time and resuspend the mRNA/beads complexes by gentle pipetting.
25. Transfer all of the suspension to a 0.2 ml PCR tube.
26. Place tubes on magnet and leave undisturbed while preparing the cDNA master mix (cDNAMM).

Preparation of a reusable cDNA library

27. Prepare cDNAMM using your standard procedures (at 10.1 μl/sample).

28. Completely remove Buffer B from the 0.2 ml PCR tubes (see Note 8).
29. Add 10 μl of cDNAMM and gently centrifuge.
30. Incubate at 42.0 °C for 1 h in a rotating hybridization oven to keep the beads in suspension (see Note 9).
31. Place PCR tubes containing the bead/RNA/cDNA mix in a PCR machine heated to 95.0 °C for 1 min (see Note 10).
32. Remove the supernatant of the first tube as quickly as possible using a 10 μl pipet tip.
33. Discard the supernatant containing tip.
34. Repeat steps 31–33 until all tubes are processed.
35. Resuspend the magnetic bead-bound cDNA in 50 μl TIT buffer and store at 4.0 °C (see Note 11).

Second strand PCR

36. Create a PCR program to generate a second strand from the bead/cDNA complex with the following settings; denaturation step 94–95.0 °C/15–20 s, annealing temperature of the primer 15 s, extension 15–30 s (based on length of amplicon and optimal temperature of the polymerase), and a final denaturation step 94–95.0 °C for 2 min.
37. Prepare a standard PCR master mix (PCRMM; 20–25 μl/sample).
38. Place a new 1.5 ml tube on a magnet (having at least two magnets simplifies this step) and a new PCR tube on a plastic rack for every sample to be analyzed.
39. Add 1.5 μl of nuclease-free water to the 1.5 ml tubes.
40. Add 10–15 μl PCRMM to the new PCR tube and set aside.
41. Briefly centrifuge the PCR tubes containing the magnetic bead-attached cDNA and place on magnet and completely remove TNT.
42. Gently resuspend beads in 10 μl PCRMM. Place tubes on PCR machine and run the second strand cycle.
43. Remove tubes immediately upon cycle completion and place on magnet.
44. Remove supernatant as quickly as possible and add to the *bottom* of the magnet-held 1.5 ml tube.
45. Immediately, transfer the second strand from the 1.5 ml tube to the PCR tube containing the PCRMM.
46. The PCR reaction is ready for amplification.
47. Wash the 1.5 ml tube with 50 μl TNT to recover any beads inadvertently transferred during the second strand recovery

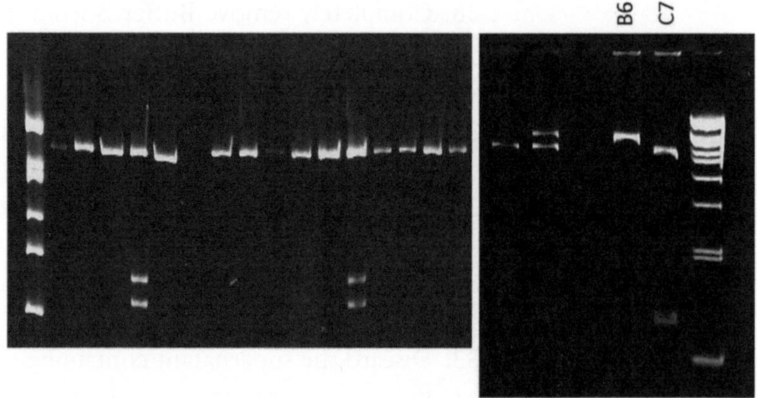

Fig. 1. Amplification of imprinted genes using morula- and early blastocyst-stage mouse embryos. Each lane contains the amplicon from an individual embryo. mRNA from C57BL/6 (B6) × B6[(CAST7)](C7) F1 hybrid embryos was extracted by affinity purification using oligo dT magnetic beads. *Left panel*—the bands are for the imprinted RNA *Kcnq1ot1* (not a polyadenylated transcript). *Right panel*—the bands are for the imprinted RNA *H19* (polyadenylated transcript). Samples were restricted with a restriction enzyme that differentiates between the parental alleles to determine the allele the RNA was transcribed form. In the gels the top band represents one of the alleles and bottom bands are the second (restricted) allele. When the RNA is expressed from only one of the alleles it is referred to as monoallelic expression. In the case where RNA is transcribed from both parental alleles the expression is said to be biallelic. Amplicons were resolved in a 7% polyacrylamide gel. (**b**) and (**c**) represent strain controls for expression and restriction enzyme digestion.

and return to the original tube containing the sample's cDNA/bead complex.

48. Store at 4.0 °C.

    Figure 1 shows examples of imprinted gene expression analysis of single mouse preimplantation embryos.

3.3. DNA Isolation

1. Add 2.5 volumes of ice cold absolute ethanol to the tube containing the supernatant from step 17 of Subheading 2.1.
2. Gently mix and incubate at −20 °C for 1 h (see Note 12).
3. Centrifuge at full speed (14,000 rpm/~18,000 × *g*) for 15 min.
4. Carefully remove half of the supernatant without disturbing the pellet (see Notes 13 and 14).
5. Centrifuge again at full speed for 5 min.
6. Place the pipet tip on the opposite side of the tube where the pellet is and remove the supernatant
7. Air dry the DNA pellet for ~5 min (see Note 15).
8. Resuspend the DNA in 21 μl nuclease free water.
9. If the isolated DNA is intended for methylation studies, proceed to the bisulfite mutagenesis procedure immediately.

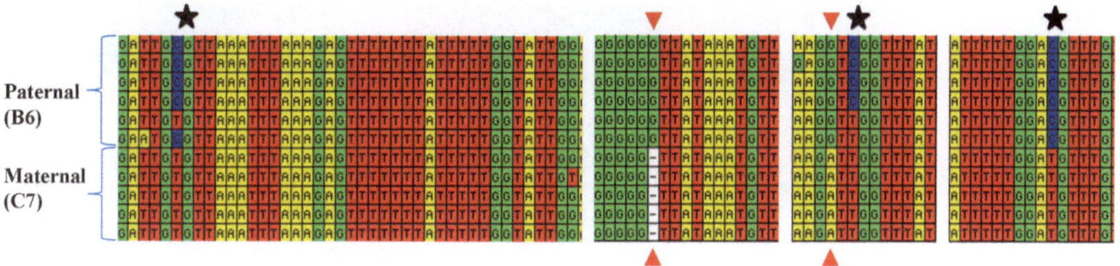

Fig. 2. Sequencing of bisulfite converted DNA of a single blastocyst-stage B6xC7 F1 hybrid embryo. DNA was isolated, bisulfite converted, and amplified by nested PCR using bisulfite converted DNA-specific primers. The amplicons were cloned and sequenced. The region amplified is the *H19/IGF2* ICR (4, 6). This ICR is normally unmethylated on the maternal allele and methylated one the paternal allele. This figure shows bisulfite sequencing information of six paternal and five maternal alleles. DNA sequence polymorphisms between B6 and Casteneus strains of mice are shown by arrowheads. *Stars* denote CpGs. A CG on the sequence means that that cytosine was methylated while a TG means that CG was unmethylated.

3.4. Bisulfute Mutagenesis

Bisulfite conversion is the selective deamination of unmethylated cytosines to uracils by treatment with sodium bisulfite. This procedure is performed using the commercially available Imprint DNA Modification Kit (Sigma; Cat. No. MOD50). See Note 16.

1. To the freshly isolated DNA add
   (a) 1 μl of a 7 ng carrier RNA solution.
   (b) 1 μl balance solution from the kit.
2. Refer to the two-step method for bisulfite mutagenesis (from the Sigma Imprint kit manual) and follow the protocol as per manufacturer's directions (see Notes 17–19).

   Figure 2 shows an example of bisulfite sequencing data obtained from one blastocyst.

4. Notes

1. Bisulfite mutagenesis is a lengthy procedure and it may not be possible to mutagenize the DNA on the same day the RNA isolation is performed. We have obtained good quality bisulfite converted DNA months after RNA collection by simply preserving the DNA in solution in a 70% EtOH (refer to step 1 of Subheading 3.3) at −80 °C for several months.
2. Care should be taken to prevent somatic cell contamination of the sample.
3. Care should be taken during the entire procedure to pipet from the front of the tube (i.e., away from the beads which will be pulled to the back of the tube by the magnet) so as not to accidentally aspirate the beads.

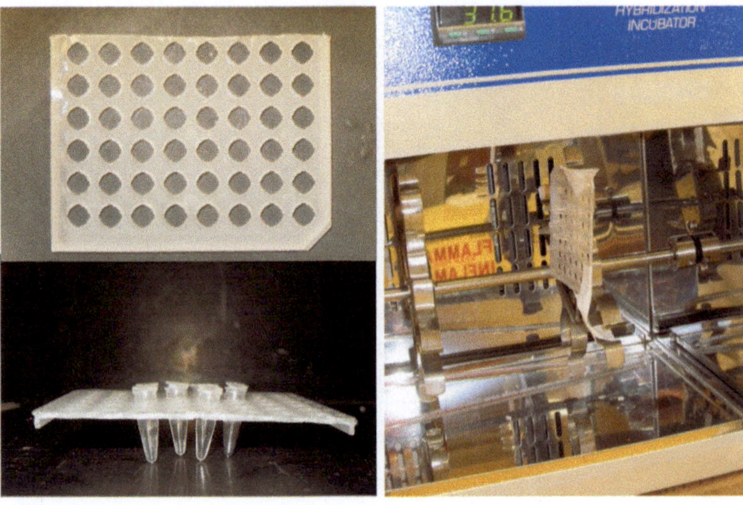

Fig. 3. Plastic rack used to facilitate the handling of samples during the cDNA synthesis step. The rack with the tubes is placed sideways in a hybridization oven and fastened to the rotating arm with tape.

4. A foam adapter for the vortexer that holds microcentrifuge tubes makes this task very simple.
5. Avoid disturbing/pipetting the RNA-bound beads to prevent RNA loss
6. The supernatant contains the sample's DNA.
7. Sometimes the supernatant does not become fully clear during the Buffer B washes (some of the Dynabeads/mRNA complexes will not readily be attracted by the magnet). When this occurs, gently aspirate the supernatant with a pipette and very slowly add back to the tube. This usually helps clear the fluid.
8. Process no more than three samples at a time. We have noticed that if the beads dry up while waiting to be resuspended in cDNAMM they may "jump" out of the tube and are lost.
9. In order to make this task simple we use a plastic rack (Fig. 3).
10. Process no more than six samples at a time to prevent reannealing.
11. We have successfully used single embryo cDNA libraries 10–20 times. The TIT may evaporate over long periods of storage. Add more TIT as necessary.
12. May be stored at −80 °C for several months with good results.
13. It should be noted that the pellet can be hardly seen at the bottom of the tube.
14. Use the smallest pipet tip possible
15. Care should be taken not to have any ethanol left on the tube, nor overdry the pellet.
16. Use a PCR machine for all incubations.

17. Please note, once the DNA Modification Mixture is added, all samples are treated as being light sensitive.
18. We have found that the following modifications to the protocol improve our success
    (a) Half-way through the 90 min incubation at 65.0 °C, remove the samples from the PCR machine, briefly mix by inverting the tubes, centrifuge, and return to the PCR machine for the remainder of the incubation.
    (b) We perform all manual-described centrifugations for 1 min instead of 20 s.
    (c) Use 23 μl of sterile water to elute the bisulfite-converted DNA from the column.
    (d) Since bisulfite-converted DNA is single-stranded, storage at −80.0 °C prevents further degradation.
    (e) In addition, aliquoting the samples before storing prevents freeze–thaw damage.
19. Other information: We have noticed that restricting the genomic DNA with an enzyme that does not have a recognition site within the region of interest prior to bisulfite mutagenesis improves the success rate of PCR amplification after conversion.

Acknowledgments

The original procedures for RNA isolation from single mouse embryos were first published by Dr. Melissa Mann (6) while in the laboratory ofDr. Marisa Bartolomei at the University ofPennsylvania. Dr. John Huntriss, University of Leeds, provided information of commercial kit for isolation of DNA from single embryos. The data used in Fig. 1 was generated by Mr. Franklin Echevarria.

References

1. Nagy A, Gertsenstein M, Vintersten K, Behringer R (2003) Manipulating the mouse embryo: a laboratory manual, 3rd edn. Cold Spring Harbor Laboratory Press, Cold Spring Harbor, NY
2. Morison IM, Ramsay JP, Spencer HG (2005) A census of mammalian imprinting. Trends Genet 21:457–465
3. Verona RI, Mann MR, Bartolomei MS (2003) Genomic imprinting: intricacies of epigenetic regulation in clusters. Annu Rev Cell Dev Biol 19:237–259
4. Rivera RM, Stein P, Weaver JR, Mager J, Schultz RM, Bartolomei MS (2008) Manipulations of mouse embryos prior to implantation result in aberrant expression of imprinted genes on day 9.5 of development. Hum Mol Genet 17:1–14
5. Doherty AS, Bartolomei MS, Schultz RM (2002) Regulation of stage-specific nuclear translocation of Dnmt1o during preimplantation mouse development. Dev Biol 242: 255–266
6. Mann MR, Lee SS, Doherty AS, Verona RI, Nolen LD, Schultz RM, Bartolomei MS (2004) Selective loss of imprinting in the placenta following preimplantation development in culture. Development 131:3727–3735

Chapter 14

Generation of cDNA Libraries from RNP-Derived Regulatory Noncoding RNAs

Mathieu Rederstorff

Abstract

Next-generation sequencing of noncoding RNA (ncRNA) libraries has become an essential tool for the profiling of ncRNAs and the identification of novel ncRNA species. Here, we describe the generation of a ncRNA-derived complementary DNA (cDNA) library by 3′-tailing of ncRNAs by CTP and poly(A) polymerase, followed by 5′-adapter ligation by T4 RNA ligase and reverse transcription of ncRNAs with an oligo-d(G) anchor primer. Preliminary selection of ncRNAs from ribonucleoprotein particles (RNPs) enables a strong enrichment of the generated libraries with functional regulatory ncRNAs compared to classical approaches.

Key words: Noncoding RNAs, Ribonucleoprotein particle, cDNA library, High-throughput sequencing (or next-generation sequencing NGS)

1. Introduction

More than 90% of the human genome appears to be actively transcribed, although only a minor portion of transcripts code for proteins (1). Therefore, it has been proposed that most of the transcripts correspond to regulatory noncoding RNAs (ncRNAs) (2–8). Over the last decade, the importance of ncRNAs has increased constantly (9–13). NcRNAs play key regulatory functions in multiple biological pathways, from DNA replication and chromosome maintenance, to RNA modification or regulation of translation (14–24). NcRNAs are also involved in regulating the transcriptional machinery, controlling transcription factors directly or mediating epigenetic changes through the interaction with proteins that modify DNA or histones (25). NcRNAs involved in epigenetic regulation are diverse. They comprise the shortest ncRNAs yet described, namely, short interfering RNAs (siRNAs, 20–22

Nora Engel (ed.), *Genomic Imprinting: Methods and Protocols*, Methods in Molecular Biology, vol. 925,
DOI 10.1007/978-1-62703-011-3_14, © Springer Science+Business Media, LLC 2012

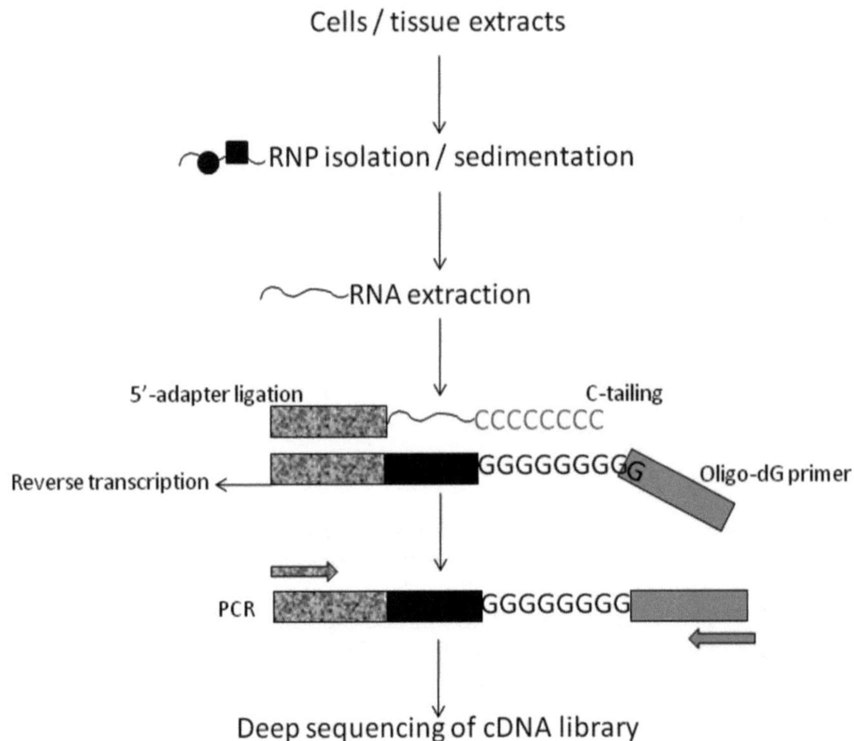

Fig. 1. Overview of the library generation approach.

nucleotides), acting within the RNA-induced transcriptional silencing (RITS) complex (26, 27), as well as some of the longest ncRNAs known to date, such as Xist (19 kb) (28) or Air (108 kb) (29), which recruit Polycomb group (PcG) complexes, like the Polycomb repressive complex 2 (PRC2), to mediate silencing (30).

To identify novel ncRNAs, a method of choice consists in the generation and sequencing of cDNA libraries, which has been considerably improved owing to the development of next-generation sequencing techniques. Previously, cDNA libraries were generated using RNA that was size-separated on denaturing gels (12, 31, 32) leading mostly to the cloning and sequencing of RNA species corresponding to ribosomal RNAs (rRNAs) or other known ncRNA species (33). We developed an alternative procedure, based on the size-selection of ribonucleoprotein (RNP) particles on glycerol gradients prior to RNA extraction, which enables a strong enrichment of the libraries with functional ncRNAs species (Fig. 1) (34, 35). NcRNAs are next 3′-tailed with CTP and poly(A) polymerase, followed by 5′-adapter ligation by T4 RNA ligase and reverse transcription with an oligo-d(G) anchor primer. Libraries obtained are compatible with most high-throughput sequencing techniques available (36).

2. Materials

Diethylpyrocarbonate (DEPC)-treated RNase-free water should be used to prepare all solutions, as well as to dissolve RNA pellets after ethanol precipitation. To prepare DEPC-treated RNase-free water, mix double-distilled water with 0.01% (vol/vol) DEPC in RNase-free bottles. Let stand overnight and autoclave. RNase-free water can be stored at room temperature (20 °C) for several months.

2.1. RNP Sedimentation

1. Prepare nuclear or cytoplasmic extracts according to Dignam et al. (35, 37).
2. L7-65 ultracentrifuge (Beckman).
3. SW41 rotor (Beckman).
4. Polyallomer centrifuge tubes: thin wall, 13.2 ml, 14×89 mm (Beckman 331372).

2.2. RNA Preparation

1. Phenol–Chloroform–Isoamyl alcohol: 25/24/1 solution (see Note 1).
2. Chloroform.
3. Sodium acetate 3 M (pH 5.2).
4. Ethanol.
5. Glycoblue™: 15 mg/ml glycogen (Ambion AM 9516) (see Note 2).

2.3. cDNA Library Generation

1. RiboLock™ RNAse Inhibitor (Fermentas Life Sciences EO0381) (see Note 3).
2. A-Plus™ poly(A) Polymerase kit, including 1× Tailing buffer: 50 mM Tris–HCl (pH 8.0), 250 mM NaCl, 10 mM $MgCl_2$ (Epicentre Biotechnologies PAP5104H).
3. Cytidine triphosphate, CTP.
4. TAP: Tobacco Acid Pyrophosphatase, including 1× TAP buffer: 50 mM sodium acetate (pH 6.0), 1 mM EDTA, 1% β-mercaptoethanol, 0.01% Triton® X-100 (Epicentre Biotechnologies T19250).
5. T4 RNA Ligase, including 1× RNA Ligation buffer: 50 mM Tris–HCl (pH 7.8), 10 mM $MgCl_2$, 10 mM DTT and 1 mg/ml BSA, 10 mM ATP (Fermentas Life Sciences EL0021).
6. 5′-Adapter (Ribonucleotides underlined): 5′-GTC AGC AAT CCC TAA C GAG-3′ (Pharmacon) (see Note 4).
7. SuperScript™ II Reverse Transcriptase, including 5× First strand buffer: 250 mM Tris–HCl (pH 8.3), 375 mM KCl, 15 mM $MgCl_2$ (Life technologies Invitrogen 100004925).
8. Deoxynucleotide triphosphate, dNTP.

9. Anchor primer (for reverse transcription): 5′-AGG AGC CAT CGT ATG TCG GGG GGG GH-3′ (Microsynth) (see Note 4).
10. *Taq* DNA Polymerase I, including 1× *Taq* buffer: 10 mM Tris–HCl (pH 9.0), 50 mM KCl, 0.1% Triton® X-100 (Life Technologies Invitrogen 10342020) (see Note 5).
11. 5′ PCR primer: 5′-(sequencing tag*)-GTC AGC AAT CCC TAA CGA G-3′ (Microsynth).
12. 3′ PCR primer: 5′-(sequencing tag*)-AGG AGC CAT CGT ATG TCG-3′ (Microsynth).

*The sequencing tags depend on the high-throughput sequencing technology chosen (see Note 4).

3. Methods

3.1. RNP Sedimentation

1. Layer the nuclear, cytoplasmic, or total extract onto a 10–30% glycerol gradient (see Note 6).
2. Spin the gradient at 198,000 × *g* (40,000 rpm) for 13 h at 4 °C in a Beckman SW41 rotor.
3. Collect the fractions of interest (see Note 7).

3.2. RNA Preparation

1. Extract RNA from each fraction with phenol–chloroform (1:1) (see Note 1).
2. Thoroughly vortex the tubes for 5 min at 20 °C to completely dissociate RNPs and extract RNA.
3. Centrifuge the tubes for 5 min at 20 °C at 10,000 × *g* to separate the organic (lower) and water (upper) phases.
4. Collect the RNA containing upper phase (see Note 8).
5. Mix the sample with chloroform (1/1).
6. Vortex the tubes for 5 min at 20 °C to extract remaining traces of phenol from the sample.
7. Centrifuge the tubes for 5 min at 20 °C at 10,000 × *g*.
8. Collect the RNA-containing upper phase.
9. Mix the sample with 3 M sodium acetate (pH 5.2) (1/10) and 100% ethanol (3/1) to precipitate RNA (see Note 2).
10. Gently vortex the tubes and precipitate RNA for 30 min at −80 °C (see Note 9).
11. Centrifuge the tubes for 15 min at 10,000 × *g* at 4 °C.
12. Remove the supernatant.
13. Gently wash the RNA pellet with 70% DEPC-ethanol (see Note 10).
14. Centrifuge the tubes for 5 min at 10,000 × *g* at 4 °C.

15. Discard the supernatant and air dry the RNA pellet (see Note 11).
16. Dissolve the RNA pellet in 20 μl DEPC-treated water.
17. Estimate RNA concentration using a spectrophotometer.

3.3. cDNA Library Generation

C-Tailing of ncRNAs (see Note 12)

1. Tail 150 ng to 1 μg of RNA in a final reaction volume of 20 μl containing 1× Tailing buffer, 4 U of A-Plus™ poly(A) polymerase, 20 U of RiboLock™ RNAse Inhibitor and 100 mM CTP (see Note 13).
2. Incubate the tube for 90 min at 37 °C.
3. Fill up the volume to 200 μl with DEPC-treated water and stop the reaction by phenol–chloroform extraction (steps 1–8 Subheading 3.2).
4. Precipitate RNA (steps 9–15 Subheading 3.2).
5. Dissolve the RNA pellet in 7 μl DEPC-treated water (see Note 14).

Decapping of ncRNAs (see Note 15)

6. Add 1× TAP buffer, 10 U of TAP and 40 U of RiboLock™ RNAse Inhibitor in a final reaction volume of 10 μl.
7. Incubate the tube for 60 min at 37 °C.
8. Fill up the volume to 200 μl with DEPC-treated water.
9. Phenol–chloroform extract and precipitate RNA (steps 1–15 Subheading 3.2).
10. Dissolve the RNA pellet in 10 μl DEPC-treated water.

5′-Adapter Ligation

11. Add 2 μl 5′-adapter (100 pmol/μl).
12. Incubate the tube for 5 min at 65 °C.
13. Add 1× RNA Ligation buffer, 1 mM ATP, 0.1 μg/μl BSA, 40 U of RiboLock™ RNAse Inhibitor and 10–20 U of T4 RNA Ligase in a final reaction volume of 20 μl.
14. Incubate the tube overnight at 4 °C (see Notes 16 and 17).
15. Phenol–chloroform extract and precipitate RNA (steps 1–15 Subheading 3.2).
16. Dissolve the RNA pellet in 10 μl DEPC-treated water.

Reverse-Transcription

17. Add 1 μl Anchor primer (100 pmol/μl) and 1 μl dNTP (10 mM each).
18. Denaturate the mixture for 5 min at 65 °C and immediately cool on ice.

19. Add 4 μl 5× First strand buffer, 2 μl DTT (100 mM), 40 U of RiboLock™ RNAse Inhibitor.
20. Incubate the tube for 2 min at 42 °C.
21. Add 200 U of SuperScript™ II Reverse Transcriptase.
22. Incubate the tube for 60 min at 42 °C.

cDNA Amplification

23. To 1/5 (4 μl) of the cDNA obtained, add 1× *Taq* buffer, 1 pmol/μl of both PCR primers, 0.25 mM $MgCl_2$, 0.1 mM dNTP and 1 U of *Taq* DNA Polymerase I in a final reaction volume of 50 μl (see Note 5).
24. Denature the sample for 2 min at 94 °C in a thermal cycler.
25. Amplify the cDNA with 20–25 cycles of PCR as follows: 1 min at 94 °C, 1 min at 54 °C, 1 min at 72 °C.
26. Finalize amplification for 5 min at 72 °C (see Notes 18 and 19).
27. The library can be sequenced with the chosen high-throughput sequencing platform (see Note 4).

4. Notes

1. The pH of the phenol–chloroform mixture used for the first extraction can be acidic (below 7, ideally 4.5 or 5) in order to denature DNA that may possibly be present as a contamination. At pH 7, DNA and RNA will appear in the water phase. At pH 5, DNA is partly denatured and will be directed to the organic phase, while RNA remains in the water phase.
2. For the first precipitation step, a carrier, e.g., glycogen or Glycoblue™, can be added since the amounts of material retrieved after the gradient may be rather low.
3. Employing RNAse inhibitors throughout the procedure in addition to using DEPC-treated water significantly reduces RNA degradation, especially during long incubation periods.
4. Additional specific sequences may have to be introduced in the 5′-adapter and anchor primer as well. These tags enable the cDNAs to bind on the sequencing slides. Regarding the 3′ extremity of ncRNAs, alternative protocols describe ligation of a 3′ adapter; however, we found it always more efficient and productive to proceed with C-tailing of the 3′ extremity.
5. The PCR step can also be carried out with a homemade *Taq* polymerase if available. However, high-fidelity enzymes are recommended to avoid too many errors.

6. The glycerol content of the sample should not exceed 10% to stay on top of the gradient when loaded. Dilute sample in an adequate buffer to reach a 10% final glycerol concentration.
7. We recommend collecting fractions starting from the bottom of the tube to avoid perturbation of the gradient, using a peristaltic pump or a syringe pump if available. Very bottom and top fractions (1 ml each) should be discarded. These fractions contain rRNA contaminations and various RNA degradation products, respectively. Remaining fractions can be pooled if no particular RNP size (10–30S) is required. One can decide to work only with specific subfractions (e.g., 15S).
8. Steps 1–4 can be repeated once.
9. Alternatively, RNA can be precipitated overnight at −20 °C.
10. To avoid washing out of small RNAs (e.g., miRNAs), pellets should be extremely gently washed throughout the protocol.
11. A pellet that is too dry may be harder to dissolve.
12. Since ncRNAs do not contain a poly(A) tail (as do mRNAs), ncRNAs are, in a first step, tailed with CTP and poly(A) polymerase (which also uses CTP as a substrate).
13. Poly(A) polymerase has a higher affinity for ATP than for CTP, therefore, increasing CTP concentration in the reaction buffer to 100 mM (10 mM for ATP), strongly improves the yield of the C-tailing reaction.
14. It is not necessary to quantify RNA here and after the forthcoming precipitation steps. Pellets are now dissolved in the adequate volume of DEPC-treated water and all the material is used for the next step.
15. This step enables all RNAs to feature a 5′ phosphorylated terminus, which is necessary for 5′-adapter ligation.
16. For the best efficiency of overnight ligation, samples should be incubated in melting ice (water + ice), rather than in crushed ice only. To avoid having the ice melt too fast, the incubation should take place in a cold room.
17. After overnight ligation at 4 °C, adding 10 more units of T4 RNA ligase for 30 min at 37 °C will increase the yield of the reaction.
18. PCR product can be purified on a native polyacrylamide gel. The user thus has the possibility of performing a size selection, e.g., products of sizes comprised between 20 and 500 nucleotides (nt), prior to high-throughput sequencing.
19. Quality control (agarose gel electrophoresis, cloning, and Sanger sequencing to check the presence of the deep-sequencing adaptors) can be performed but is generally done by the company/platform in charge of the deep-sequencing.

References

1. Birney E et al (2007) Identification and analysis of functional elements in 1% of the human genome by the ENCODE pilot project. Nature 447:799–816
2. Brosius J (2005) Waste not, want not—transcript excess in multicellular eukaryotes. Trends Genet 21:287–288
3. Carninci P et al (2005) The transcriptional landscape of the mammalian genome. Science 309:1559–1563
4. Cheng J et al (2005) Transcriptional maps of 10 human chromosomes at 5-nucleotide resolution. Science 308:1149–1154
5. Kampa D et al (2004) Novel RNAs identified from an in-depth analysis of the transcriptome of human chromosomes 21 and 22. Genome Res 14:331–342
6. Mattick JS, Makunin IV (2005) Small regulatory RNAs in mammals. Hum Mol Genet 14(1):R121–R132
7. Mattick JS, Makunin IV (2006) Non-coding RNA. Hum Mol Genet 15(1):R17–R29
8. Willingham AT, Gingeras TR (2006) TUF love for "junk" DNA. Cell 125:1215–1220
9. Couzin J (2002) Breakthrough of the year. Small RNAs make big splash. Science 298:2296–2297
10. Dennis C (2002) Small RNAs: the genome's guiding hand? Nature 420:732
11. Dennis C (2002) The brave new world of RNA. Nature 418:122–124
12. Huttenhofer A, Brosius J, Bachellerie JP (2002) RNomics: identification and function of small, non-messenger RNAs. Curr Opin Chem Biol 6:835–843
13. Rederstorff M, Huttenhofer A (2010) Small non-coding RNAs in disease development and host–pathogen interactions. Curr Opin Mol Ther 12:684–694
14. Ambros V (2001) microRNAs: tiny regulators with great potential. Cell 107:823–826
15. Bachellerie JP, Cavaille J, Huttenhofer A (2002) The expanding snoRNA world. Biochimie 84:775–790
16. Gottesman S (2002) Stealth regulation: biological circuits with small RNA switches. Genes Dev 16:2829–2842
17. Kiss T (2002) Small nucleolar RNAs: an abundant group of noncoding RNAs with diverse cellular functions. Cell 109:145–148
18. Lau NC et al (2001) An abundant class of tiny RNAs with probable regulatory roles in Caenorhabditis elegans. Science 294:858–862
19. Lee RC, Ambros V (2001) An extensive class of small RNAs in Caenorhabditis elegans. Science 294:862–864
20. Rhoades MW et al (2002) Prediction of plant microRNA targets. Cell 110:513–520
21. Storz G (2002) An expanding universe of noncoding RNAs. Science 296:1260–1263
22. Tuschl T (2002) Expanding small RNA interference. Nat Biotechnol 20:446–448
23. Tuschl T (2003) Functional genomics: RNA sets the standard. Nature 421:220–221
24. Volpe TA et al (2002) Regulation of heterochromatic silencing and histone H3 lysine-9 methylation by RNAi. Science 297: 1833–1837
25. Barrandon C, Spiluttini B, Bensaude O (2008) Non-coding RNAs regulating the transcriptional machinery. Biol Cell 100:83–95
26. Morris KV et al (2004) Small interfering RNA-induced transcriptional gene silencing in human cells. Science 305:1289–1292
27. Verdel A et al (2004) RNAi-mediated targeting of heterochromatin by the RITS complex. Science 303:672–676
28. Maenner S et al (2010) 2-D structure of the A region of Xist RNA and its implication for PRC2 association. PLoS Biol 8:e1000276
29. Braidotti G et al (2004) The Air noncoding RNA: an imprinted cis-silencing transcript. Cold Spring Harb Symp Quant Biol 69:55–66
30. Margueron R, Reinberg D (2011) The Polycomb complex PRC2 and its mark in life. Nature 469:343–349
31. Huttenhofer A et al (2001) RNomics: an experimental approach that identifies 201 candidates for novel, small, non-messenger RNAs in mouse. EMBO J 20:2943–2953
32. Huttenhofer A, Vogel J (2006) Experimental approaches to identify non-coding RNAs. Nucleic Acids Res 34:635–646
33. Jochl C et al (2008) Small ncRNA transcriptome analysis from Aspergillus fumigatus suggests a novel mechanism for regulation of protein synthesis. Nucleic Acids Res 36:2677–2689
34. Rederstorff M et al (2010) RNPomics: defining the ncRNA transcriptome by cDNA library generation from ribonucleo-protein particles. Nucleic Acids Res 38:e113
35. Rederstorff M, Huttenhofer A (2011) cDNA library generation from ribonucleoprotein particles. Nat Protoc 6:166–174
36. Metzker ML (2010) Sequencing technologies - the next generation. Nat Rev Genet 11:31–46
37. Dignam JD, Lebovitz RM, Roeder RG (1983) Accurate transcription initiation by RNA polymerase II in a soluble extract from isolated mammalian nuclei. Nucleic Acids Res 11: 1475–1489

Chapter 15

Co-Immunoprecipitation of Long Noncoding RNAs

Victoria A. Moran, Courtney N. Niland, and Ahmad M. Khalil

Abstract

It is now estimated that the human genome encodes thousands of long noncoding (lnc)RNAs. These novel molecules are causing a paradigm shift in the field of molecular biology as a number of lncRNAs have been shown to be involved in a wide range of biological functions including regulation of gene expression. Also, misregulation of lncRNAs has been observed in human diseases such as cancer and neurological disorders. These findings have spurred a huge interest in elucidating the functions and mechanisms of lncRNAs; and therefore, the need for new methods to do so. In this chapter, we discuss RIP-Seq, a method that is utilized to discover the lncRNA partners of a specific protein. This procedure involves immunoprecipitation of a protein from cross-linked cell lysate followed by reverse-cross-linking, isolation, and deep sequencing of RNAs, leading to the identification of all lncRNAs that are associated with a specific protein complex.

Key words: Long noncoding RNA, Large noncoding RNA, lncRNA, lincRNA, Ribonucleic protein complex, RNP, RNA coimmunoprecipitation, RIP, RNA–protein interactions

1. Introduction

One of the most surprising findings of the International Human Genome Sequencing Consortium was that only about 2% of the entire human genome is protein-coding (1), and is accounted for by the 20,000 or so protein-coding genes (1). It is has also been suggested that as much as 90% of the genome can be transcribed (2). While some believe that a majority of this disparity can be attributed to so-called "transcriptional noise" (3), other evidence points to a large number of genes encoding functional noncoding RNAs. RNA has long been known to play a myriad of different functions distinct from its classical protein-coding role. In this way, the "Central Dogma of Molecular Biology" has been at best an oversimplification and at worst, a scientific falsity for decades. The functional roles of ribosomal (r)RNAs, transfer (t)RNAs, snRNAs, and snoRNAs have long been elucidated. Recent advances in

Nora Engel (ed.), *Genomic Imprinting: Methods and Protocols*, Methods in Molecular Biology, vol. 925,
DOI 10.1007/978-1-62703-011-3_15, © Springer Science+Business Media, LLC 2012

technology such as tilling arrays and deep sequencing of RNAs have led to the discovery of additional classes of noncoding RNAs including small noncoding RNAs (e.g., microRNAs and piRNAs) and long (or large) noncoding RNAs (e.g., NAT and lincRNAs) (2, 4–13). Large noncoding RNAs (lncRNAs) are especially abundant in mammalian genomes, with some groups estimating the number of lncRNA encoding genes at approximately 23,000 (13).

The recent stir of this novel class of RNAs prompted a universal definition. A lncRNA is defined as any endogenous RNA that is over 200 nucleotides in length, but lacks a sensible open reading frame capable of producing a protein product, usually defined as an ORF of more than 100 amino acids (14, 15). Many new lncRNA molecules are being discovered and characterized at a rapid pace, and new functions and mechanisms are being elucidated as well. However, at this time, a relatively small number of lncRNAs have been studied in an extensive manner. Of these, arguably the best known and studied is *Xist* (X inactive specific transcript) which has an essential role in the initiation of X inactivation in mammalian female somatic cells as a means of dosage compensation (16, 17). In addition to *Xist* function in dosage compensation, other biochemical roles for lncRNAs have also been identified. So far, lncRNAs have been shown to have a prominent role in epigenomic control, genomic imprinting, transcriptional control, nuclear compartmentalization, as well as stem cell proliferation and differentiation (4). In addition, their misregulation has been observed in a host of human disorders, including cancer metastasis, neurological disease, cardiac disease, and others (18, 19). While lncRNAs function in obviously different cellular pathways, many lncRNAs function as a part of ribonucleoprotein complexes (RNPs).

RNPs are defined as any complex in which RNAs associate with protein partners to carry out a function cooperatively. These can be further broken down into classes, based on the biochemical role of the RNA itself. While outside the scope of this chapter, it is important to note a few classical examples of noncoding (nc)RNPs. Ribosomes are the most extensively studied, where the ribosomal RNA serves as the catalytic core of the ncRNP complex. Other examples of classical ncRNPs include eukaryotic telomerase and the spliceosomal snRNPs. This chapter focuses on the newly emerging field of lncRNAs and their various protein partner complexes. Although lncRNAs function in different manners, patterns are emerging based on their relationship to their protein partners. These patterns can be separated into discrete functional classes, which are highlighted below.

1.1. Ribonucleic Acids Act to Target Protein Partners to Specific Genomic Loci

One of the biggest gaps in our collective knowledge of chromatin-modifying complexes is how they are recruited to specific regions of the genome. Many chromatin-modifying complexes are capable of adding or removing histone modifications (20, 21), but many of these complexes lack DNA binding domains. Multiple recent studies have shown that some lncRNAs, which bind to chromatin-modifying complexes, are capable of recruiting them to specific genomic targets. For example, during the process of mammalian X inactivation, *Xist* recruits polycomb repressive complex 2 (PRC2) to initiate transcriptional repression (22). Also, the lncRNA HOTAIR, which is transcribed from the HOXC locus, was shown to target PRC2 initially to HOXD genes (23), and more recently, to many regions throughout the genome (24, 25). This trend of lncRNAs acting as "traffic controllers" of repressive chromatin complexes is likely to be more widespread. It has recently been shown that as many as 40% of intergenic lncRNAs (lincRNAs) are capable of directly associating with several chromatin-modifying complexes including PRC2, CoREST, and SMCX (10).

LncRNAs have also been shown to recruit chromatin-modifying complexes to imprinted genes in an allele-specific manner. Somatic cells have two copies of each autosome, one is inherited paternally and the other maternally. At most gene loci, both alleles are expressed. However, some loci are under imprinted regulation, that is, those specific loci are expressed from either the maternal or paternal allele, depending on the specific locus. Recently, it has been shown that lncRNAs may have a role in the recruitment of chromatin-modifying complexes to imprinted loci. One such example is the lncRNA *Air*. *Air* is expressed from a promoter in the antisense direction to the Igf2r gene locus, and it is only expressed from the paternal copy (26). *Air* associates with the chromatin-modifying complex G9a, a repressive histone methyltransferase, and it targets G9a in cis to the *Igf2r* locus which results in the silencing of several imprinted genes (27). Another example of a RNP acting to facilitate imprinting has been observed in the *Kcnq1* gene locus. Another lncRNA, *Kcnq1ot1*, is expressed only from the paternally inherited chromosome and antisense to the *Kcnq1* gene. *Kcnq1ot1* silences adjacent genes via the recruitment of chromatin repressor complexes. Similar to *Air*, *Kcnq1ot1* has been shown to interact with G9a, and additionally, components of the PRC2 complex (28). Interestingly, studies have also indicated that *Kcnq1ot1* associates with DNMT1, a DNA methyltransferase which is responsible for the maintenance of DNA methylation patterns (29). In this way, a single lncRNA may act through multiple pathways to regulate gene expression.

1.2. LncRNAs Perform Scaffolding Functions to Bridge Distinct Protein Partners

Another function of lncRNAs in RNPs is to serve as scaffolds, bringing proteins with different catalytic domains into close proximity (19). For example, HOTAIR binds to both PRC2 and the histone demethylase LSD1, thus serving as a scaffold between these two distinct complexes (24). Tsai and colleagues found evidence that different regions of HOTAIR bind to PRC2 and LSD1, potentially coupling their catalytic functions together (24).

This mechanism of RNA serving a structural scaffolding role is by no means restricted to mammals. A notable example of lncRNAs portraying such a function can also be seen in the dosage compensation pathway in *Drosophila melanogaster*. In fruit flies, dosage compensation takes an opposite approach to that of mammals in order to equalize sex-linked gene products. In mammals, dosage compensation is achieved by silencing one X-chromosome in XX females. In flies, dosage compensation is enacted by upregulation of the single X in XY males. This upregulation requires the Male Sex Lethal Complex (MSL), which coats the single male X, promoting active chromatin state. This MSL complex is held together by the lncRNAs, roX1 and roX2 (30). In addition, the roX RNAs facilitate the spreading of the complex over the entire X-chromosome (31, 32).

1.3. LncRNAs Function as Coregulators of Transcription Machinery

In 1999, the first lncRNA was characterized that displayed a biochemical role outside dosage compensation and imprinting. The lncRNA SRA (steroid receptor hormone activator) was discovered in screens for proteins acting in the regulation pathways of hormonal nuclear receptors (33). Nuclear receptors are a unique class of transcription factors, which only actively regulate transcription in the presence of a ligand. Specifically, ligand binding induces a conformational change in the nuclear receptors, which are then activated to bind to genomic cognate response elements, as well as to potentially recruit other coregulatory factors. It is believed that SRA forms a RNP with a common coactivator complex, named SRC-1, contributing to its binding potential, and thus regulation, of nuclear receptors (34).

Another example of coregulation of transcription can be seen in the corepressive activity of another lncRNA, *Gas5*. *Gas* (growth arrest specific) 5 is highly expressed in cells that have arrested growth function, due to the underavailability of necessary nutrients. It has been shown to be a negative regulator of glucocorticoid receptors, a specific class of nuclear receptors (35). While operating on receptor molecules as does SRA, the mechanism behind *Gas5* interaction is radically different. Gas5 interacts directly with the receptor's DNA binding domain, preventing it from binding to its DNA response element and effectively acting as a molecular decoy (35). In the absence of binding of the glucocorticoid receptor (GR), expression of GR target genes are severely downregulated. This mechanism clearly shows the possibility of a direct role in transcriptional regulation by lncRNA in a protein complex context.

1.4. RIP-Seq

The examples of RNPs discussed above only scratch the surface of the numerous potential lncRNA–protein interactions. As noted above, as many as 40% of lincRNAs are capable of directly associating with several chromatin-modifying complexes (i.e. PRC2, CoREST, and SMCX) (10). According to this observation, vast numbers of interactions of lncRNA to chromatin-modifying complexes, transcriptional regulators, and other protein complexes are yet to be characterized (19). RIP-Seq (RNA coimmunoprecipitation followed by deep sequencing) is a method that can be utilized to discover all lncRNAs that interact with a specific protein of interest genome-wide. Briefly, RNA is first cross-linked to proteins by formaldehyde, protein of interest is immunoprecipitated by a specific antibody, then the cross-linking is reversed, and RNA is isolated and prepared for deep sequencing. The RIP method described here is an adaptation from previous protocols with modifications that we worked out in our laboratory to optimize the technology (24, 36). The method described here utilizes low percentage formaldehyde (~0.3%) for cross-linking RNA to proteins prior to cell lysis, thus avoiding potential interactions that may occur after cell lysis (37). However, recent studies suggest that native (no cross-linking) RIP can also be used to identify lncRNA–protein interactions with high specificity and reproducibility (10, 38).

1.5. RIP-Seq Protocol

This technology depends highly on the quality of the antibody used; therefore a first step is to obtain a high quality antibody and test it in your laboratory. Also, all reagents and materials must be RNase-free to prevent RNA degradation. It is critical to point out that many factors play a role in the success of this procedure including antibody quality, cell type used, and successful preparation of the cell lysate; therefore, this protocol is a good starting point, and notes at each step in the following protocol can help optimize your results on a case-by-case basis.

2. Materials

1. Protein A/G magnetic beads.
2. Antibody against protein of interest, and IgG Antibody.
3. RIPA Buffer: 150 mM NaCl, 1.0% NP-40, 0.5% Sodium deoxycholate, 0.1% SDS, 50 mM Tris–HCl (pH 7.4), 1.0 mM EDTA.
4. Cells grown to 80–90% confluency.
5. Dulbecco's Modified Eagle's Media, supplemented with 10% Fetal Bovine Serum.
6. Cold 1× PBS.
7. 37% Formaldehyde.
8. 1.25 M Glycine in PBS.
9. Protease Inhibitors; 10× stock solution.
10. RNAse Inhibitor.

11. High Salt RIPA Buffer: 1.0 M NaCl, 1.0% NP-40, 0.5% Sodium deoxycholate, 0.1% SDS, 50 mM Tris–HCl (pH 7.4), 1.0 mM EDTA.
12. Buffer C: 150 mM NaCl, 50 mM Tris–HCl (pH 7.4), 5 mM EDTA, 10 mM DTT, 1.0% SDS.
13. Proteinase K.

3. Methods

3.1. Prepare Beads for Immunoprecipitation

1. Take 2 aliquots of 50 μL each of protein A/G magnetic beads, and wash two times with 300 μL RIPA buffer (see Note 1).
2. Remove wash supernatant with a magnet and resuspend beads in 500 μL of RIPA buffer.
3. Add 8 μg of antibody against your protein of interest to the first aliquot. To the second, add 8 μg of nonspecific IgG antibody (see Note 2).
4. Incubate beads with antibodies for 2 h at 4 °C with gentle rotation.

3.2. Lysate Preparation and Cross-linking

1. Grow cells to ~90% confluency. Six 15 cm plates are usually sufficient for most cell types; larger cells may require more plates to start with (see Note 3).
2. Harvest cells by trypsinization, adding an equivalent amount of media to quench the reaction. Collect cells in a 15 mL conical tube and pellet by centrifuging at 500 × *g* for 10 min.
3. Wash cells twice with PBS and collect cells after each wash by pelleting at 500 × *g* for 5 min.
4. Resuspend cellular pellet in 10 mL of PBS. Using a hemocytometer, or other quantification method, calculate the concentration of cells. Dilute suspension to two million cells per mL.
5. To perform cross-linking, add 37% formaldehyde to a final concentration of 0.3%. Incubate for 10 min, with gentle rotation at room temperature (see Note 4).
6. Quench the cross-linking reaction by adding 1.25 M glycine to a final concentration of 0.125 M. Incubate at room temperature for 5 min (see Note 5).
7. Pellet cells by centrifuging at 500 × *g* for 5 min.
8. Wash cells twice, each time with 10 mL of 1× PBS. Spin and pellet as before.
9. Resuspend pellet in 2.2 mL of RIPA buffer, supplemented with protease and RNAse inhibitors.

10. Incubate at 37 °C for 30 min. Vortex every 5 min for 30 s intervals for the duration of the incubation.
11. Homogenize the sample using a dounce homogenizer. Make certain to use enough force to disrupt the cellular membranes (see Note 6).
12. Centrifuge the lysate at maximum speed (≥10,000×*g*) using a microcentrifuge for 10 min. Collect the supernatant, discarding the pellet.
13. *Optional*: Perform pre-clearing of the lysate. Wash 30 μL of protein A/G beads two times in RIPA buffer. Remove wash buffer and add the supernatant from step 16. Incubate for 1 h with gentle rotation at 4 °C. Recover supernatant and proceed to step 18 (see Note 7).
14. Wash beads from step 4 twice with 500 μL of RIPA buffer. Discard supernatant.
15. Set aside a 100 μL aliquot of lysate to be used as an Input sample. Divide the rest of the lysate between the two antibody-mounted beads (1 mL each).
16. Incubate for 4 h or overnight at 4 °C with gentle rotation (see Note 8).
17. Remove supernatant by pelleting beads using a magnet and wash four times with high salt RIPA buffer (see Note 9).
18. Wash one time with 1× PBS.

3.3. For Protein Analysis

At this point it is necessary to ensure that the antibody used works in IP experiments, therefore, western blot analysis of the protein is critical to ensure that your antibody immunoprecipitates your protein of interest efficiently.

1. Resuspend beads in 100 μL of Laemmeli Lysis Buffer. Also add 100 μL of Laemmeli buffer to Input.
2. Incubate at 95 °C for 5 min to denature proteins and to reverse the cross-linking.
3. Run on a denaturing SDS-PAGE Gel, and perform western blot analysis on input, the IP using an antibody against your protein of interest, and IP with IgG antibody as a negative control. You should see a correct size band for your protein of interest in your input and specific antibody IP but not in your IgG IP. You should also include a positive control on your gel, such as a total cell extract with no treatment.

3.4. For RNA Analysis

1. Resuspend beads in 100 μL of Buffer C. Additionally, add 10 μg of proteinase K. Also, add 100 μL of Buffer C and proteinase K to the input sample (input volume is now 200 μL).
2. Incubate for 30 min at 42 °C, for proteinase K digestion.
3. Incubate for 4 h at 65 °C to reverse the formaldehyde cross-links.

4. Perform RNA Isolation and synthesize cDNA using random hexamers. Proceed to qPCR to analyze retrieval efficiency of known lncRNA that bind to your protein of interest. If none are known, then proceed to deep sequencing of RNA on an Illumina platform according to manufacturer's protocol (see Note 10).

4. Notes

1. Sepharose or Agarose beads can easily be used in place of magnetic beads. Since these beads tend to have higher background than magnetic beads, the preclearing step 17 should not be skipped. Additionally, protein A or protein G beads may be more optimal depending on your particular antibody.
2. Depending on the affinity of your antibody to your RNP component of interest, the mass of antibody added can be altered. While 8 μg is a good starting point, a range from 6 to 12 μg can be used.
3. Be sure to use enough starting material to produce the extract. Many lncRNAs are expressed at relatively low levels in cells; thus, it is important to certify that sufficient extract is produced. A good starting point is about ten million cells, then increase or decrease from that starting point.
4. Depending on your RNA of interest, more or less formaldehyde may be added for optimal co-IP results. While 0.3% has worked best in most cases, a concentration from 0.1 to 1% may be used. Incubation time may be altered to vary from 5 to 60 min as well.
5. If higher concentrations of formaldehyde are used, such as 1%, increase final glycine concentration to 0.25 M to ensure complete quenching.
6. Keep in mind that the more formaldehyde used to cross-link, the more rigid the cells will be. More force may be necessary to disrupt cellular membranes. You can increase the incubation and vortexing time in step 14 to 1 h to remedy this or use sonication. It is critical to point out that sonication will result in some RNA degradation which may affect the results of your experiment.
7. This step may not be necessary, depending on the specificity of your antibodies and the type of beads that are used. It is a good idea to do a preclear if using sepharose or agarose beads, due to the high potential of nonspecific interactions with these beads. If using magnetic beads, this may not be necessary.

8. It is a good idea to incubate the input along with the IPs. This way, if degradation should somehow occur, it will be equalized among all the samples.
9. If performing protein pull-down analysis, an aliquot of supernatant may be saved to determine the IP efficiency.
10. In such a procedure where so many steps can be optimized specific to each cell type and protein of interest, it is important to verify the efficiency of the pulldown for a protein by western blotting analysis.

References

1. International Human Genome Sequencing Consortium (2004) Finishing the euchromatic sequence of the human genome. Nature 431: 931–945
2. Kapranov P, Cheng J, Dike S, Nix DA, Duttagupta R, Willingham AT, Stadler PF, Hertel J, Hackermuller J, Hofacker IL et al (2007) RNA maps reveal new RNA classes and a possible function for pervasive transcription. Science 316:1484–1488
3. Wang J, Zhang J, Zheng H, Li J, Liu D, Li H, Samudrala R, Yu J, Wong GK (2004) Mouse transcriptome: neutral evolution of 'non-coding' complementary DNAs. *Nature* 431: 1 p following 757; discussion following 757
4. Nagano T, Fraser P (2011) No-nonsense functions for long noncoding RNAs. Cell 145: 178–181
5. Guttman M, Donaghey J, Carey BW, Garber M, Grenier JK, Munson G, Young G, Lucas AB, Ach R, Bruhn L et al (2011) lincRNAs act in the circuitry controlling pluripotency and differentiation. Nature 477:295–300
6. Clark MB, Mattick JS (2011) Long noncoding RNAs in cell biology. Semin Cell Dev Biol 22:366–376
7. Cabili MN, Trapnell C, Goff L, Koziol M, Tazon-Vega B, Regev A, Rinn JL (2011) Integrative annotation of human large intergenic noncoding RNAs reveals global properties and specific subclasses. Genes Dev 25:1915–1927
8. Orom UA, Derrien T, Beringer M, Gumireddy K, Gardini A, Bussotti G, Lai F, Zytnicki M, Notredame C, Huang Q et al (2010) Long noncoding RNAs with enhancer-like function in human cells. Cell 143:46–58
9. Khalil A, Huarte M, Rinn J (2010) The emerging non-coding RNA World. In: Slack F (ed) MicroRNAs in development and cancer, vol 1. Imperial College Press, London, pp 17–44
10. Khalil AM, Guttman M, Huarte M, Garber M, Raj A, Rivea Morales D, Thomas K, Presser A, Bernstein BE, van Oudenaarden A et al (2009) Many human large intergenic noncoding RNAs associate with chromatin-modifying complexes and affect gene expression. Proc Natl Acad Sci U S A 106:11667–11672
11. Guttman M, Amit I, Garber M, French C, Lin MF, Feldser D, Huarte M, Zuk O, Carey BW, Cassady JP et al (2009) Chromatin signature reveals over a thousand highly conserved large non-coding RNAs in mammals. Nature 458:223–227
12. Amaral PP, Dinger ME, Mercer TR, Mattick JS (2008) The eukaryotic genome as an RNA machine. Science 319:1787–1789
13. Carninci P, Hayashizaki Y (2007) Noncoding RNA transcription beyond annotated genes. Curr Opin Genet Dev 17:139–144
14. Ponting CP, Oliver PL, Reik W (2009) Evolution and functions of long noncoding RNAs. Cell 136:629–641
15. Lipovich L, Johnson R, Lin CY (2010) MacroRNA underdogs in a microRNA world: evolutionary, regulatory, and biomedical significance of mammalian long non-protein-coding RNA. Biochim Biophy 1799:597–615
16. Pontier DB, Gribnau J (2011) Xist regulation and function explored. Hum Genet 130: 223–236
17. Augui S, Nora EP, Heard E (2011) Regulation of X-chromosome inactivation by the X-inactivation centre. Nat Rev Genet 12: 429–442
18. Wapinski O, Chang HY (2011) Long noncoding RNAs and human disease. Trends Cell Biol 21:354–361
19. Khalil AM, Rinn JL (2011) RNA-protein interactions in human health and disease. Semin Cell Dev Biol 22(4):359–365
20. Kouzarides T (2007) Chromatin modifications and their function. Cell 128:693–705
21. Khalil AM, Boyar FZ, Driscoll DJ (2004) Dynamic histone modifications mark sex

chromosome inactivation and reactivation during mammalian spermatogenesis. Proc Natl Acad Sci U S A 101:16583–16587

22. Zhao J, Sun BK, Erwin JA, Song JJ, Lee JT (2008) Polycomb proteins targeted by a short repeat RNA to the mouse X chromosome. Science 322:750–756
23. Rinn JL, Kertesz M, Wang JK, Squazzo SL, Xu X, Brugmann SA, Goodnough LH, Helms JA, Farnham PJ, Segal E et al (2007) Functional demarcation of active and silent chromatin domains in human HOX loci by noncoding RNAs. Cell 129:1311–1323
24. Tsai MC, Manor O, Wan Y, Mosammaparast N, Wang JK, Lan F, Shi Y, Segal E, Chang HY (2010) Long noncoding RNA as modular scaffold of histone modification complexes. Science 329:689–693
25. Gupta RA, Shah N, Wang KC, Kim J, Horlings HM, Wong DJ, Tsai MC, Hung T, Argani P, Rinn JL et al (2010) Long non-coding RNA HOTAIR reprograms chromatin state to promote cancer metastasis. Nature 464:1071–1076
26. Sleutels F, Zwart R, Barlow DP (2002) The non-coding Air RNA is required for silencing autosomal imprinted genes. Nature 415:810–813
27. Nagano T, Mitchell JA, Sanz LA, Pauler FM, Ferguson-Smith AC, Feil R, Fraser P (2008) The air noncoding RNA epigenetically silences transcription by targeting G9a to chromatin. Science 322:1717–1720
28. Pandey RR, Mondal T, Mohammad F, Enroth S, Redrup L, Komorowski J, Nagano T, Mancini-Dinardo D, Kanduri C (2008) Kcnq1ot1 antisense noncoding RNA mediates lineage-specific transcriptional silencing through chromatin-level regulation. Mol Cell 32:232–246
29. Mohammad F, Mondal T, Guseva N, Pandey GK, Kanduri C (2010) Kcnq1ot1 noncoding RNA mediates transcriptional gene silencing by interacting with Dnmt1. Development 137: 2493–2499
30. Meller VH, Gordadze PR, Park Y, Chu X, Stuckenholz C, Kelley RL, Kuroda MI (2000) Ordered assembly of roX RNAs into MSL complexes on the dosage-compensated X chromosome in Drosophila. Curr Biol 10:136–143
31. Kelley RL, Meller VH, Gordadze PR, Roman G, Davis RL, Kuroda MI (1999) Epigenetic spreading of the Drosophila dosage compensation complex from roX RNA genes into flanking chromatin. Cell 98:513–522
32. Park Y, Kelley RL, Oh H, Kuroda MI, Meller VH (2002) Extent of chromatin spreading determined by roX RNA recruitment of MSL proteins. Science 298:1620–1623
33. Lanz RB, McKenna NJ, Onate SA, Albrecht U, Wong J, Tsai SY, Tsai MJ, O'Malley BW (1999) A steroid receptor coactivator, SRA, functions as an RNA and is present in an SRC-1 complex. Cell 97:17–27
34. Watanabe M, Yanagisawa J, Kitagawa H, Takeyama K, Ogawa S, Arao Y, Suzawa M, Kobayashi Y, Yano T, Yoshikawa H et al (2001) A subfamily of RNA-binding DEAD-box proteins acts as an estrogen receptor alpha coactivator through the N-terminal activation domain (AF-1) with an RNA coactivator, SRA. EMBO J 20:1341–1352
35. Kino T, Hurt DE, Ichijo T, Nader N, Chrousos GP (2010) Noncoding RNA gas5 is a growth arrest- and starvation-associated repressor of the glucocorticoid receptor. Sci Signal 3:ra8
36. Niranjanakumari S, Lasda E, Brazas R, Garcia-Blanco MA (2002) Reversible cross-linking combined with immunoprecipitation to study RNA-protein interactions in vivo. Methods 26:182–190
37. Mili S, Steitz JA (2004) Evidence for reassociation of RNA-binding proteins after cell lysis: implications for the interpretation of immunoprecipitation analyses. RNA 10: 1692–1694
38. Zhao J, Ohsumi TK, Kung JT, Ogawa Y, Grau DJ, Sarma K, Song JJ, Kingston RE, Borowsky M, Lee JT (2010) Genome-wide identification of polycomb-associated RNAs by RIP-seq. Mol Cell 40:939–953

Part VI

Imprinting in Plants

Chapter 16

Specialized Technologies for Epigenetics in Plants

Wenyan Xiao

Abstract

Plants are excellent systems for discovering and studying epigenetic phenomena, such as transposon silencing, RNAi, imprinting, and DNA methylation. Imprinting, referring to preferential expression of maternal or paternal alleles, plays an important role in reproduction development of both mammals and plants. DNA methylation is critical for determining whether the maternal or paternal alleles of an imprinted gene is expressed or silenced. In flowering plants, there is a double fertilization event in reproduction: one sperm fertilizes the egg cell to form embryo and a second sperm fuses with the central cell to give rise to endosperm. Endosperm is the tissue where imprinting occurs in plants. *MEDEA* (*MEA*), a SET domain Polycomb group gene, was the first plant gene shown to be imprinted in endosperm, and its maternal expression is controlled by DNA methylation and demethylation. Recently there has been significant progress in identifying imprinted genes as well as understanding molecular mechanisms of imprinting in plants. Up to date, approximately 350 genes were found to have differential parent-of-origin expression in plant endosperm (*Arabidopsis*, corn, and rice). In *Arabidopsis*, many imprinted genes are regulated by the DNA METHYLTRANSFERASE1 (MET1) and the DNA-demethylating glycosylase DEMETER (DME), and/or their chromatin states regulated by Polycomb group proteins (PRC2). There are also maternally expressed genes regulated by unknown mechanisms in endosperm. In this protocol, we describe in detail how to perform a genetic cross, isolate the endosperm tissue from seed, determine the imprinting status of a gene, and analyze DNA methylation of imprinted genes by bisulfite sequencing in *Arabidopsis*.

Key words: Imprinting, Epigenetics, DNA methylation, Bisulfite sequencing, Parent-of-origin, Endosperm, *Arabidopsis*, Plant

1. Introduction

Plants have been used as model organisms for studying genetics and epigenetics since Gregor Mendel did his series of seminal genetic experiments using the garden pea (*Pisum sativum*) in 1856–1863 and Barbara McClintock discovered transposable elements in maize (*Zea mays*) half a century ago (1, 2). Plant research also provided the earliest evidence for RNAi. For example, silencing of a homologous endogenous gene by the transgene in petunia

Nora Engel (ed.), *Genomic Imprinting: Methods and Protocols*, Methods in Molecular Biology, vol. 925,
DOI 10.1007/978-1-62703-011-3_16, © Springer Science+Business Media, LLC 2012

is now viewed as the earliest evidence of gene silencing by RNAi (3). Studies using potato spindle tuber viroid were the earliest evidence that RNA is an intermediate in transcriptional gene silencing (4). *Arabidopsis thaliana* is another excellent model organism for studying epigenetic mechanisms, such as imprinting and DNA methylation. The loss-of-function null mutant of DNA methyltransferase is viable in plants, thus providing a system to study how loss of DNA methylation in a genome affects imprinting, growth and development.

Imprinting is an epigenetic phenomenon that is involved in growth and development of plants and mammals. Although genomic imprinting occurs in flowering plants and mammals, it evolved independently (5). One of the widely accepted theories of genomic imprinting is the parental conflict theory (5–7), which hypothesizes that female and male parents have different interests in fitness of their progeny when one female can mate with multiple males. The female wants all her offspring to survive and thus prefers expression of genes that allocate resources equally to all her offspring, whereas a male is only interested in his progeny, therefore favoring expression of genes that maximize allocation of resources to his offspring. Thus, different interests of female and male parents result in parent-of-origin effects on expression of parental alleles of imprinted genes.

In mammals, fertilization is resulted from fusion of two haploid cells that are direct products of the preceding meiosis, whereas in plants there is a haploid gametophytic growth stage, i.e., postmeiotic cell divisions before fertilization. In *Arabidopsis*, a haploid megaspore at the end of meiosis undergoes three mitotic divisions to form an eight-nucleus, seven-cell female gametophyte containing the egg, central, synergid, and antipodal cells; the fusion of two haploid nuclei makes the nucleus of the diploid central cell. In the male gametophyte, a haploid microspore at the end of meiosis undergoes a mitotic cell division to give rise to one vegetative and one generative nucleus; then the generative cell undergoes a mitotic cell division to result in two generative nuclei (two haploid sperm cells). There is also a double fertilization event that underlies gene imprinting in flowering plants. Fertilization of the egg cell by a sperm cell gives rise to a diploid embryo that ultimately generates the organs, tissues, and meristems of the plant. Fertilization of the central cell by a second sperm generates the triploid endosperm that supports embryo or seedling growth by producing storage proteins, lipids, and starch, and by mediating the transfer of maternal-derived nutrients to be absorbed by the embryo. Imprinting mainly occurs in the endosperm tissue in plants, whereas imprinting can occur in embryos and in different adult tissues in mammals (8, 9).

Through classical experimental approaches, 11 imprinted genes (eight maternally expressed and three paternally expressed) were identified in *Arabidopsis* endosperm (10), two maternally expressed imprinted genes (*fie1*, *mee1*) in maize (11–13) and one maternal expressed imprinted gene (*OsFIE1*) in rice (14). By using extensive sequencing of cDNA libraries as well as examination of allele-specific expression of candidate genes by RT-PCR (10), Hsieh and colleagues discovered an additional 112 maternally expressed and 9 paternally expressed genes in *Arabidopsis* endosperm. These imprinted genes have various functions in plants and include transcription factors, components of hormone signaling, proteins involved in the ubiquitin-mediated protein degradation, and genes regulating DNA methylation, histone modifications, and small RNA pathways. By a similar approach of deep sequencing RNA profiles of F1 seeds of *Arabidopsis* Col-0 and Bur-0 ecotypes, Wolff and colleagues found that more than 60 genes (six maternally expressed genes overlapping those discovered by Hsieh and colleagues) that have potential parent-of-origin specific expression (15). In rice seeds, Luo and colleagues uncovered 262 putative imprinted loci in endosperm and 3 in the embryo (168 genic and 97 nongenic) via deep sequencing of cDNA libraries (16). Mechanistically, genomic imprinting is complex and is involved in many aspects of growth and development in plants. *MEA* and *PHE1*, representing maternal and paternal expressed genes, respectively, are the best characterized at the molecular level. For example, MET1 is responsible for maintaining CG methylation at the *MEA* locus (17), whereas DME excises 5-methyl cytosine of the maternal *MEA* allele, thus activating its expression (18). Interestingly, silencing of the paternal *MEA* allele is not controlled by DNA methylation. Rather, the maternally expressed Polycomb group proteins, including MEA, maintain the paternal *MEA* allele silencing (18). *PHE1* is maternally silenced and paternally expressed. Silencing of the maternal *PHE1* allele requires the FERTILIZATION INDEPENDENT SEED (FIS) Polycomb group complex as well as direct tandem repeats located downstream of the *PHE1* locus. It has been hypothesized that the differential methylation of the tandem repeats regulates the *PHE1* imprinting (19, 20).

No imprinted gene has been detected in *Arabidopsis* embryo (10). However, *mee1* and *Os10g05750* are preferentially maternally expressed in both embryo and endosperm in maize and rice, respectively (11, 16). By deep sequencing of RNA transcripts and genetic analyses of early embryogenesis in *Arabidopsis*, it has been shown that there is genome-wide dominance of maternal transcripts at the 2–4 cell stage of embryo, and the relative paternal transcripts increase at the globular stage due to a gradual activation of the paternal genome (21). Furthermore, the authors identify two antagonistic maternal epigenetic pathways that regulate parental contributions in plant embryos. It seems that imprinting has evolved independently in *Arabidopsis* and rice since imprinted loci identified through genomic approaches do not have extensive sequence conservation (10, 16).

The principle of bisulfite sequencing is that unmethylated cytosine will be converted to uracil due to hydrolytic deamination by high concentration of sodium bisulfite at pH 5.0 which will be amplified as thymine in PCR product, while 5-methyl cytosine will not be modified by sodium bisulfite and remain to be cytosine after PCR amplification (22, 23).

In this protocol, the procedures for determining if a candidate gene is imprinted in *Arabidopsis* endosperm and determining the DNA methylation status of the imprinted genes in endosperm are described. The major procedures involved in the protocol are: to perform genetic crossing (emasculating the female parent and pollinating the pistil with the male parent); to isolate endosperm tissues; to carry out RNA isolation and RT-PCR; and to perform bisulfite sequencing of the genomic DNA (bisulfite treatment, designing strand-specific primers, PCR amplification, and sequence analysis).

2. Materials

Prepare all solutions using sterile deionized distilled water and analytical grade reagents unless indicated. Prepare all reagents at room temperature unless indicated. All DNA and RNA manipulations follow standard procedures unless indicated otherwise.

2.1. Reagents

1. 70 % Ethanol: Add 30 mL water to 70 mL anhydrous ethanol.
2. 95 % Ethanol: Add 5 mL water to 70 mL anhydrous ethanol.
3. 100 % Ethanol: 100 mL anhydrous ethanol.
4. 0.3 M Sorbitol and 5 mM MES—pH 5.7: Dissolve 5.47 g Sorbitol (Aldrich #: 240850) and 98 mg MES (Sigma #: M5287, anhydrous) in 90 mL water. Adjust pH to 5.7 and bring the final volume to 100 mL.
5. Cetyltrimethyl Ammonium Bromide (CTAB) buffer: 2 % CTAB (Sigma), 100 mM Tris–HCl (pH 8.0), 20 mM EDTA (pH 8.0), 1.4 M NaCl, 1 % PVP (polyvinyl-polypyrrolidone, Mr 40,000). After making the buffer, autoclave and store it.
6. Chloroform: Sigma, >99.8 %.
7. Restriction Enzymes: XhoI, NdeI, and PstI or HindIII for the *MEA* promoter.
8. 3 M NaOH: Dissolve 1.2 g NaOH in 10 mL water.
9. 6.3 M NaOH: Dissolve 2.52 g NaOH in 10 mL water.
10. 6.24 M Urea/4 M Sodium Bisulfite: Dissolve 7.5 g of urea in 10 mL of sterile distilled water; slowly add 7.6 g of sodium metabisulfite (Sigma) over 1–2 h with heating (see Note 15); adjust the pH to 5.0 with freshly made 10 M NaOH; add sterile distilled water to a final volume to 20 mL (see Note 16).

11. Sterile distilled H_2O.
12. 10 mM Hydroquinone: Dissolve 0.011 g 99 % Hydroquinone in 10 mL water.
13. Wizard DNA Clean-Up System: Promega.
14. 10 M NH_4OAc: Dissolve 7.7 g NH_4OAc in 10 mL water.
15. 20 μg/μl tRNA: Dissolve 2 mg tRNA in 100 μL water.
16. TE buffer: 10 mM Tris–Cl, pH 7.5. 1 mM EDTA, pH 8.0. To make 1 L TE buffer: 10 mL of 1 M stock of Tris–Cl (pH 7.5), 2 mL of 500 mM stock of EDTA (pH 8.0), and 988 mL water.
17. The TOPO TA Cloning Kit: Invitrogen's TOPO® TA cloning Kit with the pCR® 2.1 vector.

2.2. Supplies and Equipment

1. Dissecting Microscope.
2. Scissors.
3. Fine Tip Forceps.
4. Jewelry Tag.
5. Plant Stakes.
6. String or Twist-Ties.
7. 4" × 2" × 8" Polyethylene Bags.
8. 3" × 1" × 1.0 mm Microscope Slides.
9. Liquid Nitrogen.
10. Liquid Nitrogen Containers.
11. Heat Block.
12. PCR Tubes.
13. Thermocycler.
14. Microcentrifuge Tubes.
15. Microcentrifuge.
16. Gel Electrophoresis facility.
17. Different *Arabidopsis thaliana* ecotypes (Columbia-0, Landsberg *erecta*, RLD, *Ws*, or *Cvi*).
18. Total RNA isolation kits.
19. NanoDrop spectrometer.

3. Methods

3.1. Genetic Crossing

1. *Emasculating the female parent.* In order to distinguish the maternal and paternal alleles of the imprinted *MEA* gene, two different ecotypes, e.g., Columbia-0 (Col-0) and RLD, will be chosen as female and male parents (see Note 1). One can emasculate the female parent by using a dissecting microscope,

magnifying visor, or naked eye (see Note 2). Locate stage-12 flowers and remove any flowers or siliques above and below them by clipping the base of the pedicel with the scissors. Sterilize forceps by dipping the base of the tip gently into a beaker of 95 % ethanol, which will remove any pollen grains on the forceps and kill the pollen as well. Gently pry apart the flower buds using forceps, and gently remove four sepals, four petals, and six stamens, leaving the pistil bare and intact (see Note 3).

2. *Picking the pollen donor and carrying out pollination.* Choose an open flower from a different ecotype plant as a male (see Note 4). Grab the flower at the base and just above the pedicel with forceps, which will cause the flower to spread open. Dust the stigma of the prepared female pistil with the anther (see Note 5). After the pollination, the stigma will be covered with yellow pollen, which can be easily observed under a microscope.
3. *Labeling the cross.* After pollinating all the emasculated pistils on a plant, we label the cross with a jewelry tag and write down information of female and male parents and date on the tag. Place a stake in the soil close to the plant, use a string to tie the stem of the inflorescence to the stake, and cover the pollinated pistils with a plastic bag (see Note 6).

3.2. Isolation of Endosperm Tissues in Arabidopsis

1. *Preparing materials.* Get necessary materials ready before harvesting the endosperm and embryo tissues: a liquid nitrogen tank, liquid nitrogen, a dissecting microscope, two new pairs of fine-tip forceps (5 INOX. FST by Dumont biology, Switzerland), microscope glass slides (3" × 1" × 1.0 mm), and a pH 5.7 solution of 0.3 M sorbitol and 5 mM MES.
2. *Collecting siliques.* At 7 or 8 day-after-pollination (DAP), the seeds are ready to be harvested at the mid- to late torpedo stage of embryogenesis and dissected for endosperm and embryo (see Note 7). Locate the crossed silique and cut the silique pedicel with small scissors.
3. *Isolating endosperm and embryo.* Put an 8-DAP silique on a glass slide under a microscope, use a pair of forceps to hold the silique pedicel and use the tip of another pair of forceps to slide open the silique in the margin where two carpels fuse (see ref. 24). Use a pair of forceps to pick up one seed onto pH 5.7 solutions of 0.3 M sorbitol and 5 mM MES (see Note 8), make a small cut at the micropyle end (see Fig. 1a) to slide out embryo (see Fig. 1b), and then carefully separate endosperm (see Fig. 1c) from seed coat (see Fig. 1d). Put the embryo and endosperm into separate microtubes in liquid nitrogen. Continue this until accumulating enough embryo or endosperm from 15 to 20 siliques and then store the tube in a −80 °C freezer (see Note 9).

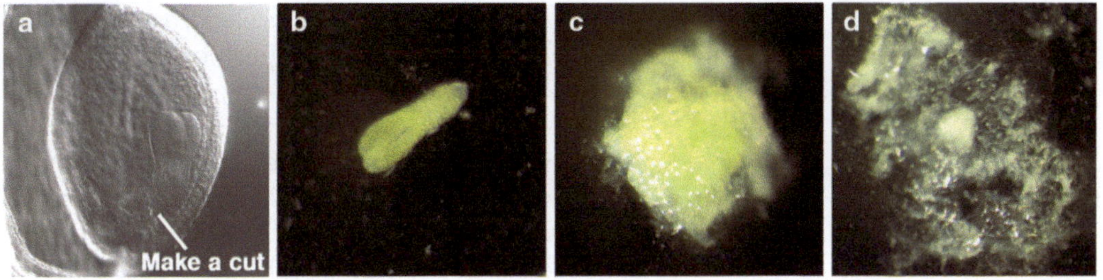

Fig. 1. An *Arabidopsis* seed and dissected embryo, endosperm, and seed coat. (**a**) A seed at the late heart stage is ready to dissect out embryo, endosperm, and seed coat. A dissected embryo (**b**), endosperm (**c**), and seed coat (**d**) from an *Arabidopsis* seed at the torpedo stage.

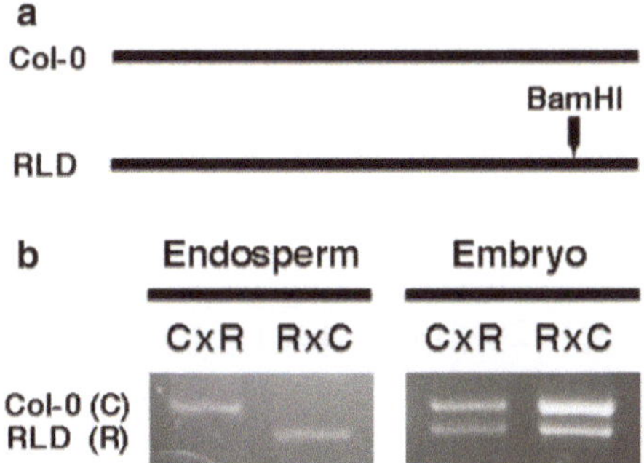

Fig. 2. Determining the imprinting status of *MEA*. (**a**) Shows a sequence polymorphism in the exon 17 of *MEA* between Col-0 and RLD that can be used to convert to a dCAPS marker. After PCR amplification of the fragment, digestion of the PCR fragment with BamHI will give a 239-bp band in Col-0 and two bands (207 and 32 bp) in RLD. (**b**) The maternal *MEA* allele is specifically expressed in the reciprocal crosses between Col-0 and RLD (C × R and R × C) and the paternal *MEA* allele is silenced in *Arabidopsis* endosperm. Biallelic *MEA* expression is detected in *Arabidopsis* embryo.

3.3. RNA Isolation and RT-PCR

1. Isolate total RNA from the endosperm tissue collected above (see Note 10).
2. Adjust RNA concentration to 100 ng/μL and use total 1 μg total RNA (10 μL of 100 ng/μL) for reverse transcription (RT) reaction (see Note 11).
3. Identify polymorphisms between two ecotypes within the coding region of your candidate gene. Design a dCAPS marker. For *MEA*, there is a polymorphism between Col-0 (or Ler) and RLD at the end of the coding region (exon 17) that can be used to design a dCAPS marker to distinguish Col-0 and RLD alleles (see Fig. 2a and Note 12).
4. Use 1 μL of RT reaction in a total volume of 20 μL PCR reaction.
5. Perform restriction enzyme digestion after the PCR reaction. In the PCR reaction for amplifying *MEA* fragments, perform BamHI reaction for 6–12 h.
6. Run a 3 % agarose gel for 1 h to distinguish the Col-0 and RLD allele (see Fig. 2b and Note 13). The maternal *MEA*

allele is expressed and the paternal allele is silenced in *Arabidopsis* endosperm, whereas biallelic *MEA* expression is detected in embryo (see Fig. 2b).

3.4. Bisulfite Sequencing

1. *Preparing required reagents.* Cetyltrimethyl ammonium bromide (CTAB) for isolating genomic DNA, restriction enzymes, 3 M NaOH (Freshly made), 6.42 M urea/4 M sodium bisulfite (2 M sodium metabisulfite, Sigma-Aldrich, S9000, $Na_2S_2O_5$, Molecular Weight: 190), 10 mM hydroquinone, a DNA purification kit (Promega Wizard DNA Clean-up System, Cat. # A7280), TE buffer, 6.3 M NaOH (freshly made), 10 M NH_4OAc, 20 μg/μL tRNA, and 100 % ethanol.
2. *Treating genomic DNA with the sodium bisulfite.*
   (a) Perform genetic cross (see Subheading 3.1) and collect endosperm tissue (see Subheading 3.2) as described.
   (b) Isolate genomic endosperm DNA using a CTAB procedure (see Note 14).
   (c) Digest 100 ng to 2 μg of endosperm genomic DNA in 20–100 μL total volume with restriction enzymes that cut outside the region to be analyzed. For the *MEA* promoter, we use XhoI, NdeI, and PstI or HindIII.
   (d) Denature the restriction enzymes by boiling the DNA for 5 min after the restriction enzyme digestion and then quench on ice.
   (e) Add 1/9 volume (2.2 μL for 20 μL digested DNA) of 3 M NaOH and incubate at 37 °C for 15 min.
   (f) Transfer the solution to a 250 μL PCR tube.
   (g) Dissolve 7.5 g of urea in 10 mL of sterile distilled water; slowly add 7.6 g of sodium metabisulfite (Sigma Cat. # S900-1KG) over 1–2 h and heating usually helps dissolving (see Note 15); adjust the pH to 5.0 with freshly made 10 M NaOH; add sterile distilled water to a final volume to 20 mL. This is 6.42 M urea/4 M sodium bisulfite solution (see Note 16).
   (h) Add 6.42 M urea/4 M sodium bisulfite solution to a final concentration of 5.36 M and 3.44 M, respectively (see ref. 25). For example, add 208 μL of 6.42 M urea/4 M sodium bisulfite solution to the above 20 μL of starting genomic DNA reaction.
   (i) Add 10 mM hydroquinone to the DNA at a final concentration of 0.5 mM (12 μL for 20 μL digestion).
   (j) Conduct bisulfite treatment in a PCR machine: 30 cycles of 55 °C for 15 min and 95 °C for 30 s (see Note 17).
   (k) Desalt the bisulfite treated DNA using the Wizard DNA Clean-Up System from Promega and follow up the protocol (see Note 18).

Table 1
Amounts of solutions being added depends on the volume of TE recovered

TE recovered (μL)	44	46	48	50	52	54
6.3 M NaOH (μL)	2.2	2.3	2.4	2.5	2.6	2.7
10 M NH_4Ac (μL)	21	21.95	22.9	23.86	24.82	25.77
100 % EtOH (μL)	206	215.9	224.7	234	243	252.4

(l) Measure the exact volume of TE recovered from the column after desalting. Add 6.3 M NaOH to a final concentration of 0.3 M. Incubate at 37 °C for 15 min (see Table 1 and Note 19).

(m) Add 10 M NH_4OAc (pH 7.0) to a final concentration of 3 M, 2 μL of 20 μg/μL tRNA, and 3 volumes of 100 % ethanol and then mix (see Table 1 and Note 19). Centrifuge for 15 min at 16,800 × *g*.

(n) Wash the pellet once with 70 % ethanol, do a short centrifuge, and remove extra ethanol.

(o) Dry pellet in a speedvac for 5–10 min and resuspend in 25–100 μL TE buffer depending on starting amount of DNA. The sodium bisulfite-treated DNA is now ready for PCR analysis.

3.5. PCR Amplification

1. To sequence the 4-kb *MEA* promoter, we designed many sets of primers and amplified 14 overlapping fragments to cover the entire region (see Note 20).
2. To sequence the top-strand (see Fig. 3a), in designing a forward primer, (1) choose a G (guanine)-rich region in order to have a higher annealing temperature without extra long nucleotides in the primers; (2) change C (cytosine) to Y (pyrimidine) at CG and CNG contexts and change the remaining C to T (thymine). In designing a reverse primer, (1) choose a C-rich region (5′–3′ strand direction); (2) change G to R (purine) at CG and CNG contexts and change the remaining G to A (adenine).
3. To sequence the bottom-strand (see Fig. 3b), in designing a forward primer, (1) choose a C-rich region (5′–3′ strand direction); (2) change G to R at CG and CNG contexts and change the remaining G to A. In designing a reverse primer, (1) choose a G-rich region (5′–3′ strand direction); (2) change C to Y (pyrimidine) at CG and CNG contexts and change the remaining C to T.

```
a Forward primer
5'-GGAAGATTGTTAAATGTTAAATATTTAATTYGAA-3'
5'-GGAAGATTGTTAAATGTCAAATATTTAATTCGAA...GAGGATGGTCCATCAGCCGGCTGTGTT-3'
3'-CCTTCTAACAATTTACAGTTTATAAATTAAGCTT...CTCCTACCAGGTAGTCGGCCGACACAA-5'
                                      3'-CTCCTACCAAATARTCRRCCAACACAA-5'
                                                      Reverse primer

b Forward primer
5'-CRRCRRATAAACTTAACCTCCCCATTCRT-3'
5'-CGGCGGATAGACTTAACCTCCCCATTCGT...CTATTTCCTGCTAGGAATTTAACCCGTATATATGTCACA-3'
3'-GCCGCCTATCTGAATTGGAGGGGTAAGCA...GATAAAGAACGATCCTTAAATTGGGCATATATACAGTGT-5'
                                3'-AGAATGATTTTTAAATTGGGYATATATATAGTGT-5'
                                                        Reverse primer
```

Fig. 3. General rule in designing primers for bisulfite sequencing. (**a**) Strategy for designing forward and reverse primers to sequence the top-strand in bisulfite sequencing. In designing a forward primer, choose a G (guanine)-rich region in order to have a higher annealing temperature without extra long nucleotides in the primers; change C (cytosine) to Y (pyrimidine) at CG and CNG contexts and change the remaining C to T (thymine). In designing a reverse primer, choose a C-rich region; change G to R (purine) at CG and CNG contexts and change the remaining G to A (adenine). (**b**) Strategy for designing forward and reverse primers to sequence the bottom-strand in bisulfite sequencing. In designing a forward primer, choose a C-rich region; change G to R at CG and CNG contexts and change the remaining G to A. In designing a reverse primer, choose a G-rich region; change C to Y (pyrimidine) at CG and CNG contexts and change the remaining C to T.

4. We usually use 1–2 μL of the sodium bisulfite-treated DNA as a template for each PCR amplification (see Note 21). The PCR product needs to be analyzed by gel electrophoresis to confirm the correct size of the fragment, then to be gel purified and cloned into the TOPO TA cloning vector pCR2.1 (Invitrogen) as an insert. A single colony is picked up and cultured; plasmid DNA extracted and sent for sequencing.
5. *Sequence analysis*. After obtaining the sequencing result, we compare it with the strand-specific template that is used for PCR amplification (see Fig. 4a). If a cytosine residue in the template reads as a thymine in the sequencing result, it indicates that the cytosine is not methylated (see Fig. 4b). If a cytosine residue in the template remains a cytosine in the sequencing, it means that the cytosine is methylated (see Fig. 4b). After sequencing many colonies, one can compile the bisulfite sequencing result together as shown in Fig. 5 (see Note 22).

4. Notes

1. For choosing a female parent for genetic crossing, it is important to choose a healthy plant to emasculate. Young plants with large un-open flower buds are easier to emasculate for beginners.
2. In emasculation, different people can use different approaches. For beginners, the emasculation procedure should be done

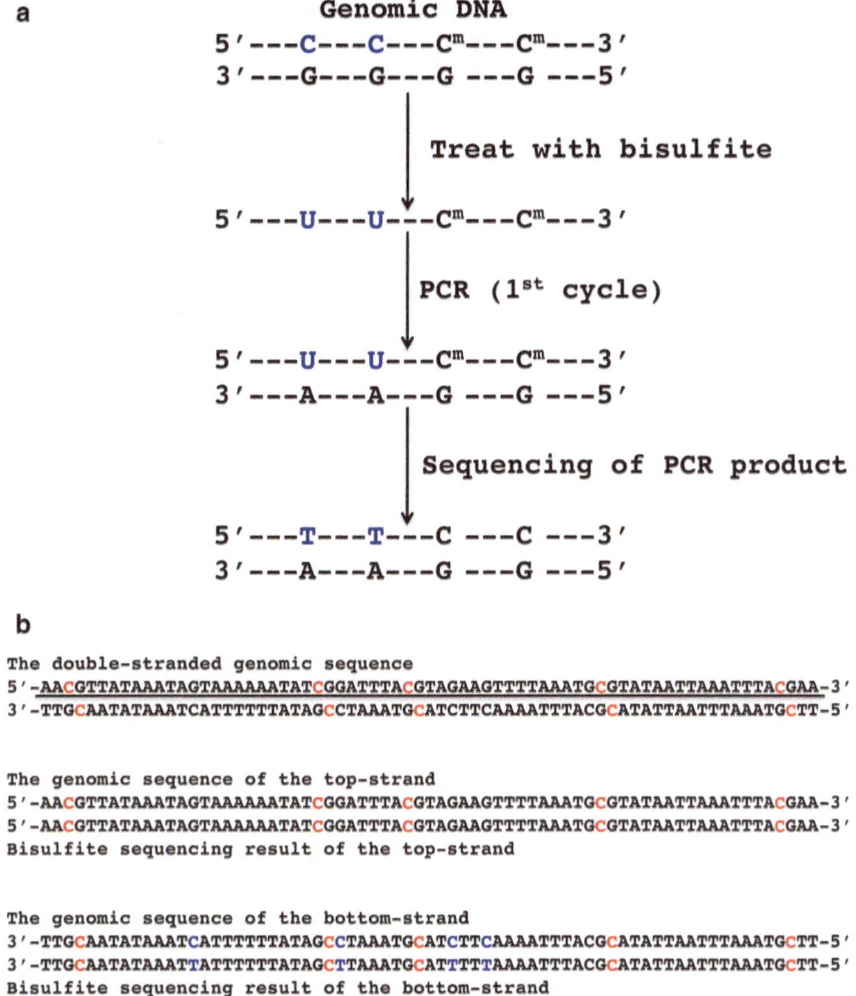

Fig. 4. Strategy for bisulfite sequencing. (**a**) Shows an outline of the major bisulfite sequencing procedures. (**b**) Shows the genomic sequence of the *MEA* promoter between –587 and –519 bp. There are five methylated cytosine at five CG sites in the promoter region. Fig. 4b shows how one can compare the bisulfite sequencing result with the original genomic sequence and deduce whether a cytosine residue is methylated or not.

under a dissecting microscope to make sure that all six stamens are removed from the flower. A very experienced person can do emasculation with naked eyes. One common problem for beginners is the damage made to the plant during emasculation. Thus, it is critical to gently grasp or hold the plant stems without damaging them.

3. Try to avoid damage to the carpels during this process. Another option is that you wait 24–48 h to allow ovules from emasculated plants to reach maturity and synchronize to the same growth stage before performing pollination.

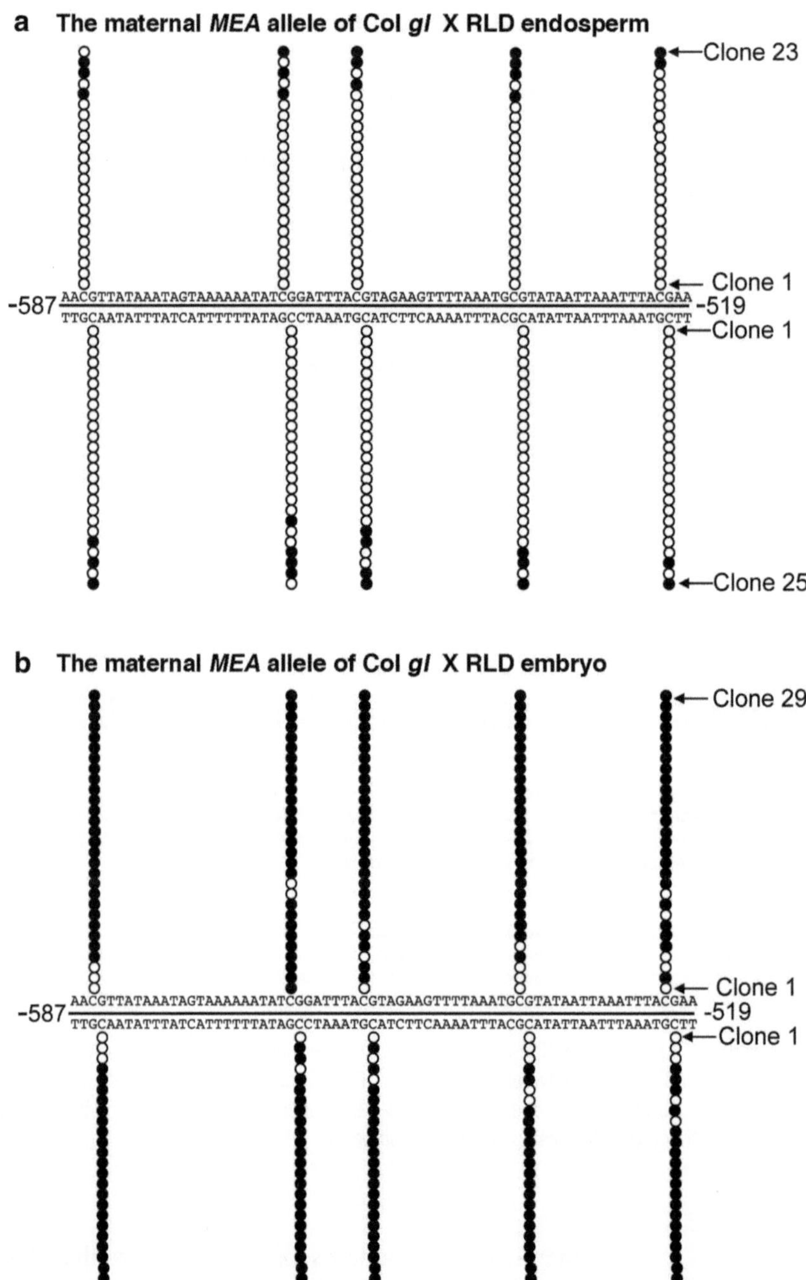

Fig. 5. The methylation status of the maternal *MEA* allele in −500 bp region of the *MEA* promoter. (**a**) and (**b**) show the methylation status of five methylated CpG sites in the sequenced clones of (Col *gl* X RLD) endosperm and embryo, respectively. *Black filled circles* and *white unfilled circles* indicate methylated and unmethylated cytosines, respectively. Number of sequences is relative to the translation start site of *MEA* (Xiao et al., 2003; Gehring et al., 2006).

4. The best pollen donors are anthesed open flowers at the stage 14 with the petals extending at a 90° angle to the pistil (26), in which a lot of pollen is shedding. For examining *MEA* imprinting status, we have used reciprocal crosses between Col-0 and RLD, both crosses work.
5. Pollination should be done under a microscope because it is almost impossible to see whether enough pollen has been put onto the stigma with naked eyes.
6. Covering the pollinated pistils with a plastic bag has two purposes: one is to avoid cross-pollination contamination and the other is to avoid water evaporation from the emasculated pistil. The emasculated pistil can easily wilted if not covered with a plastic bag.
7. Sometimes, it might take nine DAP if the emasculated flowers are too young. Actual time depends on the plant growth condition. If it is too early, the endosperm is not cellularized at all but fluid. If it is too late, the endosperm is completely cellularized, the endosperm tissue might not be optimal for imprinting analysis.
8. When embryo and endosperm are located in the buffer solution, sometimes it is not easy to pick them up and put into the eppendorf tube in the liquid nitrogen tank. One can accumulate endosperm tissues from several seeds before putting them into the eppendorf tube in the liquid nitrogen tank. Another optional is not using the buffer solution and isolating the embryo and endosperm directly on the top of a slide. After dissecting embryo and endosperm from one seed, immediately put into the eppendorf tube in the liquid nitrogen tank. This isolation method needs to reduce exposure of dissected endosperm or embryo in the air to the minimum before freezing them in liquid nitrogen in order to avoid degradation of RNA.
9. It is relatively easy to separate embryo from endosperm and seed coat, but it is tedious to separate endosperm from seed coat, especially for seed at the heart stage of embryogenesis. Since seed coat only contributes a very small amount of tissue, for some genes, e.g., *MEA* and *FWA*, we do not have to separate endosperm from seed coat. That means that we can isolate RNAs from a mixture of endosperm and seed coat tissues, check expression of the maternal and paternal alleles of a gene, compare with expression in embryo, and determine the imprinting status of the gene.
10. Total RNA can be isolated using your own method. We have used kits from Ambion and Qiagen and they both work fine.
11. Total RNA amount used for RT reaction is variable. We have used any amount between 500 ng and 2 μg depending expression levels of your candidate gene and total amount RNA

available. We have used RT kits from Ambion and New England Biolab and both are good.

12. For *MEA*, we use the following two primers to amplify the PCR fragments to distinguish Col-0 and RLD alleles: *MEA-L1:* 5′-GACCTAACTGCTACGCCAAG-3′; *MEA-R2d*: 5′-AAGGACTGCTTGAATTGCTGCTTCTCCTCGGATC-3′ (27).
13. A 3 % agarose gel is very easy to solidify after dissolving. Get gel box and comb ready and immediately pour the gel after dissolving. If you only need to determine imprinting status of your candidate gene, you are done after you run a gel as shown in Fig. 2b.
14. Isolate genomic DNA using CTAB procedure is fine (28). Make sure to thoroughly clean the bench to avoid contamination before starting DNA isolation.
15. It is not easy to dissolve sodium bisulfite. Start with a little bit sodium bisulfite until it is dissolved and gradually add the rest of it over a 1–2 h period.
16. Another way to make sodium bisulfite solution is: dissolve 40.5 g of sodium bisulfite (Fisher Cat. #S654-500) in 80 mL distilled water with slow stirring to avoid aeration. Adjust pH to 5.1 with freshly made 10 M NaOH. Add 3.3 mL of 20 mM Hydroquinone. Adjust the final volume to 100 mL (29).
17. Different labs have used different protocol to conduct bisulfite treatment. We have tried different bisulfite treatments in a PCR machine in order to make the PCR amplification working. Another bisulfite treatment condition is: 55 °C for 16 h in the dark with a jolt to 95 °C for 5 min every 3 h (29). Recently, several plant biology labs have leaded the field of bisulfite sequencing and sequenced the whole methylome at single-base resolution in *Arabidopsis* and human by combining the bisulfite treatment with deep sequencing technology (30–35). Several commercial kits have been used successfully, which provide easier alternatives to the classical bisulfite treatment (Dr. Hsieh's personal communication). Another important development is that a higher concentration of ammonium bisulfite solution (10 M) has been shown to be more effective in converting cytosine to uracil than traditional sodium bisulfite solution (4–5 M) (36). These kits are: The Imprint® DNA Modification Kit (Sigma-Aldrich), the EpiTect Bisulfite Kits (Qiagen), the MethylEasy *Xceed* kit (Human Genetic Signatures, NSW, Australia) (34), the EZ DNA Methylation-Gold™ Kit (Zymo Research), and the MethylCode kit (Invitrogen, Life Technologies).
18. For desalting the bisulfite treated DNA, just follow Promega protocol of the Wizard DNA Clean-Up System (Cat. #: 288742).

19. The exact amount of 6.3 M NaOH, 10 M NH_4Ac and 100 % ethanol solutions being added depends on the volume of TE recovered (see Table 1).

20. Since unmethylated cytosines are converted to uracil by bisulfite-treatment, it is difficult to amplify a large fragment using the bisulfite-treated DNA as a template. Thus, we usually design primers to amplify a product no longer than 500 bp (17).

21. Sodium bisulfite-treatment causes damage to DNA. It is challenging to amplify your expected fragment from a bisulfite-treated genomic DNA for beginners and this can be the most frustrating part in the bisulfite sequencing. After restriction enzyme digestion, the genomic DNA needs to be completely denatured into single-stranded since the bisulfite deamination almost does not work on the cytosine residues in double-stranded DNA structures. Another key is that bisulfite treatment needs to be complete. If a known unmethylated cytosine is not converted to thymine in the bisulfite sequencing, it indicates that the bisulfite treatment is not conducted properly. Alternatively, if many cytosine residues in an unknown region are not changed to thymine in the bisulfite sequencing, it is most likely due to amplification of unconverted DNA rather than indication of all methylated cytosine residues in the region. For bisulfite sequencing, the designed primers need to be strand-specific. If it is impossible to amplify an expected fragment from the bisulfite-treated genomic DNA due to damaged DNA template during sodium bisulfite treatment, we usually try different sets of PCR primers in different regions and shorter fragments. When you are sequencing an unknown region and you do not know whether a particular cytosine is methylated or not, you can use a degenerate nucleotide (**Y** for C and T; **R** for G and A) in a primer in order to increase the chance of pairing between the primer and template. For example, as shown in Fig. 3a, in designing a forward primer to sequence the top-strand, you can change both C to Y in the forward primer. Finally, if your bisulfite sequencing is still not working after you have tried all of the above, you can use commercially available bisulfite conversion kits (see Note 17). These kits are easier to use, less time-consuming, and have a higher successful rate.

22. Sodium bisulfite sequencing can precisely reveal whether a particular cytosine residue is methylated or not in a genome if the experiment is conducted correctly (37). Combining the sodium bisulfite sequencing with recent high-throughput sequencing techniques, such as the Illumina Genome Analyzer high-throughput sequencing platform, one can map the epigenome at single-base resolution in plants and mammals (31–35).

Acknowledgments

The author thanks colleagues in the lab for discussion and Dr. Tzung-Fu Hsieh for critical reading of the manuscript. This work is supported by startup fund from Saint Louis University and National Institutes of Health grants 1R15GM086846-01 and 3R15GM086846-01S1.

References

1. McClintock B (1951) Chromosome organization and genic expression. Cold Spring Harbor Symp Quant Biol 16:13–47
2. McClintock B (1965) The control of gene action in maize. Brookhaven Symp Biol 18: 162–184
3. Napoli C, Lemieux C, Jorgensen R (1990) Introduction of a chimeric chalcone synthase gene into petunia results in reversible co-suppression of homologous genes in trans. Plant Cell 2:279–289
4. Wassenegger M, Heimes S, Sanger HL (1994) An infectious viroid RNA replicon evolved from an in vitro-generated non-infectious viroid deletion mutant via a complementary deletion in vivo. EMBO J 13:6172–6177
5. Feil R, Berger F (2007) Convergent evolution of genomic imprinting in plants and mammals. Trends Genet 23:192–199
6. Haig D, Westoby M (1989) Parent-specific gene expression and the triploid endosperm. Am Nat 134:147–155
7. Moore T, Haig D (1991) Genomic imprinting in mammalian development: a parental tug-of-war. Trends Genet 7:45–48
8. Gregg C et al (2010) High-resolution analysis of parent-of-origin allelic expression in the mouse brain. Science 329:643–648
9. Constancia M, Kelsey G, Reik W (2004) Resourceful imprinting. Nature 432:53–57
10. Hsieh TF et al (2011) Regulation of imprinted gene expression in Arabidopsis endosperm. Proc Natl Acad Sci U S A 108:1755–1762
11. Jahnke S, Scholten S (2009) Epigenetic resetting of a gene imprinted in plant embryos. Curr Biol 19:1677–1681
12. Hermon P, Srilunchang KO, Zou J, Dresselhaus T, Danilevskaya ON (2007) Activation of the imprinted Polycomb Group Fie1 gene in maize endosperm requires demethylation of the maternal allele. Plant Mol Biol 64:387–395
13. Danilevskaya ON et al (2003) Duplicated fie genes in maize: expression pattern and imprinting suggest distinct functions. Plant Cell 15:425–438
14. He G et al (2010) Global epigenetic and transcriptional trends among two rice subspecies and their reciprocal hybrids. Plant Cell 22: 17–33
15. Wolff P et al (2011) High-resolution analysis of parent-of-origin allelic expression in the arabidopsis endosperm. PLoS Genet 7:e1002126
16. Luo M, Taylor JM, Spriggs A, Zhang H, Wu X, Russell S, Singh M, Koltunow A (2011) A genome-wide survey of imprinted genes in rice seeds reveals imprinting primarily occurs in the endosperm. PLoS Genet 7:e1002125
17. Xiao W et al (2003) Imprinting of the MEA Polycomb gene is controlled by antagonism between MET1 methyltransferase and DME glycosylase. Dev Cell 5:891–901
18. Gehring M et al (2006) DEMETER DNA glycosylase establishes MEDEA polycomb gene self-imprinting by allele-specific demethylation. Cell 124:495–506
19. Makarevich G, Villar CB, Erilova A, Kohler C (2008) Mechanism of PHERES1 imprinting in Arabidopsis. J Cell Sci 121:906–912
20. Villar CB, Erilova A, Makarevich G, Trosch R, Kohler C (2009) Control of PHERES1 imprinting in Arabidopsis by direct tandem repeats. Mol Plant 2:654–660
21. Autran D et al (2011) Maternal epigenetic pathways control parental contributions to Arabidopsis early embryogenesis. Cell 145: 707–719
22. Frommer M et al (1992) A genomic sequencing protocol that yields a positive display of 5-methylcytosine residues in individual DNA strands. Proc Natl Acad Sci U S A 89: 1827–1831
23. Clark SJ, Harrison J, Paul CL, Frommer M (1994) High sensitivity mapping of methylated cytosines. Nucleic Acids Res 22:2990–2997
24. Rea M et al (2011) Determination of DNA methylation of imprinted genes in

Arabidopsis endosperm. J Vis Exp 47, http://www.jove.com/index/Details.stp?ID=2327, doi: 10.3791/2327

25. Paulin R, Grigg GW, Davey MW, Piper AA (1998) Urea improves efficiency of bisulphite-mediated sequencing of 5′-methylcytosine in genomic DNA. Nucleic Acids Res 26: 5009–5010
26. Smyth DR, Bowman JL, Meyerowitz EM (1990) Early flower development in Arabidopsis. Plant Cell 2:755–767
27. Kinoshita T, Yadegari R, Harada JJ, Goldberg RB, Fischer RL (1999) Imprinting of the MEDEA polycomb gene in the Arabidopsis endosperm. Plant Cell 11:1945–1952
28. Rogers SO, Bendich AJ (1988) Extraction of DNA from plant tissues. Plant Mol Biol Manual A6:1–10
29. Jacobsen SE, Sakai H, Finnegan EJ, Cao X, Meyerowitz EM (2000) Ectopic hypermethylation of flower-specific genes in Arabidopsis. Curr Biol 10:179–186
30. Gehring M, Bubb KL, Henikoff S (2009) Extensive demethylation of repetitive elements during seed development underlies gene imprinting. Science 324:1447–1451
31. Cokus SJ et al (2008) Shotgun bisulphite sequencing of the Arabidopsis genome reveals DNA methylation patterning. Nature 452: 215–219
32. Hsieh TF et al (2009) Genome-wide demethylation of Arabidopsis endosperm. Science 324: 1451–1454
33. Lister R et al (2008) Highly integrated single-base resolution maps of the epigenome in Arabidopsis. Cell 133:523–536
34. Lister R et al (2009) Human DNA methylomes at base resolution show widespread epigenomic differences. Nature 462:315–322
35. Lister R et al (2011) Hotspots of aberrant epigenomic reprogramming in human induced pluripotent stem cells. Nature 471:68–73
36. Hayatsu H, Negishi K, Wataya Y (2009) Progress in the bisulfite modification of nucleic acids. Nucleic Acids Symp Ser 53:217
37. Henderson IR, Chan SR, Cao X, Johnson L, Jacobsen SE (2010) Accurate sodium bisulfite sequencing in plants. Epigenetics 5:47–49

Part VII

Evolution of Imprinted Genes

Chapter 17

Computational Studies of Imprinted Genes

Martina Paulsen

Abstract

Computational studies on imprinted genes can have very different purposes: one major aim of these studies is the identification of DNA elements that distinguish imprinted genes from biallelically expressed genes. Comparative studies may help to identify imprinting regulatory elements and to understand common mechanisms of imprinted gene regulation in mammalian species. To date, the continuously growing number of genomic and epigenetic data sets makes detailed, genome-wide analyses on imprinted genes feasible. However, imprinted genes are characterized by genomic features that can influence statistics and can make such studies difficult. Hence, comparative computational studies can get very complex and require a tight interaction between bioinformaticians and biologists. Furthermore, analyses of raw data that are generated by micro-array hybridization and high-throughput sequencing technologies require computational approaches that have been designed especially for the epigenetic field. This chapter gives an overview about databases and software that is suitable for analyses of imprinted genes. Furthermore, possible difficulties that are typical for computational and statistical analyses of imprinted genes are described.

Key words: Imprinting, Bioinformatics, Comparative genomics

1. Introduction

In mammalian genomes most genes are active on both parental chromosomes. However, quite a number of genes are silenced on one of the two chromosomes, i.e., they are mono-allelically expressed. Among these genes are genes on the X chromosomes in females, olfactory receptor genes, and imprinted genes. In all these cases, epigenetic modifications such as DNA methylation and histone modifications appear to be the key players in silencing one of the gene copies. Imprinted genes are particularly interesting since the parental origin decides which of the two gene copies remains active and which is repressed. Silencing of one of the two gene copies is initiated in one of the parental germ lines, and is maintained after fertilization.

Nora Engel (ed.), *Genomic Imprinting: Methods and Protocols*, Methods in Molecular Biology, vol. 925,
DOI 10.1007/978-1-62703-011-3_17,

Although allele-specific DNA methylation and histone modification patterns are prerequisites for epigenetic regulation of imprinted genes, the genetic basis for this regulatory mechanism is not understood, i.e., we do not know what DNA elements in imprinted regions attract the epigenetic modification machinery to these genes in one germ line but keep them free of the respective modifications in the other germ line. Similarly, it is unclear how such parental origin-specific modifications survive remodeling of the epigenome after fertilization. Hence, during the last years many studies aimed at identifying the DNA elements that are responsible for allele-specific epigenetic modifications (1–6). Furthermore, it has been questioned if the mono-allelic regulation of these genes is associated with specific expression patterns and functions in distinct tissues or developmental stages (7, 8). Such questions can be asked for individual imprinted genes or can also address the features of imprinted genes in general. In both cases biocomputational studies can help to identify specific DNA elements, cell type-specific expression patterns, or putative functions. In turn, the design of experimental studies can be optimized, or hypotheses or models on the functional or regulatory particularities of imprinted genes can be developed.

We are currently facing a continuously growing number of data sets comprising genomic sequences, epigenome, and transcriptome data in a genome-wide scale. In parallel, computational hardware capacities have been expanded and software tools have been developed that allow the fast processing and annotation of raw data and subsequently complex statistical analyses. Hence, computational analyses of imprinted genes will benefit from this growing amount of available data. However, analyses of this data will require a combination of bioinformaticians who provide the computational tools for such analyses, statisticians for statistical support, and biologists who are experienced in the field of genomic imprinting for the initial design of such studies and interpretation of the obtained results. The success of such interdisciplinary projects depends to a high degree on the interaction of these scientists and on a basic knowledge of the particularities of imprinted genes and computational analyses and statistics of all participants. Therefore, this review aims to help strategies for computational analyses on imprinted genes and tries to point out possible difficulties that are related to the particular genetic and epigenetic properties of these genes.

2. Materials

2.1. Databases of Imprinted Genes

During the past two decades much effort has been invested in the identification of imprinted genes. Screens for imprinted genes are based on different methodologies. The first imprinted genes were discovered by investigations of genomic regions in mouse and human, where mutations show a parental origin-specific inheritance pattern for the phenotype. Further imprinted genes have been identified in vicinity of already known genes. Furthermore, computational approaches have been used for the prediction of candidate genes that are likely to be imprinted. For some of these candidate genes imprinting was verified by additional experiments in the lab. Collections of imprinted genes are provided by several databases that focus on different aspects of genomic imprinting and are therefore shortly described in the following.

2.1.1. The Catalogue of Parent of Origin Effects

The database at the University of Otago (New Zealand) provides lists of imprinted genes in various mammalian species (http://igc.otago.ac.nz/home.html) (9). In addition to defined genes, genomic regions are listed for which diseases or phenotypes with parental-specific biases in inheritance patterns have been observed. The database comprises currently more than 450 entries for the human and mouse genomes. The information on imprinted genes and phenotypes is derived from screens of the available literature, i.e., numerous different studies, and is regularly updated. For each entry the database provides a summary on features such as the expressing parental allele, imprinting status of the respective gene in other species, chromosomal location, tissue-specific imprinting effects, literature references, etc. The Internet surface allows searches by species, chromosome location, and name of the gene.

2.1.2. MRC Harwell Database on Imprinted Genes in the Mouse

This database focuses on imprinted genes in mouse (http://www.har.mrc.ac.uk/research/genomic_imprinting/index.html) (10). The information provided is structured similarly to the Otago database and is complemented by graphic maps on the chromosomal locations of imprinted genes.

2.1.3. Geneimprint Database

This database is a service of the Jirtle Laboratory at Duke University (USA) (http://www.geneimprint.com/). In addition to experimentally identified imprinted genes, the database also provides lists of genes that have been predicted to be imprinted by computational approaches (4, 5). For each gene information on cDNA and genomic sequences are given. In addition, the genes are linked to the SNP, Gene, and PubMed databases at the NCBI and to the UCSC genome browser.

2.1.4. The ncRNAimprint Database

The database is run by the Key Laboratory of Gene Engineering of Ministry of Education at Sun Yat-Sen University (China) (http://rnaqueen.sysu.edu.cn/ncRNAimprint/index.php) (11). The database focuses on imprinted noncoding RNAs and encompasses a comprehensive collection of microRNAs, snoRNAs, antisenseRNAs, etc. that are located in imprinted regions.

2.1.5. Transcriptome Data Sets on Imprinted Gene Expression

In the past few years, systematic experimental screens that cover the entire human or mouse genome have been applied and computational approaches have helped to identify further imprinted genes. One frequently used approach is the evaluation of monoallelic gene expression on the basis of SNP's occurrence in cDNA sequences. Subjects of these studies are usually public EST libraries derived from human or mouse tissues. The EST populations of individual genes are investigated for biases towards one sequence variant at heterozygous SNP positions. Such biases can be caused either by parental imprinting or by genetic variability. By combining the information on allelic biases from different EST libraries genetic effects can often be distinguished from parental imprinting effects. Another interesting approach is the comparison of gene expression levels in androgenetic and parthenogenetic embryos. Approximately 2,000 genes show differential expression patterns in these embryos. However, of the genes that were identified as possibly imprinted genes by transcriptome analyses, only very few have been verified as being indeed imprinted. Hence, the lists of candidate genes may contain rather high numbers of false positives. Some studies that provide lists of imprinted candidate genes are listed below:

- Candidates for imprinting based on SNP distributions in EST libraries in the human (12).
- Two data sets on transcriptome sequences derived from brain samples of F1 mice (13, 14).
- Differentially methylated regions in reciprocal human uniparental disomy samples (15).
- Differentially expressed genes in parthenogenetic and androgenetic mouse embryos (16, 17).

2.1.6. Computationally Predicted Imprinted Genes

A substantial number of imprinted genes have been predicted for the human and mouse genomes based on the special densities and distribution of repetitive elements in imprinted regions. The predictions were performed by statistical classifiers that had been trained on the special distributions of repetitive elements (4, 5) and epigenetic properties of imprinted regions (18). Similar to the transcriptome data sets, imprinted expression has been proven for only a few of these candidate genes.

2.2. Databases for Information on Genomes and DNA Sequences of Different Species

For quite a number of mammalian genomes assembled versions exist that have been made available by the UCSC Genome Browser, and the ENSEMBL Genome Browser. For these genomes, both browsers annotate a high number of different genome features that can be downloaded. The annotated features encompass information on genome organization such as gene organization, CpG islands, repetitive elements, etc. Information on different aspects of sequence conservation is available such as the annotation of

SNPs and highly conserved elements. Furthermore, information on gene expression and epigenetic modifications is given.

The NCBI provides a broad platform (http://www.ncbi.nlm.nih.gov/sites/genome) for the access to additional genome resources. This includes links to information on genomes that are not annotated in the UCSC or ENSEMBL genome browsers, for example, for the genome of species that are not considered common model organisms in medicine or evolutionary research. In addition, the NCBI provides information on the current status of genome sequencing projects that are still in progress (see Note 1).

2.3. Databases for Information on Gene Expression and Epigenetic Modifications

2.3.1. Gene Expression Data

The Genevestigator database (https://www.genevestigator.com/gv/index.jsp) is a joint project of the the NEBION company and the ETH Zuerich, Switzerland (19). Currently, the database includes gene expression data for 14 species. The Genevestigator homepage offers several options for doing some basic statistical analyses.

2.3.2. Epigenome Data

Epigenome data are to some extent already annotated in the UCSC genome browser. More data sets and information from ongoing epigenome projects can be accessed via the browser of the Epigenome Roadmap project of the NIH Roadmap Epigenomics Mapping Consortium (http://www.roadmapepigenomics.org/) (20).

2.4. Software for the Annotation of Genomic Features

The appearance of newly sequenced genomes may make it necessary to annotate genomic features, and even in case of genomes that are available in the UCSC or ENSEMBL genome browsers, it may sometimes be necessary to annotate genomic features anew. Newly released genomes or sequences that have been generated by the investigators' personal project may require special annotations. A selection of useful software for the annotation of genomic features of imprinted genes is listed below:

Identification of repetitive elements:

- RepeatMasker (http://www.repeatmasker.org/): Detection of (retro-)transposed elements.
- Tandem Repeats Finder (http://tandem.bu.edu/trf/trf.html): Identification of tandem repeat arrays (21).

Identification of CpG-rich regions:

- EMBOSS CpG plot: http://www.ebi.ac.uk/Tools/emboss/cpgplot/.
- CpG island searcher: http://cpgislands.usc.edu/ (22).
- CpG cluster: http://genius-index.com/cluster.aspx (23).
- More comprehensive overviews on the performance of available software for CpG island identification are given in two recent reviews (24, 25).

Identification of transcription factor binding sites:

- Transfac database: http://www.gene-regulation.com/pub/databases.html.
- An overview of the computational analyses and discovery of transcription factor binding sites is given in a recent review published in Methods in Molecular Biology (26).

Epigenetic features:

- Epigraph: Relationships between DNA sequence and epigenetic modifications (http://epigraph.mpi-inf.mpg.de/WebGRAPH/), (27).
- BIQ Analyzer HAT: Processing and statistical analyses of bisulfite sequencing data (http://biq-analyzer-ht.bioinf.mpi-inf.mpg.de/) (28).
- Methvisual package: Visualization and statistical analyses of bisulfite sequencing data (http://methvisual.molgen.mpg.de/) (29).
- The Methdb database (http://www.methdb.de/links.html) provides a comprehensive list of computational tools that are specifically designed for applications in epigenetics (30).

Data organization and statistical analyses:

- Bioconductor provides the framework for sequence retrieval, annotation, and statistical analyses (http://www.bioconductor.org/).
- R package: Open source package for statistics (http://www.r-project.org/).

3. Methods

Genome-wide studies on imprinted genes consist usually of four phases that will be the subject of the following paragraphs:

1. Selection of gene sets
2. Sequence retrieval—Evaluation of sequence quality
3. Annotation of genomic and epigenomic features
4. Statistical analyses of different parameters

3.1. Selection of Gene Sets

3.1.1. Collecting Gene Sets of Suitable Size

If studies on a rather broad group of imprinted genes are planned the selection of sets of imprinted genes can be hampered by various factors that are to some extent specific for this special class of genes.

The major problem with selecting a set of imprinted genes is the danger of too small a sample size with a gene number below hundred, or even below 50. In order to ensure a sound statistical analysis gene numbers below 30 should be avoided.

3.1.2. Manually Collected Data Sets

In most cases the selection of imprinted genes is driven by the need to get a set of genes whose imprinting status is out of question. As a primary source, public databases on imprinted genes are usually adequate. For consistent selection, several parameters can be used as criteria, such as a minimal number of cited publications, conservation of imprinting in at least two species, application of at least two different methods for detection of allele-specific expression or allele-specific modifications, successful tests for imprinted expression in a distinct number of samples or individuals, etc. Examples of such manually collected lists can be found in (8, 31, 32).

3.1.3. Selection of Imprinted Candidate Genes from Transcriptome Data Sets

Another option can be the usage of one of the transcriptome data sets on imprinted expression described in Subheading 2.1.5. This has the advantage that the imprinted candidate genes are all discovered by the same method. From such a data set imprinted genes can be chosen by setting a threshold for a score for parental allele-specific expression that has to be passed. This procedure has the advantage of being free of any biases caused by manual curation of gene lists. However, there are indications that such a procedure might be biased as well. For example, data sets on allele-specific gene expression are biased towards strongly expressed genes that achieve higher scores for allele-specific expression than weakly expressed genes (33).

3.1.4. Imprinted Genes Can Be Divided into Protein-Encoding Genes and RNA Genes

Imprinted regions in the human and mouse genomes contain many genes that encode untranslated RNAs. Among these are long non-spliced transcripts that appear to be involved in epigenetic silencing and spliced noncoding RNAs such as the *H19* RNA whose function in imprinted gene regulation is still an enigma (34, 35). Especially the microRNA and snoRNA clusters of imprinted regions are famous for their high numbers of small RNAs (36–38). For studies that aim to identify features that distinguish imprinted genes from non-imprinted genes one should consider analyzing imprinted protein-coding genes and untranslated RNA genes separately since genes of noncoding RNAs and protein-encoding genes differ strongly in terms of their exon/intron structure.

3.2. Sequence Retrieval: Evaluation of Sequence Quality

The next step in computational analysis is usually the download of genomic DNA sequences. Usually, this will be done via the server of the UCSC or Ensembl Genome browser or via the Genome project homepage of the NCBI.

Due to the differences in progress of the numerous sequencing projects, genomic sequences can show pronounced differences in quality as is nicely described for comparison of the *Gnas* locus in three species (39). Especially if several unfinished genomes are to be compared, an eye should be kept on parameters such as read coverage and contig length of the chosen genomes, and it might be a good idea to select for comparative studies only genomes with similar average contig length or sequence coverage.

In past studies (own unpublished data) we have seen that the average contig length of assembled sequences can differ significantly between the genomes of different species. We also observed differences in contig lengths between imprinted genes and other autosomal genes. For example repetitive elements may cause assembly problems. Hence, sequence and assembly quality can cause biases in statistical analyses on features such as the density of repetitive elements, and CpG and G+C content.

3.3. Annotation of Sequence Features

3.3.1. Organization of a Systematic Annotation

Before starting with the actual annotation of DNA sequences it is always very useful to develop a plan for the consistent annotation of the physical structure of all genes. This means that the investigators need to define the starts and ends of transcriptional units and the extent of intergenic regions. This sounds rather trivial but can be difficult in case of overlapping genes, gene with more than one promoter, etc. Another typical problem is the multiple annotation of transcripts of the same gene. Such a phenomenon can be caused by the presence of more than one reference sequence for one gene. For example, if this affects mostly strongly expressed genes, such multiple annotations may bias studies on gene expression towards strongly expressed genes.

As already mentioned above, many features are annotated in the UCSC or Ensembl genome browsers. Positions, scores, etc. can be downloaded in table formats and can subsequently be used for statistical comparisons of imprinted genes vs. control genes. In rare cases, for example if newly sequenced genomes are part of the analyses such sequences have to be annotated by the investigator. Some useful software suites are listed above (see Subheading 2.3).

Especially when DNA sequences need to be converted to different formats or if basic statistical analyses on general sequence features are planned, the Bioconductor package can be very useful (http://www.bioconductor.org/). The Bioconductor package is an open source package. It provides tools for DNA sequence analyses, investigations on gene expression patterns, and also software for epigenetic analyses. The incorporated statistics packages use R as statistical programming language.

3.3.2. Annotation of DNA Methylation Patterns

Although the information derived from some of the epigenetic data sets has been incorporated into the Ensembl and UCSC browsers, a lot of information is only available in the form of raw data. One example for a software that helps to determine DNA methylation patterns using bisulfite sequencing raw data is the BiQAnalyzer HT package. The recently updated version is available as a download and allows the processing of large sets of raw data as they are typically generated by high-throughput sequencing (28). Implemented into the package are some useful tools for the visualization of methylation data and tools for statistical analyses of DNA methylation patterns. An alternative for statistical analyses and the visual presentation of bisulfite data is the Methvisual package (http://methvisual.molgen.mpg.de/) (29). This software package is implemented in an R/Bioconductor environment.

3.3.3. How to Identify Genomic Features that are Associated with Epigenetic Modifications

A common question in epigenetics is if a specific pattern of epigenetic modifications is associated with special features of DNA sequence (see Note 2). For example investigators might be interested if differentially methylated regions in imprinted regions show specific DNA features. For such questions the EpiGraph suite provides useful computational tools for the identification and statistical evaluation of DNA features that are associated with a specific type of epigenetic modifications (27). Furthermore, the tool allows the prediction of further genomic regions that are likely to show the same epigenetic modification patterns as the test samples.

3.4. Statistical Analyses

3.4.1. Planning Statistical Tests

After annotation of sequence features, the major task of genome-wide studies is to compare imprinted vs. non-imprinted genes or to compare the imprinted genes of different species. Usually tables of annotated elements are taken as a basis for the subsequent statistical analyses. Due to the large amount of available genome data statistics can become very complex. Therefore, it is recommended to use from the beginning a professional software such as the R package (http://www.r-project.org/).

Especially researchers with little experiences with statistical analyses should keep in mind that not every statistical test can be applied to every question or every sample. Hence, it is usually helpful to check first if the chosen statistical test can indeed be applied to the selected samples. A frequent mistake is for example the application of t-tests to samples that do not show a normal distribution.

Often a study encompasses not just one but several similarly structured comparisons of different data sets. In such cases the consistency of the entire study can be improved if a statistical test is chosen that can be used for all planned comparisons.

3.4.2. Statistical Problems with Sample Sizes

Rarely do comparative studies on imprinted genes encompass more than 50 genes. In special cases such as differentially methylated regions or retrotansposed imprinted genes numbers drop to less than 20 or even 10. In such cases, most statistical methods are not suitable.

Another statistical problem can be the comparison of a moderately sized group of imprinted genes (approx. 50) to a large control group, for example all other autosomal genes (approx. 20,000). Due to the extremely large number of control genes, such comparisons produce easily significant p-values. One option to show that an observed difference is unlikely to occur by chance is background testing, i.e., multiple comparison of a randomly selected, similarly sized group of genes to all other genes (8).

4. Notes

Mammalian genomes consist of segments with different G+C content, the so-called isochores. Different isochores are characterized by differences in gene densities and gene activities. Isochores are typical examples of how the global structure of a genomic region can influence gene activity. Thus, general features such as G+C content, repetitive elements, etc. have a major input on the organization of imprinted regions. Such differences in genomic features can influence comparative studies on imprinted genes, and it is therefore sensible to keep a close eye on possible interdependencies of genomic features and species-specific effects.

1. Species-specific differences

   Studies on the identification of imprinted genes focus on human and mouse, and information on imprinted gene expression in other species is scarce. However, imprinted gene expression is apparently well conserved among different mammals: a survey on orthologous protein-encoding genes that have been investigated in somatic human and mouse tissues revealed that 87 % were imprinted in both species (8). Despite the conservation of imprinted gene expression in human tissues and the murine embryo proper, imprinted expression in the placenta might differ considerably between the two species. For the mouse, quite a number of genes have been identified that show a bias towards maternal expression in the placenta but not in the embryo proper. Although this phenomenon has been known for quite a while, for many of these genes it is still unclear if the maternal bias is indeed caused by a true imprinting effect, or rather by contamination with maternal tissues (40).

   Furthermore, not only in imprinted genes, but also in the entire genome, features such as G+C and CpG contents and repetitive elements can differ considerably between different species. For example, the mouse possesses fewer and shorter CpG islands than the human (24). In addition, different species possess different types of repetitive elements, for example *Alu* elements are primate-specific repetitive elements and are absent in other mammals.

2. Correlations between different genomic features

   As already mentioned above, genome-wide comparative studies on imprinted genes can be hampered by technical biases in association with sequence quality. Genome-wide studies should take into account that quite a number of interdependencies exist between different genomic parameters. For example, differences in intron length between imprinted and control genes may have some impact on the intragenic repetitive element or

CpG island content (32). Hence, statistical analyses should be normalized against the lengths of intragenic and intergenic regions. A second parameter that can have a pronounced influence on the analyses of genomic features is the G+C content. Imprinted genes show highly variable G+C contents. On average, the G+C content of imprinted genes is slightly but not significantly elevated in comparison to other autosomal genes. This slight increase, nevertheless, may be associated with an elevated CpG content, and may therefore have some impact on CpG island identification. Hence, an elevated CpG island content might simply be the result of an elevated G+C content as is might be also seen in other (non-imprinted) regions of the genome.

In the human, many Alu elements contain CpG islands. Hence, in human genomes there are many CpG islands within repetitive elements. As a consequence, the Alu depletion that is typical for human imprinted genes influences the CpG island content of these genes in a species-specific manner (2). One rather simple option to avoid such complications is to apply repeat masking to sequences and to analyze single copy sequences and repetitive elements separately.

Nevertheless, the described interdependencies between different parameters call for a more detailed analysis using different approaches such as multivariate statistics.

Acknowledgments

Computational studies in the Paulsen group are supported by the Deutsche Forschungsgemeinschaft (DFG grant PA750/3-1).

References

1. Greally JM (2002) Short interspersed transposable elements (SINEs) are excluded from imprinted regions in the human genome. Proc Natl Acad Sci USA 99:327–332
2. Hutter B et al (2006) Tandem repeats in the CpG islands of imprinted genes. Genomics 88: 323–332
3. Kobayashi H et al (2006) Bisulfite sequencing and dinucleotide content analysis of 15 imprinted mouse differentially methylated regions (DMRs): paternally methylated DMRs contain less CpGs than maternally methylated DMRs. Cytogenet Genome Res 113:130–137
4. Luedi PP et al (2007) Computational and experimental identification of novel human imprinted genes. Genome Res 17:1723–1730
5. Luedi PP et al (2005) Genome-wide prediction of imprinted murine genes. Genome Res 15: 875–884
6. Ke X et al (2002) The distinguishing sequence characteristics of mouse imprinted genes. Mamm Genome 13:639–645
7. Varrault A et al (2006) Zac1 regulates an imprinted gene network critically involved in the control of embryonic growth. Dev Cell 11:711–722
8. Steinhoff C et al (2009) Expression profile and transcription factor binding site exploration of imprinted genes in human and mouse. BMC Genomics 10:15
9. Morison IM et al (2005) A census of mammalian imprinting. Trends Genet 21:457–465

10. Williamson CM et al. (2011), MRC Harwell, Oxfordshire. World Wide Web Site - Mouse Imprinting Data and References - http://www.har.mrc.ac.uk/research/genomic_imprinting/
11. Zhang Y et al (2010) ncRNAimprint: a comprehensive database of mammalian imprinted noncoding RNAs. RNA 16:1889–1901
12. Seoighe C et al (2006) Maximum likelihood inference of imprinting and allele-specific expression from EST data. Bioinformatics 22:3032–3039
13. Wang X et al (2008) Transcriptome-wide identification of novel imprinted genes in neonatal mouse brain. PLoS One 3:e3839
14. Gregg C et al (2010) High-resolution analysis of parent-of-origin allelic expression in the mouse brain. Science 329:643–648
15. Nakabayashi K et al (2011) Methylation screening of reciprocal genome-wide UPDs identifies novel human-specific imprinted genes. Hum Mol Genet 20(16):3188–3197
16. Mizuno Y et al (2002) Asb4, Ata3, and Dcn are novel imprinted genes identified by high-throughput screening using RIKEN cDNA microarray. Biochem Biophys Res Commun 290:1499–1505
17. Nikaido I et al (2003) Discovery of imprinted transcripts in the mouse transcriptome using large-scale expression profiling. Genome Res 13:1402–1409
18. Brideau CM et al (2010) Successful computational prediction of novel imprinted genes from epigenomic features. Mol Cell Biol 30:3357–3370
19. Hruz T et al (2008) Genevestigator v3: a reference expression database for the meta-analysis of transcriptomes. Adv Bioinformatics 2008:420747
20. Bernstein BE et al (2010) The NIH roadmap epigenomics mapping consortium. Nat Biotechnol 28:1045–1048
21. Benson G (1999) Tandem repeats finder: a program to analyze DNA sequences. Nucleic Acids Res 27:573–580
22. Takai D, Jones PA (2002) Comprehensive analysis of CpG islands in human chromosomes 21 and 22. Proc Natl Acad Sci USA 99:3740–3745.2
23. Hackenberg M et al (2006) CpGcluster: a distance-based algorithm for CpG-island detection. BMC Bioinformatics 7:446
24. Hutter B et al (2009) Identifying CpG islands by different computational techniques. Omics-a Journal of Integrative Biology 13:153–164
25. Zhao Z, Han L (2009) CpG islands: algorithms and applications in methylation studies. Biochem Biophys Res Commun 382:643–645.14
26. Ladunga I (2010) An overview of the computational analyses and discovery of transcription factor binding sites. Methods Mol Biol 674: 1–22
27. Bock C et al (2010) Web-based analysis of (Epi-) genome data using EpiGRAPH and Galaxy. Methods Mol Biol 628:275–296
28. Lutsik P et al (2011) BiQ Analyzer HT: locus-specific analysis of DNA methylation by high-throughput bisulfite sequencing. Nucleic Acids Res 39:W551–W556
29. Zackay A, Steinhoff C (2010) MethVisual - visualization and exploratory statistical analysis of DNA methylation profiles from bisulfite sequencing. BMC Res Notes 3:337.20
30. Negre V, Grunau C (2006) The MethDB DAS server: adding an epigenetic information layer to the human genome. Epigenetics 1:101–105
31. Hutter B et al (2010) Divergence of imprinted genes during mammalian evolution. BMC Evol Biol 10:10
32. Hutter B et al (2010) Imprinted genes show unique patterns of sequence conservation. BMC Genomics 11:649
33. Nothnagel M et al (2011) Statistical inference of allelic imbalance from transcriptome data. Hum Mutat 32:98–106
34. Sleutels F, Zwart R, Barlow DP (2002) The non-coding Air RNA is required for silencing autosomal imprinted genes. Nature 415:810–813
35. Gabory A et al (2009) H19 acts as a trans regulator of the imprinted gene network controlling growth in mice. Development 136: 3413–3421
36. Seitz H et al (2003) Imprinted microRNA genes transcribed antisense to a reciprocally imprinted retrotransposon-like gene. Nat Genet 34:261–262
37. Kircher M, Bock C, Paulsen M (2008) Structural conservation versus functional divergence of maternally expressed microRNAs in the Dlk1/Gtl2 imprinting region. BMC Genomics 9:346
38. Cavaillé J et al (2000) Identification of brain-specific and imprinted small nucleolar RNA genes exhibiting an unusual genomic organization. Proc Natl Acad Sci USA 97: 14311–14316
39. Hart EA et al (2007) Lessons learned from the initial sequencing of the pig genome: comparative analysis of an 8 Mb region of pig chromosome 17. Genome Biol 8:R168
40. Proudhon C, Bourc'his D (2010) Identification and resolution of artifacts in the interpretation of imprinted gene expression. Brief Funct Genomics 9:374–384

Chapter 18

Insights on Imprinting from Beyond Mice and Men

Andrew Pask

Abstract

Genomic imprinting is an epigenetic phenomenon that results in the silencing of alleles, dependent on their parent of origin. Within vertebrates, this phenomenon is restricted only to the mammals and has been identified in eutherians and marsupials but not in the egg-laying monotremes. Many hypotheses have been put forward to explain why genomic imprinting evolved, most of which are centered on the regulation of nutrient provisioning from parent to offspring. The three different mammalian lineages have adopted very different modes of reproduction and, as a result, vary widely in the amount of nutrient provisioning to the conceptus. Examining imprinting across the three mammal groups enables us to test hypotheses on the origin of this phenomenon in mammals and also to investigate changes in the genome coincident with its evolution.

Key words: Genomic imprinting, Eutherian, Marsupial, Monotreme, Genome evolution

1. Introduction

The class Mammalia is divided up into three extant lineages, the eutherians, marsupials, and monotremes. Eutherian mammals last shared a common ancestor with marsupials around 160 million years ago and monotremes around 180 million years ago (Fig. 1) (1). Each lineage is characterized by a unique mode of reproduction. Eutherian mammals typically give birth to well-developed young after an extended gestation period. Fetal growth is supported by a large and invasive placenta, and pregnancies place large physiological demands on the mother (Fig. 2). Marsupials, like eutherian mammals, have a fully functional placenta to deliver nutrition to their developing fetus, but it is less invasive and short-lived. Most marsupial species give birth to very small young that place a minimal demand on the mother's resources during preg-nancy. Maternal contribution to pregnancy is further reduced in the egg-laying monotremes. Even so, the early stages of

Nora Engel (ed.), *Genomic Imprinting: Methods and Protocols*, Methods in Molecular Biology, vol. 925,
DOI 10.1007/978-1-62703-011-3_18, © Springer Science+Business Media, LLC 2012

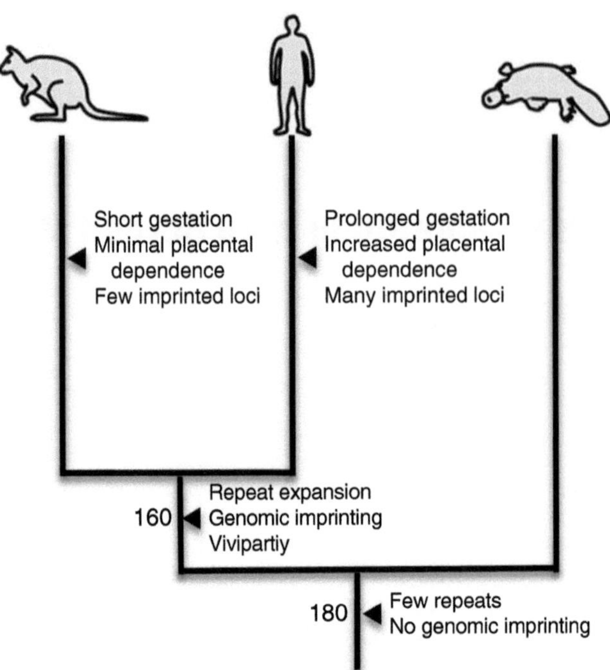

Fig. 1. Evolutionary tree of the three extant mammalian lineages. Numbers indicate divergence times in millions of years. *Arrows* indicate the acquisition of various features associated with genomic imprinting.

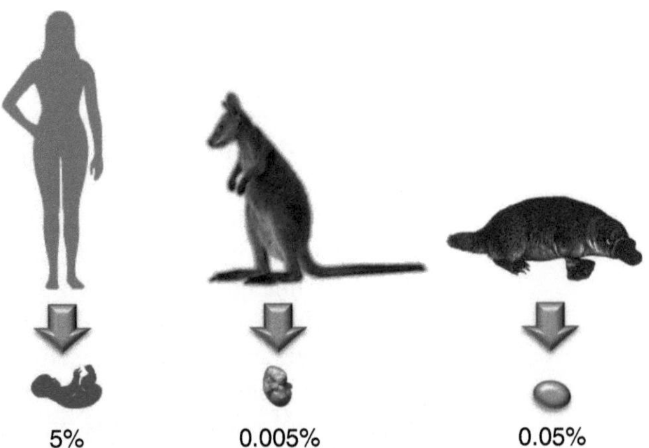

Fig. 2. Relative maternal contributions to offspring for each mammalian lineage (Eutherians—left, marsupials—center and monotremes—right). Numbers indicate the average weight of the offspring as a percentage of the maternal weight.

development and nutritional supplies for the egg require maternal resource provisioning, but this places little demand on the mother (2). Each of these strategies results in different maternal contributions of nutrients to the offspring in utero and during postnatal care, making them especially useful models for examining the origins of genomic imprinting.

Genomic imprinting is the epigenetic silencing of certain alleles dependant on their parent of origin. As a result, the eutherian conceptus is haploinsufficient for around 100 different genes, many of which are essential regulators of growth and development (3). This presents a perplexing evolutionary strategy since any mutations in the single expressed copy of each of these genes could lead to disease. In fact, we see many disease states in humans arising from mutations in imprinted loci (4). Many different theories have been put forward to explain what the evolutionary advantage might be of adopting much a mechanism, the most supported of which is the parental conflict/kinship hypotheses (5, 6). Briefly, these theories attempt to explain the different investments of the male and female genomes in the fitness of the offspring. Paternally expressed genes that favor nutrient provisioning and growth of the offspring will result in greater genetic fitness of the father, even if this occurs at the mother's expense. The genetic fitness of the mother is increased by restricting the nutrients given to any one offspring, so that she may have many successive pregnancies. In line with these hypotheses, the vast majority of imprinted genes identified in mice and humans are expressed in the placenta or fetus and affect growth and nutrition in utero (7). Thus, the parental conflict/kinship hypotheses would have very different predictions for the prevalence of genomic imprinting across the three extant mammalian lineages. One would expect that the evolutionary forces that have favored the evolution of imprinting in eutherians would be significantly reduced in the marsupials and monotremes.

Here I review how studies in marsupials and monotremes have contributed to our understanding of why and how genomic imprinting evolved. Such evolutionary studies have been fundamental to our understanding of the mechanisms governing genomic imprinting and the evolutionary forces selecting for it.

2. The Absence of Genomic Imprinting in the Nonmammalian Vertebrates

Genomic imprinting is presumed not to exist in nonmammalian vertebrates since all lineages are capable of producing viable parthenogenotes (embryos created from only maternal chromosomal contributions). This finding suggests that the genes must not be differently silenced between the sexes. In contrast, parthenogenetic mouse embryos show early embryonic lethality, due to the loss of critical gene function, and the few mammalian parthenogenetic embryos that were able to make it to post implantation stages of development show stunted placental growth (8).

Despite the presumed absence of genomic imprinting in the genomes of nonmammalian vertebrates, only a few studies have actually verified this assumption. Two highly conserved genes that show monoallelic expression in mice in humans (*IGF2* and *IGF2R*) were shown to be biallelically expressed in the chicken embryo (9). Most mammalian imprinted genes exist in clusters in the genome and are often regulated by shared imprinting control regions (ICRs). While no such imprinting control regions have been identified outside of the therian (marsupial and eutherian) mammals, the genes themselves are clustered in the same highly conserved arrangements in chickens and show asynchronous replication which may have served as a prelude to genomic imprinting (10).

3. Insights on the Evolution of Genomic Imprinting from the Monotreme Genome

Monotremes, like all oviparous animals, provide some maternal nutrients to the developing embryo before the egg is laid (2). However, this contribution is often minimal, and would not have as major an impact on maternal resources as an extended pregnancy (Fig. 2). Therefore, based on the parental conflict/kinship hypotheses, one might expect a lack of genomic imprinting associated with genes regulating fetal nutrition in monotremes, and this is indeed the case. Imprinting has not been detected in any of the ten eutherian imprinted genes so far investigated in the platypus (11–21) (Table 1). However, due to the protected status of this species, the imprint status for each of these genes has only been examined in adult material and not in the developing young or fetal membranes. Nevertheless, it is generally concluded that imprinting does not occur in monotremes, making them ideal comparative models for examining the evolution of this epigenetic phenomenon.

Many hypotheses have been put forward that attempt to explain what the evolutionary advantage of imprinting might be, but it is equally important to examine how such a mechanism arose. The host defense hypothesis suggests that genomic imprinting evolved from endogenous mechanisms that silence transposable elements and invading foreign DNA within the genome (44). This hypothesis is supported by the observation that most imprinted genes in eutherians are associated with a high density of repeat sequences and endogenous retroviruses that could have attracted silencing to the region (45). As such, the monotreme (platypus) genome (46) provides a unique resource to compare with marsupials (opossum (47) and tammar (48)) and eutherians (49, 50), to determine what changes were coincident with the evolution of imprinting. These analyses revealed that most imprinted genes were highly conserved at the nucleotide level in all mammals (51). Furthermore, most genes resided in similar clusters, suggesting their spatial arrangement predated the evolution of imprinting (51). The regions of the platypus genome surrounding all the

Table 1
A list of the eutherian imprinted loci that have been investigated in marsupials and monotremes

Location (human)	Gene	Eutherian	Marsupial	Monotreme
6q25	*IGF2R*	Yes (22)	Yes (13)	No (13)
	Air	Yes (23)	No (20)	
7q21.3	*SGCE*	Yes (24)	No (17)	No (17)
	PEG10	Yes (25)	Yes (17)	Absent (17)
	PPP1R9A	Yes (26)	No (17)	
	ASB4	Yes (27)	No (17)	
7q32.2	*MEST*	Yes (28)	Yes (18)	
11p15	*H19*	Yes (29)	Yes (16)	
	IGF2	Yes (30)	Yes (31)	No (14)
	INS	Yes (32)	Yes (33)	
	CDKN1C	Yes (34)	No (35)	
	PHLDA2	Yes (36)	No (19)	
14q32	*DLK1*	Yes (37)	No (11)	No (11)
	MEG3	Yes (38)	Absent (20, 21)	
	RTL1	Yes (39)	Absent (11)	
	DIO3	Yes (40)	No (11)	No (11)
15q11–12	*SNURF-SNRPN*	Yes (41)	No (15)	No (15)
	UBE3A	Yes (42)	No (15)	No (15)
20q11.23	*NNAT*	Yes (43)	Absent (12)	

The genes are grouped into clusters based on their arrangement in the human genome. The imprint status for each gene is indicated for each mammalian lineage: Yes, denotes that the gene is imprinted; No, denotes the gene is not imprinted; Absent, denotes that the gene is not present in that lineage

imprinted gene orthologues contained significantly less long-terminal repeats (LTRs) and DNA transposable elements than in the therian orthologous regions (51). In addition, other classes of transposable elements (such as SINEs and LINEs) were seen to expand in certain therian imprinted gene clusters (51). An expansion in repeat elements is also seen across the entire mammalian genome after the divergence of monotremes (46). Thus, it appears that imprinting evolved coincident with repeat expansion in the mammalian genome. This is consistent with the host defense hypothesis (44), and suggests that repeats attracted epigenetic silencing to many regions throughout the genome. Where this silencing caused an evolutionary advantage it was selected for, and maintained, and the gene became imprinted. If the silencing had no effect, or a deleterious effect, it was lost (51). While the platypus genome analyses provided important data in support of the host defense hypothesis, the first direct evidence of such a mechanism came from imprinted gene analyses in marsupials.

4. The Birth of Genomic Imprinting in Marsupial Mammals

Marsupials are viviparous and give birth to relatively small young, placing minimal demand on maternal resources (Fig. 2) (52). Under the predictions of the parental conflict/kinship hypotheses we would expect to see reduced selection for imprinting in this lineage. This prediction holds true, and of the 19 eutherian imprinted orthologues investigated in marsupials, only six are imprinted (Table 1) (11–13, 15–21, 31, 33, 35), three of which reside in the *IGF2-H19* cluster. One of these genes, ***insulin*** (*INS*), is exclusively imprinted in the eutherian placenta but not elsewhere in the embryo (53). Similarly, *INS* is exclusively imprinted in the placental membranes in marsupials (33). Since *INS* has maintained imprinting solely in the therian placenta, evolutionary pressures in this tissue alone must have been sufficient to drive genomic imprinting (33). This provides strong support for the parental conflict/kinship hypotheses and for the placenta being a focal point for the selection of genomic imprinting.

In eutherians, imprinted genes are generally regulated by a differentially methylated region (DMR) (54). However, in marsupials, DMRs have only been found to be associated with two imprinted genes, *PEG10* (17) and *H19* (16). Interestingly, these are also the only two genes that show complete silencing from the imprinted alleles. The remaining four marsupial imprinted genes show parental bias, resulting in diminished expression from the imprinted allele, but not complete silencing (2). These data suggest that genomic imprinting does not require differential methylation, and that such a mechanism may have evolved over time, due to increased selective pressure, to further silence imprinted loci.

The accumulation of regulatory mechanisms appears to be a feature of many imprinted clusters across marsupial and eutherian mammals. For example, in eutherian mammals the *CDKN1C* gene is syntenic to *IGF2* and oppositely imprinted. Both genes are expressed in the placenta and are antagonistic in function (55). *CDKN1C* and *IGF2* are also in a syntenic arrangement in the marsupial genome and *CDKN1C* is expressed alongside *IGF2* in the tammar wallaby placenta. In spite of this conservation, only *IGF2* is imprinted and *CDKN1C* shows biallelic expression (35, 56). Thus, *CDKN1C* imprinting is not dependant on its syntenic localization with an imprinted *IGF2*. Imprinting of *CDKN1C* in mice is regulated by *KCNQ1OT1*, a long noncoding antisense RNA derived from an intron of the *KCNQ1* gene under the control of a DMR (57). Although *CDKN1C* is not imprinted in marsupials, they still produce the *KCNQ1OT1* transcript but there is no evidence of a DMR within the region. *PHLDA2* is another eutherian imprinted gene from the same region that negatively controls placental growth in mice (58). Similar to *CDKN1C*, marsupial *PHLDA2* shows a conserved syntenic arrangement and expression in the placenta, but is not imprinted (19). Together, these data suggest that placental expression and antisense transcription preceded the acquisition of imprinting within certain gene clusters.

In addition to the accumulation of imprinting control mechanisms, marsupial analyses have also shown the evolution of new imprinted regions. *Peg10* is an imprinted gene derived from the *Sushi-ichi* transposon, essential for placental development in mice (25). Analyses of *PEG10* across the three extant mammalian lineages showed that it was recently inserted into the mammalian genome in the therian ancestor after the monotreme split (17). This insertion event was coupled with methylation and the acquisition of an imprint. This imprint has remained in marsupials and *PEG10* is one of only two imprinted genes in marsupials with a DMR (17). Interestingly, in marsupials imprinting is restricted only to *PEG10* and not the surrounding loci. However, in eutherian mammals, imprinting in the *PEG10* region has spread to encompass the neighboring genes *SGCE*, *PPP1R9A* and *ASB4*, suggesting that genesis of a new DMR can lead the evolution of an entire imprinted gene cluster (2). The only other marsupial imprinted gene with an identified DMR is *H19* (16). *H19* also appears to be a recent insertion in to the therian genome, and has not yet been found in the monotremes.

Similar retrotransposition events also appear to have driven the evolution of the *SNRPN-UBE3A* (Prader–Willi/Angelman syndrome) region in mice and humans. This imprinted region is not present in marsupials, in which *SNRPN* and *UBE3A* map to different chromosomes and are both biallelically expressed. Thus, it appears that the region was formed and acquired imprinting only in the eutherian lineage from a major rearrangement event linking *SNRPN* and *UBE3A*. This rearrangement was concomitant with the insertion of retrotransposed genes and snoRNAs that regulate imprinted gene expression from the region in mice and humans (59). Similarly, genomic imprinting of the *DLK-DIO3* region is a recent evolutionary event and occurs only in eutherian mammals. While the genomic arrangement of these genes is conserved between eutherians and marsupials, both *DLK* and *DIO3* are biallelically expressed in the wallaby (11). This is in contrast to eutherian mammals where both genes are imprinted (60). Comparisons of the entire imprinted region between the mouse, human and wallaby genomes revealed the insertion of the retrotransposed gene *RTL1* and noncoding transcripts including microRNAs and snoRNAs exclusively in the eutherian lineage (11, 59).

Since the insertion of retrotransposed genes and snoRNAs are known to attract methylation, it is likely that they attracted imprinting to the regions described above (2). These findings provide direct evidence in support of the host defense hypothesis and confirm that retrotransposed genes can acquire methylation dependent imprinting in mammals. Furthermore, they show that DMR associated imprinting arose before the marsupial–eutherian split.

5. What Have Studies Beyond Mice and Men Taught Us About Genomic Imprinting?

Comparative analyses of imprinted regions across the three extant mammalian groups and in birds have enabled us to draw many conclusions on the origins and selective pressures of genomic imprinting. First, we see that the genesis of genomic imprinting in vertebrates is coincident with the evolution of viviparity. Second, we see that many of the genes that have retained imprinted expression in marsupials are expressed in the placenta. Together, these data are in strong support of the parental conflict/kinship hypotheses for explaining why imprinting has been maintained in mammals. Furthermore, it suggests that in marsupials, even though the placental attachment is short lived and maternal nutrient contribution to the developing fetus is minimal, there is sufficient evolutionary pressure to retain imprinting. Many of the eutherian imprinted gene orthologues that are not imprinted in marsupials are still expressed in the placenta. This suggests that placental expression for many loci predated the acquisition of genomic imprinting. Furthermore, we also see a reduced prevalence of genomic imprinting in marsupials compared to eutherians as predicted by the parental conflict/kinship hypotheses. However, it is possible that marsupials have their own, as yet unidentified, unique imprinted loci not present in eutherians.

Another consistent feature of imprinted regions is that they have expanded in eutherians as compared to marsupials, to encompass neighboring loci. Furthermore, we also see an increase in the complexity of regulatory mechanisms. In many instances this involves the expression of long noncoding antisense RNA transcripts such as *KCNQ1OT1* that regulates imprinting of the *CDKN1C* gene. Interestingly, *KCNQ1OT1* is conserved and expressed in marsupials, but *CDKN1C* is not imprinted suggesting further regulatory features are needed to initiate silencing of this region in eutherians.

The stringency of imprinting is also increased in eutherian mammals as compared to marsupials. The majority of marsupial imprinted genes lack differential methylation and show biased expression rather than complete silencing from one allele. This is in contrast to eutherians where DMRs are a feature of imprinted regions. Interestingly the only two marsupial imprinted genes that show complete silencing of one allele are also the only two associated with DMRs. These data suggest that not all genomic imprinting is methylation based, and that the evolution of DMRs, in at least some clusters, evolved after the initial imprint was attracted to the region. Selection for complete silencing and the evolution of DMRs was greatest in the eutherian lineage, again consistent with the parental conflict/kinship hypotheses.

In addition to providing empirical evidence explaining why genomic imprinting evolved, comparative studies in mammals have also shed light on how it evolved. Examination of imprinted gene clusters across all mammals showed that the advent of imprinting was coincident with repeat expansion and retrotransposition events in many regions. The retrotransposition of *PEG10* in therian mammals provided the first direct evidence that such an event can trigger the evolution of an entire imprinted region. This was also the first direct evidence of the host defense hypothesis to explain the origins of genomic imprinting.

These studies highlight the importance of evolutionary perspectives to understanding complex genetic mechanisms. By comparing orthologous gene clusters across the three extant mammalian groups we have gained a deep insight into the evolutionary pressures that selected for genomic imprinting, as well as the mechanisms driving it.

References

1. Luo ZX, Yuan CX, Meng QJ, Ji Q (2011) A Jurassic eutherian mammal and divergence of marsupials and placentals. Nature 476: 442–445
2. Renfree MB, Papenfuss AT, Shaw G, Pask AJ (2009) Eggs, embryos and the evolution of imprinting: insights from the platypus genome. Reprod Fertil Dev 21:935–942
3. Radford EJ, Ferron SR, Ferguson-Smith AC (2011) Genomic imprinting as an adaptative model of developmental plasticity. FEBS Lett 585:2059–2066
4. Hirasawa R, Feil R (2010) Genomic imprinting and human disease. Essays Biochem 48: 187–200
5. Haig D (2004) Genomic imprinting and kinship: how good is the evidence? Annu Rev Genet 38:553–585
6. Moore T, Haig D (1991) Genomic imprinting in mammalian development: a parental tug-of-war. Trends Genet 7:45–49
7. Fowden AL, Sibley C, Reik W, Constancia M (2006) Imprinted genes, placental development and fetal growth. Horm Res 65(Suppl 3):50–58
8. Surani MA, Barton SC, Norris ML (1984) Development of reconstituted mouse eggs suggests imprinting of the genome during gametogenesis. Nature 308:548–550
9. Smith SM, Mefford M, Sodora D, Klase Z, Singh M, Alexander N, Hess D, Marx PA (2004) Topical estrogen protects against SIV vaginal transmission without evidence of systemic effect. AIDS 18:1637–1643
10. Dunzinger U, Nanda I, Schmid M, Haaf T, Zechner U (2005) Chicken orthologues of mammalian imprinted genes are clustered on macrochromosomes and replicate asynchronously. Trends Genet 21:488–492
11. Edwards CA, Mungall AJ, Matthews L, Ryder E, Gray DJ, Pask AJ, Shaw G, Graves JA, Rogers J, Dunham I, Renfree MB, Ferguson-Smith AC (2008) The evolution of the DLK1-DIO3 imprinted domain in mammals. PLoS Biol 6:e135
12. Evans HK, Weidman JR, Cowley DO, Jirtle RL (2005) Comparative phylogenetic analysis of blcap/nnat reveals eutherian-specific imprinted gene. Mol Biol Evol 22:1740–1748
13. Killian JK, Byrd JC, Jirtle JV, Munday BL, Stoskopf MK, MacDonald RG, Jirtle RL (2000) M6P/IGF2R imprinting evolution in mammals. Mol Cell 5:707–716
14. Killian JK, Nolan CM, Stewart N, Munday BL, Andersen NA, Nicol S, Jirtle RL (2001) Monotreme IGF2 expression and ancestral origin of genomic imprinting. J Exp Zool 291: 205–212

15. Rapkins RW, Hore T, Smithwick M, Ager E, Pask AJ, Renfree MB, Kohn M, Hameister H, Nicholls RD, Deakin JE, Graves JA (2006) Recent assembly of an imprinted domain from non-imprinted components. PLoS Genet 2:e182
16. Smits G, Mungall AJ, Griffiths-Jones S, Smith P, Beury D, Matthews L, Rogers J, Pask AJ, Shaw G, VandeBerg JL, McCarrey JR, Renfree MB, Reik W, Dunham I (2008) Conservation of the H19 noncoding RNA and H19-IGF2 imprinting mechanism in therians. Nat Genet 40:971–976
17. Suzuki S, Ono R, Narita T, Pask AJ, Shaw G, Wang C, Kohda T, Alsop AE, Marshall Graves JA, Kohara Y, Ishino F, Renfree MB, Kaneko-Ishino T (2007) Retrotransposon silencing by DNA methylation can drive mammalian genomic imprinting. PLoS Genet 3:e55
18. Suzuki S, Renfree MB, Pask AJ, Shaw G, Kobayashi S, Kohda T, Kaneko-Ishino T, Ishino F (2005) Genomic imprinting of IGF2, p57(KIP2) and PEG1/MEST in a marsupial, the tammar wallaby. Mech Dev 122:213–222
19. Suzuki S, Shaw G, Kaneko-Ishino T, Ishino F, Renfree MB (2011) Characterisation of marsupial PHLDA2 reveals eutherian specific acquisition of imprinting. BMC Evol Biol 11:244
20. Weidman JR, Dolinoy DC, Maloney KA, Cheng JF, Jirtle RL (2006) Imprinting of opossum Igf2r in the absence of differential methylation and air. Epigenetics 1:49–54
21. Weidman JR, Maloney KA, Jirtle RL (2006) Comparative phylogenetic analysis reveals multiple non-imprinted isoforms of opossum Dlk1. Mamm Genome 17:157–167
22. Barlow DP, Stoger R, Herrmann BG, Saito K, Schweifer N (1991) The mouse insulin-like growth factor type-2 receptor is imprinted and closely linked to the Tme locus. Nature 349:84–87
23. Lyle R, Watanabe D, te Vruchte D, Lerchner W, Smrzka OW, Wutz A, Schageman J, Hahner L, Davies C, Barlow DP (2000) The imprinted antisense RNA at the Igf2r locus overlaps but does not imprint Mas1. Nat Genet 25:19–21
24. Muller B, Hedrich K, Kock N, Dragasevic N, Svetel M, Garrels J, Landt O, Nitschke M, Pramstaller PP, Reik W, Schwinger E, Sperner J, Ozelius L, Kostic V, Klein C (2002) Evidence that paternal expression of the epsilon-sarcoglycan gene accounts for reduced penetrance in myoclonus-dystonia. Am J Hum Genet 71: 1303–1311
25. Ono R, Nakamura K, Inoue K, Naruse M, Usami T, Wakisaka-Saito N, Hino T, Suzuki-Migishima R, Ogonuki N, Miki H, Kohda T, Ogura A, Yokoyama M, Kaneko-Ishino T, Ishino F (2006) Deletion of Peg10, an imprinted gene acquired from a retrotransposon, causes early embryonic lethality. Nat Genet 38:101–106
26. Nakabayashi K, Makino S, Minagawa S, Smith AC, Bamforth JS, Stanier P, Preece M, Parker-Katiraee L, Paton T, Oshimura M, Mill P, Yoshikawa Y, Hui CC, Monk D, Moore GE, Scherer SW (2004) Genomic imprinting of PPP1R9A encoding neurabin I in skeletal muscle and extra-embryonic tissues. J Med Genet 41:601–608
27. Mizuno Y, Sotomaru Y, Katsuzawa Y, Kono T, Meguro M, Oshimura M, Kawai J, Tomaru Y, Kiyosawa H, Nikaido I, Amanuma H, Hayashizaki Y, Okazaki Y (2002) Asb4, Ata3, and Dcn are novel imprinted genes identified by high-throughput screening using RIKEN cDNA microarray. Biochem Biophys Res Commun 290:1499–1505
28. Kaneko-Ishino T, Kuroiwa Y, Miyoshi N, Kohda T, Suzuki R, Yokoyama M, Viville S, Barton SC, Ishino F, Surani MA (1995) Peg1/Mest imprinted gene on chromosome 6 identified by cDNA subtraction hybridization. Nat Genet 11:52–59
29. Zhang Y, Shields T, Crenshaw T, Hao Y, Moulton T, Tycko B (1993) Imprinting of human H19: allele-specific CpG methylation, loss of the active allele in Wilms tumor, and potential for somatic allele switching. Am J Hum Genet 53:113–124
30. DeChiara TM, Robertson EJ, Efstratiadis A (1991) Parental imprinting of the mouse insulin-like growth factor II gene. Cell 64: 849–859
31. O'Neill MJ, Ingram RS, Vrana PB, Tilghman SM (2000) Allelic expression of IGF2 in marsupials and birds. Dev Genes Evol 210:18–20
32. Moore GE, Abu-Amero SN, Bell G, Wakeling EL, Kingsnorth A, Stanier P, Jauniaux E, Bennett ST (2001) Evidence that insulin is imprinted in the human yolk sac. Diabetes 50:199–203
33. Ager E, Suzuki S, Pask A, Shaw G, Ishino F, Renfree MB (2007) Insulin is imprinted in the placenta of the marsupial, Macropus eugenii. Dev Biol 309:317–328
34. Hatada I, Mukai T (1995) Genomic imprinting of p57KIP2, a cyclin-dependent kinase inhibitor, in mouse. Nat Genet 11:204–206
35. Ager EI, Pask AJ, Gehring HM, Shaw G, Renfree MB (2008) Evolution of the CDKN1C-KCNQ1 imprinted domain. BMC Evol Biol 8:163
36. Salas M, John R, Saxena A, Barton S, Frank D, Fitzpatrick G, Higgins MJ, Tycko B (2004) Placental growth retardation due to loss of

imprinting of Phlda2. Mech Dev 121: 1199–1210

37. Wylie AA, Murphy SK, Orton TC, Jirtle RL (2000) Novel imprinted DLK1/GTL2 domain on human chromosome 14 contains motifs that mimic those implicated in IGF2/H19 regulation. Genome Res 10:1711–1718
38. Miyoshi N, Wagatsuma H, Wakana S, Shiroishi T, Nomura M, Aisaka K, Kohda T, Surani MA, Kaneko-Ishino T, Ishino F (2000) Identification of an imprinted gene, Meg3/Gtl2 and its human homologue MEG3, first mapped on mouse distal chromosome 12 and human chromosome 14q. Genes Cells 5:211–220
39. Seitz H, Youngson N, Lin SP, Dalbert S, Paulsen M, Bachellerie JP, Ferguson-Smith AC, Cavaille J (2003) Imprinted microRNA genes transcribed antisense to a reciprocally imprinted retrotransposon-like gene. Nat Genet 34: 261–262
40. Tsai CE, Lin SP, Ito M, Takagi N, Takada S, Ferguson-Smith AC (2002) Genomic imprinting contributes to thyroid hormone metabolism in the mouse embryo. Curr Biol 12: 1221–1226
41. Leff SE, Brannan CI, Reed ML, Ozcelik T, Francke U, Copeland NG, Jenkins NA (1992) Maternal imprinting of the mouse Snrpn gene and conserved linkage homology with the human Prader-Willi syndrome region. Nat Genet 2:259–264
42. Herzing LB, Cook EH Jr, Ledbetter DH (2002) Allele-specific expression analysis by RNA-FISH demonstrates preferential maternal expression of UBE3A and imprint maintenance within 15q11–q13 duplications. Hum Mol Genet 11:1707–1718
43. Kagitani F, Kuroiwa Y, Wakana S, Shiroishi T, Miyoshi N, Kobayashi S, Nishida M, Kohda T, Kaneko-Ishino T, Ishino F (1997) Peg5/Neuronatin is an imprinted gene located on sub-distal chromosome 2 in the mouse. Nucleic Acids Res 25:3428–3432
44. Barlow DP (1993) Methylation and imprinting: from host defense to gene regulation? Science 260:309–310
45. McDonald JF, Matzke MA, Matzke AJ (2005) Host defenses to transposable elements and the evolution of genomic imprinting. Cytogenet Genome Res 110:242–249
46. Warren WC, Hillier LW, Marshall Graves JA, Birney E, Ponting CP, Grutzner F, Belov K, Miller W, Clarke L, Chinwalla AT, Yang SP, Heger A, Locke DP, Miethke P, Waters PD, Veyrunes F, Fulton L, Fulton B, Graves T, Wallis J, Puente XS, Lopez-Otin C, Ordonez GR, Eichler EE, Chen L, Cheng Z, Deakin JE, Alsop A, Thompson K, Kirby P, Papenfuss AT, Wakefield MJ, Olender T, Lancet D, Huttley GA, Smit AF, Pask A, Temple-Smith P, Batzer MA, Walker JA, Konkel MK, Harris RS, Whittington CM, Wong ES, Gemmell NJ, Buschiazzo E, Vargas Jentzsch IM, Merkel A, Schmitz J, Zemann A, Churakov G, Kriegs JO, Brosius J, Murchison EP, Sachidanandam R, Smith C, Hannon GJ, Tsend-Ayush E, McMillan D, Attenborough R, Rens W, Ferguson-Smith M, Lefevre CM, Sharp JA, Nicholas KR, Ray DA, Kube M, Reinhardt R, Pringle TH, Taylor J, Jones RC, Nixon B, Dacheux JL, Niwa H, Sekita Y, Huang X, Stark A, Kheradpour P, Kellis M, Flicek P, Chen Y, Webber C, Hardison R, Nelson J, Hallsworth-Pepin K, Delehaunty K, Markovic C, Minx P, Feng Y, Kremitzki C, Mitreva M, Glasscock J, Wylie T, Wohldmann P, Thiru P, Nhan MN, Pohl CS, Smith SM, Hou S, Nefedov M, de Jong PJ, Renfree MB, Mardis ER, Wilson RK (2008) Genome analysis of the platypus reveals unique signatures of evolution. Nature 453: 175–183
47. Mikkelsen TS, Wakefield MJ, Aken B, Amemiya CT, Chang JL, Duke S, Garber M, Gentles AJ, Goodstadt L, Heger A, Jurka J, Kamal M, Mauceli E, Searle SM, Sharpe T, Baker ML, Batzer MA, Benos PV, Belov K, Clamp M, Cook A, Cuff J, Das R, Davidow L, Deakin JE, Fazzari MJ, Glass JL, Grabherr M, Greally JM, Gu W, Hore TA, Huttley GA, Kleber M, Jirtle RL, Koina E, Lee JT, Mahony S, Marra MA, Miller RD, Nicholls RD, Oda M, Papenfuss AT, Parra ZE, Pollock DD, Ray DA, Schein JE, Speed TP, Thompson K, VandeBerg JL, Wade CM, Walker JA, Waters PD, Webber C, Weidman JR, Xie X, Zody MC, Graves JA, Ponting CP, Breen M, Samollow PB, Lander ES, Lindblad-Toh K (2007) Genome of the marsupial Monodelphis domestica reveals innovation in non-coding sequences. Nature 447: 167–177
48. Renfree MB, Papenfuss AT, Deakin JE, Lindsay J, Heider T, Belov K, Rens W, Waters PD, Pharo EA, Shaw G, Wong ES, Lefevre CM, Nicholas KR, Kuroki Y, Wakefield MJ, Zenger KR, Wang C, Ferguson-Smith M, Nicholas FW, Hickford D, Yu H, Short KR, Siddle HV, Frankenberg SR, Chew KY, Menzies BR, Stringer JM, Suzuki S, Hore TA, Delbridge ML, Mohammadi A, Schneider NY, Hu Y, O'Hara W, Al Nadaf S, Wu C, Feng ZP, Cocks BG, Wang J, Flicek P, Searle SM, Fairley S, Beal K, Herrero J, Carone DM, Suzuki Y, Sagano S, Toyoda A, Sakaki Y, Kondo S, Nishida Y, Tatsumoto S, Mandiou I, Hsu A, McColl KA, Landsell B, Weinstock G, Kuczek E, McGrath A, Wilson P, Men A, Hazar-Rethinam M, Hall A, Davies J, Wood D, Williams S,

Sundaravadanam Y, Muzny DM, Jhangiani SN, Lewis LR, Morgan MB, Okwuonu GO, Ruiz SJ, Santibanez J, Nazareth L, Cree A, Fowler G, Kovar CL, Dinh HH, Joshi V, Jing C, Lara F, Thornton R, Chen L, Deng J, Liu Y, Shen JY, Song XZ, Edson J, Troon C, Thomas D, Stephens A, Yapa L, Levchenko T, Gibbs RA, Cooper DW, Speed TP, Fujiyama A, Graves JA, O'Neill RJ, Pask AJ, Forrest SM, Worley KC (2011) Genome sequence of an Australian kangaroo, Macropus eugenii, provides insight into the evolution of mammalian reproduction and development. Genome Biol 12:R81

49. Waterston RH, Lindblad-Toh K, Birney E, Rogers J, Abril JF, Agarwal P, Agarwala R, Ainscough R, Alexandersson M, An P, Antonarakis SE, Attwood J, Baertsch R, Bailey J, Barlow K, Beck S, Berry E, Birren B, Bloom T, Bork P, Botcherby M, Bray N, Brent MR, Brown DG, Brown SD, Bult C, Burton J, Butler J, Campbell RD, Carninci P, Cawley S, Chiaromonte F, Chinwalla AT, Church DM, Clamp M, Clee C, Collins FS, Cook LL, Copley RR, Coulson A, Couronne O, Cuff J, Curwen V, Cutts T, Daly M, David R, Davies J, Delehaunty KD, Deri J, Dermitzakis ET, Dewey C, Dickens NJ, Diekhans M, Dodge S, Dubchak I, Dunn DM, Eddy SR, Elnitski L, Emes RD, Eswara P, Eyras E, Felsenfeld A, Fewell GA, Flicek P, Foley K, Frankel WN, Fulton LA, Fulton RS, Furey TS, Gage D, Gibbs RA, Glusman G, Gnerre S, Goldman N, Goodstadt L, Grafham D, Graves TA, Green ED, Gregory S, Guigo R, Guyer M, Hardison RC, Haussler D, Hayashizaki Y, Hillier LW, Hinrichs A, Hlavina W, Holzer T, Hsu F, Hua A, Hubbard T, Hunt A, Jackson I, Jaffe DB, Johnson LS, Jones M, Jones TA, Joy A, Kamal M, Karlsson EK, Karolchik D, Kasprzyk A, Kawai J, Keibler E, Kells C, Kent WJ, Kirby A, Kolbe DL, Korf I, Kucherlapati RS, Kulbokas EJ, Kulp D, Landers T, Leger JP, Leonard S, Letunic I, Levine R, Li J, Li M, Lloyd C, Lucas S, Ma B, Maglott DR, Mardis ER, Matthews L, Mauceli E, Mayer JH, McCarthy M, McCombie WR, McLaren S, McLay K, McPherson JD, Meldrim J, Meredith B, Mesirov JP, Miller W, Miner TL, Mongin E, Montgomery KT, Morgan M, Mott R, Mullikin JC, Muzny DM, Nash WE, Nelson JO, Nhan MN, Nicol R, Ning Z, Nusbaum C, O'Connor MJ, Okazaki Y, Oliver K, Overton-Larty E, Pachter L, Parra G, Pepin KH, Peterson J, Pevzner P, Plumb R, Pohl CS, Poliakov A, Ponce TC, Ponting CP, Potter S, Quail M, Reymond A, Roe BA, Roskin KM, Rubin EM, Rust AG, Santos R, Sapojnikov V, Schultz B, Schultz J, Schwartz MS, Schwartz S, Scott C, Seaman S, Searle S, Sharpe T, Sheridan A, Shownkeen R, Sims S, Singer JB, Slater G, Smit A, Smith DR, Spencer B, Stabenau A, Stange-Thomann N, Sugnet C, Suyama M, Tesler G, Thompson J, Torrents D, Trevaskis E, Tromp J, Ucla C, Ureta-Vidal A, Vinson JP, Von Niederhausern AC, Wade CM, Wall M, Weber RJ, Weiss RB, Wendl MC, West AP, Wetterstrand K, Wheeler R, Whelan S, Wierzbowski J, Willey D, Williams S, Wilson RK, Winter E, Worley KC, Wyman D, Yang S, Yang SP, Zdobnov EM, Zody MC, Lander ES (2002) Initial sequencing and comparative analysis of the mouse genome. Nature 420:520–562

50. Venter JC, Adams MD, Myers EW, Li PW, Mural RJ, Sutton GG, Smith HO, Yandell M, Evans CA, Holt RA, Gocayne JD, Amanatides P, Ballew RM, Huson DH, Wortman JR, Zhang Q, Kodira CD, Zheng XH, Chen L, Skupski M, Subramanian G, Thomas PD, Zhang J, Gabor Miklos GL, Nelson C, Broder S, Clark AG, Nadeau J, McKusick VA, Zinder N, Levine AJ, Roberts RJ, Simon M, Slayman C, Hunkapiller M, Bolanos R, Delcher A, Dew I, Fasulo D, Flanigan M, Florea L, Halpern A, Hannenhalli S, Kravitz S, Levy S, Mobarry C, Reinert K, Remington K, Abu-Threideh J, Beasley E, Biddick K, Bonazzi V, Brandon R, Cargill M, Chandramouliswaran I, Charlab R, Chaturvedi K, Deng Z, Di Francesco V, Dunn P, Eilbeck K, Evangelista C, Gabrielian AE, Gan W, Ge W, Gong F, Gu Z, Guan P, Heiman TJ, Higgins ME, Ji RR, Ke Z, Ketchum KA, Lai Z, Lei Y, Li Z, Li J, Liang Y, Lin X, Lu F, Merkulov GV, Milshina N, Moore HM, Naik AK, Narayan VA, Neelam B, Nusskern D, Rusch DB, Salzberg S, Shao W, Shue B, Sun J, Wang Z, Wang A, Wang X, Wang J, Wei M, Wides R, Xiao C, Yan C, Yao A, Ye J, Zhan M, Zhang W, Zhang H, Zhao Q, Zheng L, Zhong F, Zhong W, Zhu S, Zhao S, Gilbert D, Baumhueter S, Spier G, Carter C, Cravchik A, Woodage T, Ali F, An H, Awe A, Baldwin D, Baden H, Barnstead M, Barrow I, Beeson K, Busam D, Carver A, Center A, Cheng ML, Curry L, Danaher S, Davenport L, Desilets R, Dietz S, Dodson K, Doup L, Ferriera S, Garg N, Gluecksmann A, Hart B, Haynes J, Haynes C, Heiner C, Hladun S, Hostin D, Houck J, Howland T, Ibegwam C, Johnson J, Kalush F, Kline L, Koduru S, Love A, Mann F, May D, McCawley S, McIntosh T, McMullen I, Moy M, Moy L, Murphy B, Nelson K, Pfannkoch C, Pratts E, Puri V, Qureshi H, Reardon M, Rodriguez R, Rogers YH, Romblad D, Ruhfel B, Scott R, Sitter C, Smallwood M, Stewart E, Strong R, Suh E, Thomas R, Tint NN, Tse S, Vech C, Wang G, Wetter J, Williams S, Williams M, Windsor S, Winn-Deen E, Wolfe K, Zaveri J, Zaveri K, Abril JF, Guigo R, Campbell MJ,

Sjolander KV, Karlak B, Kejariwal A, Mi H, Lazareva B, Hatton T, Narechania A, Diemer K, Muruganujan A, Guo N, Sato S, Bafna V, Istrail S, Lippert R, Schwartz R, Walenz B, Yooseph S, Allen D, Basu A, Baxendale J, Blick L, Caminha M, Carnes-Stine J, Caulk P, Chiang YH, Coyne M, Dahlke C, Mays A, Dombroski M, Donnelly M, Ely D, Esparham S, Fosler C, Gire H, Glanowski S, Glasser K, Glodek A, Gorokhov M, Graham K, Gropman B, Harris M, Heil J, Henderson S, Hoover J, Jennings D, Jordan C, Jordan J, Kasha J, Kagan L, Kraft C, Levitsky A, Lewis M, Liu X, Lopez J, Ma D, Majoros W, McDaniel J, Murphy S, Newman M, Nguyen T, Nguyen N, Nodell M, Pan S, Peck J, Peterson M, Rowe W, Sanders R, Scott J, Simpson M, Smith T, Sprague A, Stockwell T, Turner R, Venter E, Wang M, Wen M, Wu D, Wu M, Xia A, Zandieh A, Zhu X (2001) The sequence of the human genome. Science 291:1304–1351

51. Pask AJ, Papenfuss AT, Ager EI, McColl KA, Speed TP, Renfree MB (2009) Analysis of the platypus genome suggests a transposon origin for mammalian imprinting. Genome Biol 10:R1
52. Tyndale-Biscoe CH, Renfree MB (1987) Reproductive physiology of marsupials. Cambridge University Press, Cambridge
53. Giddings SJ, King CD, Harman KW, Flood JF, Carnaghi LR (1994) Allele specific inactivation of insulin 1 and 2, in the mouse yolk sac, indicates imprinting. Nat Genet 6:310–313
54. Mann JR, Szabo PE, Reed MR, Singer-Sam J (2000) Methylated DNA sequences in genomic imprinting. Crit Rev Eukaryot Gene Expr 10:241–257
55. Obata Y, Kaneko-Ishino T, Koide T, Takai Y, Ueda T, Domeki I, Shiroishi T, Ishino F, Kono T (1998) Disruption of primary imprinting during oocyte growth leads to the modified expression of imprinted genes during embryogenesis. Development 125:1553–1560
56. Ager EI, Pask AJ, Shaw G, Renfree MB (2008) Expression and protein localisation of IGF2 in the marsupial placenta. BMC Dev Biol 8:17
57. Arima T, Kamikihara T, Hayashida T, Kato K, Inoue T, Shirayoshi Y, Oshimura M, Soejima H, Mukai T, Wake N (2005) ZAC, LIT1 (KCNQ1OT1) and p57KIP2 (CDKN1C) are in an imprinted gene network that may play a role in Beckwith-Wiedemann syndrome. Nucleic Acids Res 33:2650–2660
58. Frank D, Fortino W, Clark L, Musalo R, Wang W, Saxena A, Li CM, Reik W, Ludwig T, Tycko B (2002) Placental overgrowth in mice lacking the imprinted gene Ipl. Proc Natl Acad Sci USA 99:7490–7495
59. Runte M, Huttenhofer A, Gross S, Kiefmann M, Horsthemke B, Buiting K (2001) The IC-SNURF-SNRPN transcript serves as a host for multiple small nucleolar RNA species and as an antisense RNA for UBE3A. Hum Mol Genet 10:2687–2700
60. da Rocha ST, Edwards CA, Ito M, Ogata T, Ferguson-Smith AC (2008) Genomic imprinting at the mammalian Dlk1-Dio3 domain. Trends Genet 24:306–316

Chapter 19

Nonmammalian Parent-of-Origin Effects

Elena de la Casa-Esperón

Abstract

Chromosomes acquire different epigenetic marks during oogenesis and spermatogenesis. After fertilization, if retained and selected, these differences may result in imprinting effects. Rather than being an oddity, imprinting effects have been found in many sexually reproducing organisms. Interestingly, imprinting can result in disparate effects under different selective forces. At the same time, epigenetic mechanisms and selective pressures shared by sexually reproducing organisms could underlie common imprinting effects. Large-scale studies are revealing that parent-of-origin effects are more common than previously thought and supporting the important contribution of imprinting to many traits and diseases.

Key words: Parent-of-origin effects, Parental origin effects, Imprinting, Allelic expression, Chromosome elimination, Epigenetic reprogramming

"A chromosome which passes through the male germ line acquires an "imprint" which will result in behavior exactly opposite to the "imprint" conferred on the same chromosome by the female germ line. In other words, the "imprint" a chromosome bears is unrelated to the genic constitution of the chromosome and is determined only by the sex of the germ line through which the chromosome has been inherited"—Helen Crouse, 1960.

1. Introduction: Imprinting, a Form of Parent-of-Origin Effect

Parent-of-origin effects (also parental origin effects, POEs) comprise a broad range of phenomena that result from the different influence of each parent on the offspring; therefore, POEs are caused by sexual differences between the parents. The best-known form of POE is imprinting. The term "imprint" was coined by Helen Crouse to describe the differential marking of maternal and paternal chromosomes (1). It refers to a reversible mark of epigenetic nature that is differentially established during oogenesis and spermatogenesis, and transmitted to the offspring. By extension, the term "imprinting" has also been applied to those POEs derived

Nora Engel (ed.), *Genomic Imprinting: Methods and Protocols*, Methods in Molecular Biology, vol. 925,
DOI 10.1007/978-1-62703-011-3_19,

from the different chromatin properties of maternally and paternally inherited chromosomes, which includes POEs affecting gene expression, chromosome segregation, heterochromatinization and a variety of other chromosomal functions. These disparate POEs are called "imprinting" in their respective fields, creating some confusion across them. For instance, mammalian researchers often restrict the use of the term "imprinting" to imprinted gene expression and their associated epigenetic marks. This form of imprinting, in which expression of one of the two copies of a gene is repressed in a parental origin dependent manner, has been the focus of most imprinting studies in mammals and plants. Only a few studies have explored other imprinting phenomena that occur within the same species. In contrast, analyses of other organisms have revealed a broad diversity of effects caused by differences between the maternally and paternally inherited genomes. Interestingly, the phenomena affected by imprinting in these species can also be imprinted in mammals, and the lessons learned from imprinting studies in one species have the potential of revealing new insights about imprinting phenomena in others. Therefore, the aim of this review is to go beyond the classical imprinting studies in mammals, in order to provide an overview of the diversity and the distribution of imprinting effects, as well as to discuss their implications.

Not all POEs are imprinting effects: POEs also comprise other phenomena, such as maternal effects, which result from the influence of the maternally provided environment on the phenotype of her offspring (2). They include effects derived from maternal transcripts present in the egg, maternal nursing behavior, etc. and, therefore, they are very common. Although it is beyond the scope of this review to summarize the broad spectrum of maternal (and paternal) effects identified to date, it is important to notice that they can be confused with effects derived of imprinted gene expression. Consequently, several strategies have been proposed to distinguish diverse types of POEs (3–5).

For the purpose of this review, the term "imprinting" is restricted to the POEs associated with parent-of-origin dependent epigenetic imprints (1), while those POEs unrelated to imprinting (such as maternal effects) will not be discussed in depth. An extensive analysis reveals that a plethora of imprinting effects and variants occurs in sexually reproducing organisms, although many of them display recurrent themes. Moreover, diverse forms of imprinting can occur within the same organism. Imprinting diversity will be discussed by grouping the effects according to the affected function: gene expression, whole chromosome heterochromatinization and/or elimination, chromosomal interactions and replication, epigenetic marks and other phenotypes. Understanding imprinting diversity and complexity could, in turn, help us to approach the analysis of many complex phenotypes from a broader perspective and may unveil novel phenomena that result from the simple fact that we have a mom and a dad.

2. Imprinting Effects in Expression Can Affect Single Genes, Entire Chromosomes and Relocated Genes

Although a large body of imprinting studies (particularly in mammals and plants) have been performed on the expression of individual genes, whether they are isolated or clustered in imprinted domains (6, 7), imprinting can also affect the transcription of entire chromosomes, as well as rearranged genes and transgenes upon exposure to foreign chromatin. In insects, imprinting of whole chromosome gene expression has been observed in several species: condensation and transcriptional inactivation of the entire paternal set occurs in males of lecanoid coccids (mealybugs) and *Hypothenemus hampei* (coffee berry borer) (8, 9). In *Drosophila*, paternal transmission of a *Dp(1;f)LJ9* mini-X chromosome results in transcriptional silencing of more than a hundred genes (10, 11). Chromosome-wide imprinting is not exclusive of insects: in female mouse extraembryonic tissues (and possibly brain) and marsupial somatic cells, the paternal X chromosome is preferentially condensed and transcriptionally silenced (imprinted X-chromosome inactivation) (12–14). In addition, POEs have been observed during the X-chromosome choice process that precedes random X-inactivation in mouse embryonic cell lineages (15).

Unlike mammals and plants, in which there are many reports of endogenous imprinted genes (http://www.har.mrc.ac.uk/research/genomic_imprinting/) (7), imprinting effects on gene expression have not been found in karyotypically normal *Drosophila*. In fruit flies, imprinting effects on expression are restricted to rearranged genes and transgenes that fall under the repressive influence of heterochromatic regions (16). Spreading of this heterochromatin to nearby regions results in patchy silencing of genes located in the vicinity, a phenomenon known as position effect variegation (PEV). Silencing of the mini-X chromosome genes is an example of imprinted PEV, caused by rearrangements and deletions that relocated genes under the effect of centric heterochromatin (11). *Ultraabdominal^1* (*Uab^1*) inversion, a bithorax-complex rearrangement, results in a mutant phenotype that is only observed when paternally transmitted (17). PEVs of multiple transgene insertions in the heterochromatic Y chromosome are also subject to POEs, implying that imprinting is a general property of the *Drosophila* Y chromosome (18) (unlike other species, the Y chromosome is necessary for *Drosophila* male fertility, but it does not determine sex; therefore, it can be transmitted through both males and females). Y chromosome imprinting can also affect PEV in other chromosomes: transmission of the Y chromosome through a *mod(mgd4)* male (a mutant of the *mod(mgd4)/E(var)3-93D* enhancer of PEV) enhances eye color variegation of the X-linked *w^{m4h}* allele (a *white* locus juxtaposed to centric X heterochromatin).

However, and unlike typical imprinting, this Y chromosome effect is not reset in the germ line and persists over several generations, even when the Y chromosome is transmitted through wild type flies (19).

Imprinted transgene expression has also been observed in other organisms, such as mice and worms (20, 21). In *Caenorhabditis elegans*, paternal transmission of transgenes results in greater expression than maternal transmission. This imprint is acquired through both male (X0) and hermaphrodite (XX) spermatogenesis, resulting in similar transgene expression levels, and is reset by passage through oogenesis (21). But unlike other species in which imprinting is considered to be fully reset during gametogenesis in every generation, germ-line imprint erasure does not seem to be complete in *C. elegans*. After transmission of the transgene through the same germ line for several generations, expression resetting requires multiple passages through the opposite germ line (21).

In summary, both genome-wide and individual gene studies have shown that imprinting can affect gene expression in many different ways and species, being more common than previously thought. Moreover, expression analyses of transgenes and rearrangements in *C. elegans* and *Drosophila* have revealed that, even in the absence of endogenous imprinted genes, many chromosomal regions are capable of generating imprinted expression upon insertion of transgenes. This implies that such regions have parent-of-origin dependent chromatin properties, which may be involved in imprinting effects other than gene expression.

3. Imprinting Effects in Chromosome Segregation and Elimination

Paternal and maternal chromosomes may behave differently during mitotic and meiotic segregation. These imprinting effects can be dramatic and lead to chromosome elimination, or more subtle and result in preferential segregation (transmission ratio distortion) of specific chromosomes. Proper chromosome segregation depends on centromeric integrity and telomere organization, as well as on chromatin condensation, meiotic pairing and recombination (which will be discussed in later sections), and, as we will see, imprinting effects have been observed in all of them.

Although the term "imprinting" is often restricted to the parent-of-origin dependent monoallelic expression of genes, it was first coined by Helen Crouse (1) to explain a different phenomenon that was also influenced by the sex of the transmitting parent: in sciarid flies, she observed that paternal X chromosomes are selectively eliminated during early embryonic stages (one chromosome in females and two in males) and in germ cells (one chromosome in both sexes). In addition, all paternal chromosomes are discarded during meiosis in males (1, 22). Chromosome elimination of the entire paternal set is also observed in other invertebrates and, in several cases, such elimination is part of the sex determination process. In diaspidid coccids, sex determination depends on the elimination of the entire paternal set while, in lecanoid coccids (including mealybugs), the paternal chromosomes are heterochromatized and silenced during the cleavage stage of embryogenesis. Later on, during spermatogenesis, these paternal chromosomes are also eliminated in lecanoids (9). Similarly, paternal chromosome condensation in male somatic cells and elimination during spermatogenesis is observed in the beetle *Hypothenemus hampei*, the coffee berry borer (8). In the wasp *Nasonia vitripennis*, males are haploid and develop parthenogenetically from unfertilized eggs, while fertilized eggs render diploid females. However, males can develop from diploid eggs after abnormal paternal genome condensation and elimination caused by *Wolbachia*-induced cytoplasmic incompatibility or by the effect of an extranumerary paternal sex ratio (PSR) chromosome. *Wolbachia* is an intracellular bacterium that causes cytoplasmic incompatibility in insects; in *Nasonia* embryos of uninfected females mated with infected males, it triggers paternal chromosomal loss (23). POEs can also be caused by the presence of a PSR chromosome in *Nasonia* sperm, which results in condensation and subsequent loss of all paternal chromosomes except itself; PSR chromosomes are also present in other hymenoptera with similar POEs (24–26).

Imprinting in chromosome exclusion also occurs in *C. elegans*: in XX larvae grown in specific bacterial metabolites, a single X chromosome (preferentially the paternal one) occasionally suffers nondisjunction and is lost, resulting in X0 males (27); X chromosomal loss has been observed in larvae from X0 fathers, but not from self-fertilized hermaphrodites grown in the same conditions. However, discarding paternally inherited chromosomes is not the universal rule: several species of *Corbicula*, a mollusc, can undergo spontaneous androgenesis by female pronucleus extrusion in the zygote, thus eliminating the entire maternal chromosomal set (28, 29).

In *Drosophila*, several mutations also show POEs on chromosome elimination in zygotes and early embryonic stages: maternal chromosomes are lost in *ncd* (*nonclaret disjunctional*) mutants, whereas *pal* (*paternal loss*) and *Horka*^{D} mutants eliminate paternal ones (30–33). In addition, paternal chromosome loss due to telomere fusion has been observed in *Drosophila* embryos fathered by *k81* mutants. K81 is a telomeric protein that protects *Drosophila* telomeres during spermatogenesis and constitutes a paternal mark that is required for the functional reestablishment of the paternal chromosomes after fertilization (34). Therefore, the underlying cause of these mutation-associated POEs appears to reside in the differential contribution of the mutated gene products to the processing of the chromosomes during spermatogenesis respect to that of oogenesis.

Imprinting can have more subtle effects on chromosome segregation than chromosome loss. Transmission ratio distortion (TRD) in favor of paternal or maternal alleles has been reported in several mouse and human loci, including imprinted regions (35–37). TRD can be the result of preferential meiotic segregation or postfertilization processes, and imprinting effects may act in both (38). While chromosome loss in insects is often associated to large chromatin differences (such as chromosome heterochromatinization and condensation) between the parental sets (8, 22), it is unclear which epigenetic marks mediate imprinted TRDs. Proper chromosome segregation depends on the centromeres and the heterochromatic domains that surround them, and thus, parent-of-origin dependent marks in these regions are candidates for mediating imprinting effects in chromosome transmission (39).

4. Imprinting Effects in Replication and Chromosomal Interactions

Imprinting effects also modulate other chromosomal functions, such as replication, chromosomal interactions during interphase and meiosis (pairing and recombination), and nuclear compartmentalization. Imprinting has very interesting effects in replication: in mammalian somatic cells, all imprinted domains tested to date replicate asynchronously, with the paternal allele replicating early and the maternal one late (40). But in *Drosophila*, imprinting of the *Dp(1;f)LJ9* mini-X chromosome affects a different aspect of replication: the extent of polytenization in somatic cells. During polytenization, the DNA is replicated multiple times without division, resulting in polyploid (polytene) chromosomes. Paternally transmitted mini-X chromosomes undergo less endoreplication than maternally transmitted ones. Underreplication, in turn, reduces copy number and, consequently, the RNA levels of many adjacent genes. In this way, *Drosophila* displays a unique form of imprinting that affects both replication and expression levels (10).

During meiosis, homologous chromosomes pair and recombine. A link between imprinting and meiotic pairing has been reported in *C. elegans*: in XX embryos, the paternal X chromosome is refractory to accumulating activating histone marks present in the rest of the chromosomes. This POE is never observed past 20-cell stage in cross-progeny from X0 males, while it lasts shorter (not past 10-cell stage) in self-progeny of XX hermaphrodites. This observation, as well as the analyses of several mutants bearing one or two X chromosomes, suggest that *C. elegans* unpaired X chromosomes acquire more repressive and stable imprints than paired X chromosomes (41). In *Drosophila*, imprinting effects in gene expression are also observed within two genomic contexts in which meiotic pairing can be deficient or absent: rearrangements and

Y-chromosomes. Meiotic unpaired DNA can sometimes acquire repressive marks -e.g., *Neurospora* meiotic silencing by unpaired DNA (MSUD) that represses the expression of homologous unpaired genes (42). These observations have led Menon and Meller (16) to propose that pairing disruption might be necessary to establish *Drosophila* germ line-specific imprints. The reciprocal, imprinting effects on meiotic processes have also been reported: analyses of meiotic crossovers in the offspring of F1 hybrid mice have shown that recombination rates can be affected by the parental origin of the chromosomes (43–45).

Aside from meiosis, imprinted regions also participate in other chromosome interactions, sometimes with a clear preference for the paternal or the maternal allele. These interactions have been postulated to contribute to the parent-of-origin dependent activity of imprinted genes (46–48). For instance, high throughput analyses have revealed that imprinted loci are overrepresented among the regions that interact with the Igf2/H19 imprinting controlling region (ICR); some of these regions preferentially associate with the maternally inherited ICR (49). These observations have lead to the concept of "imprinting interactomes," which might facilitate the regulation of the epigenetic properties (gene expression, asynchronous replication, etc.) of multiple imprinted regions in *trans* (48, 49).

Additional imprinting effects on nuclear organization and localization of certain genomic sequences have been observed in insects, as well as in mammals (50, 51). In *Sciarid* male germ line, paternal and maternal chromosome sets occupy distinct compartments in prophase I nuclei; this separation may facilitate their meiotic segregation to opposite poles and the subsequent paternal chromosome set elimination (22, 52). A transitory separation of maternal and paternal chromosomes is also observed in preimplantation mouse embryos (53).

In conclusion, studies in several invertebrate species have shown links between imprinting and chromosome replication, meiotic chromosomal interactions and nuclear compartmentalization. These processes are also affected by imprinting in mammals and, although the effects are very different, taken together they indicate that parent-of-origin epigenetic differences present through nuclear space and at critical times (replication and meiosis) have allowed their selection for very diverse functions.

5. Imprinting Affects Many Phenotypes

I have presented several examples of imprinting effects that illustrate the diversity of phenomena and species affected. Imprinting can also influence other processes (Fig. 1). A few additional examples have been selected to illustrate that: (1) different imprinting

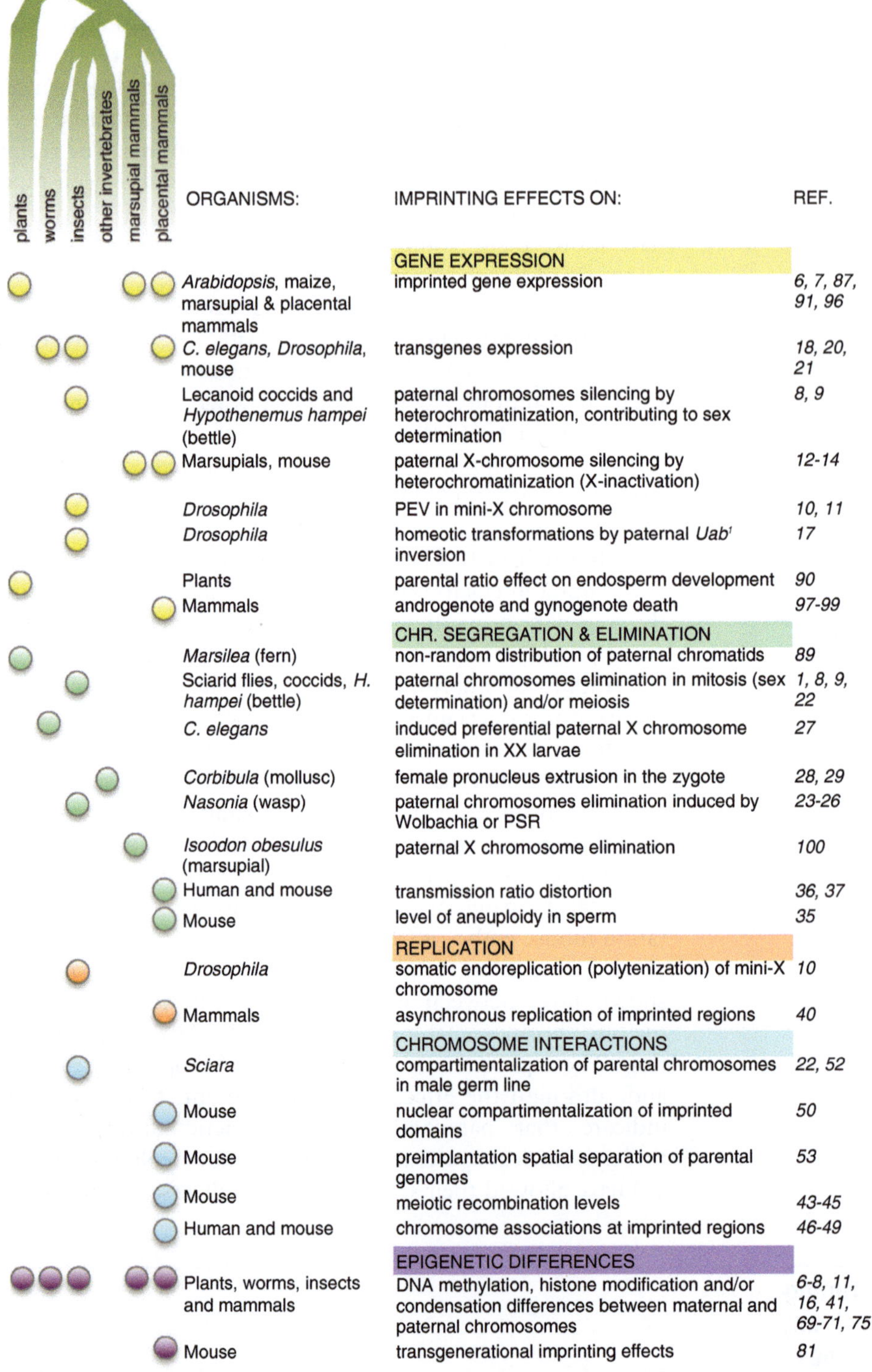

ORGANISMS:	IMPRINTING EFFECTS ON:	REF.
	GENE EXPRESSION	
Arabidopsis, maize, marsupial & placental mammals	imprinted gene expression	*6, 7, 87, 91, 96*
C. elegans, Drosophila, mouse	transgenes expression	*18, 20, 21*
Lecanoid coccids and *Hypothenemus hampei* (bettle)	paternal chromosomes silencing by heterochromatinization, contributing to sex determination	*8, 9*
Marsupials, mouse	paternal X-chromosome silencing by heterochromatinization (X-inactivation)	*12-14*
Drosophila	PEV in mini-X chromosome	*10, 11*
Drosophila	homeotic transformations by paternal *Uab*[1] inversion	*17*
Plants	parental ratio effect on endosperm development	*90*
Mammals	androgenote and gynogenote death	*97-99*
	CHR. SEGREGATION & ELIMINATION	
Marsilea (fern)	non-random distribution of paternal chromatids	*89*
Sciarid flies, coccids, *H. hampei* (bettle)	paternal chromosomes elimination in mitosis (sex determination) and/or meiosis	*1, 8, 9, 22*
C. elegans	induced preferential paternal X chromosome elimination in XX larvae	*27*
Corbibula (mollusc)	female pronucleus extrusion in the zygote	*28, 29*
Nasonia (wasp)	paternal chromosomes elimination induced by Wolbachia or PSR	*23-26*
Isoodon obesulus (marsupial)	paternal X chromosome elimination	*100*
Human and mouse	transmission ratio distortion	*36, 37*
Mouse	level of aneuploidy in sperm	*35*
	REPLICATION	
Drosophila	somatic endoreplication (polytenization) of mini-X chromosome	*10*
Mammals	asynchronous replication of imprinted regions	*40*
	CHROMOSOME INTERACTIONS	
Sciara	compartimentalization of parental chromosomes in male germ line	*22, 52*
Mouse	nuclear compartimentalization of imprinted domains	*50*
Mouse	preimplantation spatial separation of parental genomes	*53*
Mouse	meiotic recombination levels	*43-45*
Human and mouse	chromosome associations at imprinted regions	*46-49*
	EPIGENETIC DIFFERENCES	
Plants, worms, insects and mammals	DNA methylation, histone modification and/or condensation differences between maternal and paternal chromosomes	*6-8, 11, 16, 41, 69-71, 75*
Mouse	transgenerational imprinting effects	*81*

Fig. 1. Imprinting effects in sexually reproducing organisms. Although other POEs have been reported in additional species and phenotypes, it remains to be determined if they are actually caused by imprinting (see text) and, therefore, have not been included in this list.

mechanisms can affect the same phenomenon, (2) a chromosomal imprint can have multiple effects and (3) large-scale studies have the potential to unravel novel imprinted loci and imprinting effects in additional species.

Sex determination exemplifies how different imprinting effects can lead to similar outcomes. As previously discussed, sex determination is mediated by diverse imprinting effects in coccids and *sciarid* flies and, under some conditions, in *Nasonia* and *C. elegans.* The diverse mechanisms include heterochromatinization or elimination of one chromosome or an entire parental set (9, 22, 27, 54). Moreover, in *Nasonia vitripensis*, several lines of evidence have demonstrated that sex determination depends on the imprinted expression of the *transformer* gene (a member of the cascade of genes that control insect sex determination) (55). In this wasp, diploid female development depends on the transcription of the zygotic *transformer* gene (*Nvtra*). This gene appears to be maternally repressed in unfertilized eggs (which develop into haploid males); in contrast, the presence of a paternal genome triggers a peak of *Nvtra* zygotic transcription in fertilized eggs, rendering female embryos (55, 56).

Drosophila provides an example of how the imprinting of one chromosome (the Y) can have diverse effects: (1) on the expression of individual Y-linked transgenes and (2) on the entire gene expression of other chromosome. As previously discussed, studies of chromosome Y-inserted transgenes have shown differential expression according to their parental origin (18). In addition, a maternally inherited Y chromosome can rescue mutations that disrupt X chromosome dosage compensation (92). Normally, this is a mechanism that compensates for unequal X chromosome dosage between males (with one X) and females (with two X) by upregulating the transcription of the genes of the single male X chromosome. This process is mediated by the chromatin-modifying male-specific lethal (MSL) complex; mutations of some MSL components (*roX1* and *roX2*) disrupt expression upregulation and cause male-specific lethality. However, a maternally inherited Y chromosome can suppress this mutant male lethality by elevating the expression of X-linked genes (92). How imprinting of the Y-chromosome can rescue a mutant phenotype in other chromosome is very intriguing, because no endogenous imprinted genes that could mediate this effect have been reported in *Drosophila.*

Recent years have seen tremendous advances in large-scale genotyping, phenotyping and mapping studies, changing our view of POEs. Until recently, imprinted expression appeared to be limited to a few genes (http://www.har.mrc.ac.uk/research/genomic_imprinting/) (7). However, recent large-scale studies have shown that imprinted gene expression and imprinted epigenetic marks are more frequent and widespread than previously thought (57–62). This conclusion is further supported by observations of imprinting effects in methylation and expression of transgenes inserted in multiple locations in diverse species (4, 18, 20, 21, 63). Moreover, quantitative trait loci (QTL) analyses have revealed that POEs are common and can affect many traits and diseases (http://igc.otago.ac.nz/home.html; Richard Mott, personal communication) (59); these studies have also found complex patterns of POEs (64, 65). Therefore, large-scale expression studies are very promising for uncovering novel POEs; however, as previously discussed, in many cases the nature of these POEs and whether they are originated by imprinting remains to be determined. An example is found in chicken: while no conclusive evidence of imprinted gene expression has been reported in birds (66), QTL mapping has identified several POEs (67). Their nature is still unknown, but some POE candidate regions map in or close to chicken orthologues of mammalian imprinted genes (67). Moreover, these avian genes are clustered and replicate asynchronously, as observed in their mammalian orthologues (68).

In summary, while numerous imprinting effects have been known for a long time, the list is still growing. Figure 1 does not intend to be a comprehensive catalog, but illustrates the extreme diversity of imprinting and imprinting-candidate POEs and their widespread distribution in sexually reproducing organisms.

6. The Common Theme: Paternal and Maternal Chromosomes Bear Different Epigenetic Marks

Although not all imprinting effects have been studied with the same depth, most have been associated with parent-of-origin dependent epigenetic marks, such as DNA methylation and histone modifications, chromatin condensation (euchromatin vs. heterochromatin), and chromatin remodeling factors (such as noncoding RNAs and CTCF). Many of the chromatin modifications involved in mammalian or plant imprinting (7, 8) are also associated with a variety of imprinting effects in other organisms. For instance, imprinted heterochromatinization of chromosomes is observed during sex determination of lecanoid coccids and *Hypothenemus hampei*, as well as in mammalian imprinted X-inactivation (39, 40, 44, 45). In coccids, the paternal DNA is hypomethylated in both males and females. Additionally, in males, the paternal chromosomes are heterochromatized and condensed, accumulating nuclear-resistant chromatin, Heterochromatin Protein 1 (HP1) protein and repressive histone modifications (9, 69, 70). In *Sciara ocellaris*, chromosome elimination in early germ nuclei and male meiosis is accompanied by differential condensation and differential histone H3 and H4 methylation and acetylation between maternal and paternal chromosomes (71, 72). In *Drosophila*, heterochromatin imprints affect gene expression (16). Mutant analyses have shown that this imprinting depends on the products of Suppressor of variegation (Su(var)) and trithorax (trx-G) chromatin modifier genes (73). CTCF is a common player in both fruit fly and mammalian imprinting, although the role of DNA methylation in *Drosophila* (74) has not been clarified yet. These data indicate that histone methylation and acetylation, as well as other well-known mammalian epigenetic marks, contribute to the imprint of parental chromosomes also in insects.

The origin of parent-of-origin dependent epigenetic differences (and the subsequent imprinting effects) resides on a simple fact: maternal and paternal chromosomes can retain part of the chromatin configuration assembled during the formation of two extremely specialized cell types: the gametes. The epigenetic reprogramming that occurs during oogenesis and spermatogenesis has been extensively studied in mammals and plants (75, 76) and also occurs in other organisms. For instance, different patterns of histone marks (such as histone H3 lysine 9 methylation) are incorporated during spermatogenesis vs. oogenesis in coccids (mealybugs) and *C. elegans* (41, 77). In many organisms, tight chromosome condensation and replacement of histones for protamines occurs only during spermatogenesis. Still, some histones can be retained in the mature sperm and transmit imprinted epigenetic marks to the embryo, constituting a potential source of POEs (78). After fertilization, two pronuclei with very different chromatin organization encounter in the zygote. Therefore and in order to generate a totipotent embryo, many epigenetic differences between the parental genomes are removed and new ones established (79, 80). For instance, recent reports of imprinted gene expression not only in endosperm, but also in plant embryos, suggest that imprint erasure and resetting mechanisms are present in plants (7). Through these cyclic reprogramming events, the imprints are reset and established in every generation (6, 75, 76). Nevertheless, transgenerational POEs have been observed, suggesting incomplete gametic erasure of certain epigenetic marks (81).

The links between gametic epigenetic differences, imprinting and embryonic reprogramming can be illustrated with an example found in *C. elegans*: after fertilization, the paternal pronucleus lacks all tested histone modifications, but rapidly accumulates them as it decondenses (41). As a result, maternal and paternal chromosomes end up displaying similar histone marks in the early embryos. All but the X chromosome transmitted by the sperm, which, as previously discussed, remains devoid of activating histone modifications (methylated H3/Lys4 and acetylated H3) (41). This epigenetic imprint might be a direct consequence of the transcriptional and epigenetic history of the *C. elegans* X: this chromosome is devoid of spermatogenesis-specific genes and, therefore, it remains silent during sperm development, while it is actively transcribed during oogenesis (41, 82). Interestingly, the paternal epigenetic imprint of the X-chromosome is transitory: it is maintained in the embryo through several cell divisions, but histone modifications accumulate at later stages and the imprint is no longer observed past the 20-cell stage (41).

This marked, although transient, epigenetic asymmetry between the two parental genomes is also found in other organisms during early embryogenesis and has two important implications: first, reprogramming involves unique mechanisms for each of the two parental genomes (79). For instance, in paternal pronuclei tightly condensed and packed around protamines, these must be replaced by histones and the chromosomes decondensed. Preimplantation reprogramming leads to chromatin equalization compatible with embryo development, but the differences between the paternal and maternal chromosomes remodeling processes can also result in the acquisition of distinct epigenetic marks. Second, this epigenetic asymmetry manifests in a number of POEs: nuclear separation (53), pericentric heterochromatin (39) and asynchronous replication (83, 84) distinguish the two parental sets during mammalian early embryonic stages. Although these POEs fade after a few cell divisions, some POEs can remain at later stages; for instance, asynchronous replication (resulting in early replication of paternal alleles respect to maternal ones) is first observed between paternal and maternal chromosomes after fertilization, but becomes restricted to imprinted loci after a few cell divisions and then for life (84). Moreover, early embryonic POEs could contribute to the establishment of novel epigenetic imprints that may result in imprinting effects at later stages.

7. The Lessons of Comparative Analyses: Imprinting Effects are Widespread and Diverse, yet They Can Be Interrelated

In spite of the vast embryonic reprogramming, epigenetic differences between maternal and paternal chromosomes are present in somatic cells. The specificity of many of these parent-of-origin dependent marks and the key roles they play in the imprinting of crucial processes suggest that, besides being derived from gametic epigenetic marks or early embryonic POEs, they have been selected due to their functional relevance. Several theories about the selective forces operating at imprinting effects on expression have been proposed (reviewed in (85)), although none of them is sufficient to explain the evolution of all imprinted genes. Moreover, the diversity of imprinting effects suggests that different selective pressures have operated over different loci and organisms. For instance, imprinting has no known expression effects in karyotypically normal *Drosophila* genes, only on PEV of rearranged and inserted genes; therefore, *Drosophila* chromosomes are imprinted, but the role and origin of this imprinting might have nothing to do with the control gene expression as in mammals (16). Nevertheless, fruit fly and mammalian imprinting share some common features: for instance, the CTCF chromatin insulator contributes to both mammalian and *Drosophila* imprinting control (16, 73, 74, 86) and the physical size of several *Drosophila* imprinted regions (including the imprinted mini-X chromosome) is comparable to the size of some mammalian imprinted clusters (10). Besides placental and marsupial mammals, imprinted gene expression has been found in plants, but not in other animals (87); therefore, imprinting is believed to have evolved independently in those two groups (88). However, mammals and plants share several common principles of imprinting regulation (such as DNA methylation and other epigenetic marks) (88) and even other imprinting effects (89–91). Therefore, imprinting shows recurrent features among distantly related species and imprinted phenomenon: as an additional example, CTCF has been reported to contribute to asynchronous replication (92), intra- and interchromosomal interactions (47, 93) and imprinted expression regulation in diverse organisms (48, 74, 86). Furthermore, the phylogenetic distribution of parent-of-origin dependent epigenetic differences (Fig. 1) indicates that they are an ancient phenomenon in sexually reproducing organisms. This suggests that, along with specific selective pressures in different organisms, there might be common selective forces operating on paternal and maternal epigenomes (85, 94, 95).

In summary, the development of two highly differentiated cell types that participate in fertilization culminates with the zygotic encounter of two chromosome sets with dramatically different chromatin configurations. Although most epigenetic differences must be erased and reset to generate totipotent cells, natural selection may also operate to maintain some of them for particular functions (Fig. 1). Any epigenetic difference between maternally and paternally inherited chromosomes offers the possibility of an imprinting effect: the substrate is there. While different selective forces have led to a large spectrum of imprinting effects (Fig. 1), very diverse organisms, despite their different evolutionary paths, share several imprinting effects and, maybe, common selective pressures. Moreover, disparate imprinting effects are observed within the same species. It is possible that intraspecific imprinting diversity is a common occurrence and it is just our inability to detect certain effects (or our lack of curiosity or awareness) that keep them out of our sight. Future studies will expand our knowledge of imprinting in additional species, discover novel or complex imprinted phenomena, and unveil the links between diverse imprinting effects.

Acknowledgments

I apologize to colleagues whose original research papers could not be cited due to space limitations. I would like to thank Elena Becker Barroso and Jose Javier García Ramírez for their helpful comments to this manuscript. I am also grateful to Francisco R. Jiménez Díaz for technical support and to the Consejería de Educación y Ciencia (ref. PPII10-0259-4347) for financial support.

References

1. Crouse HV (1960) The controlling element in Sex chromosome behavior in sciara. Genetics 45:1429–1443
2. Wolf JB, Wade MJ (2009) What are maternal effects (and what are they not)? Philos Trans R Soc Lond B Biol Sci 364:1107–1115
3. Hager R, Cheverud JM, Wolf JB (2008) Maternal effects as the cause of parent-of-origin effects that mimic genomic imprinting. Genetics 178:1755–1762
4. Wang AD, Sharp NP, Spencer CC, Tedman-Aucoin K, Agrawal AF (2009) Selection, epistasis, and parent-of-origin effects on deleterious mutations across environments in Drosophila melanogaster. Am Nat 174:863–874
5. Wittkopp PJ, Haerum BK, Clark AG (2006) Parent-of-origin effects on mRNA expression in Drosophila melanogaster not caused by genomic imprinting. Genetics 173:1817–1821
6. Bartolomei MS, Ferguson-Smith AC (2011) Mammalian genomic imprinting. Cold Spring Harb Perspect Biol 3(7). pii: a002592
7. Raissig MT, Baroux C, Grossniklaus U (2011) Regulation and flexibility of genomic imprinting during seed development. Plant Cell 23:16–26
8. Brun LO, Stuart J, Gaudichon V, Aronstein K, French-Constant RH (1995) Functional haplodiploidy: a mechanism for the spread of insecticide resistance in an important international insect pest. Proc Natl Acad Sci USA 92: 9861–9865
9. Khosla S, Mendiratta G, Brahmachari V (2006) Genomic imprinting in the mealybugs. Cytogenet Genome Res 113:41–52
10. Anaka M, Lynn A, McGinn P, Lloyd VK (2009) Genomic imprinting in Drosophila has properties of both mammalian and insect imprinting. Dev Genes Evol 219:59–66
11. Lloyd VK, Sinclair DA, Grigliatti TA (1999) Genomic imprinting and position-effect variegation in Drosophila melanogaster. Genetics 151:1503–1516
12. Deakin JE, Chaumeil J, Hore TA, Marshall Graves JA (2009) Unravelling the evolutionary origins of X chromosome inactivation in mammals: insights from marsupials and monotremes. Chromosome Res 17:671–685
13. Takagi N, Sasaki M (1975) Preferential inactivation of the paternally derived X chromosome in the extraembryonic membranes of the mouse. Nature 256:640–642
14. Wang X, Soloway PD, Clark AG (2010) Paternally biased X inactivation in mouse neonatal brain. Genome Biol 11:R79
15. Chadwick LH, Willard HF (2005) Genetic and parent-of-origin influences on X chromosome choice in Xce heterozygous mice. Mamm Genome 16:691–699
16. Menon DU, Meller VH (2010) Germ line imprinting in Drosophila: Epigenetics in search of function. Fly (Austin) 4:48–52
17. Kuhn DT, Packert G (1988) Paternal imprinting of inversion Uab1 causes homeotic transformations in Drosophila. Genetics 118: 103–107
18. Maggert KA, Golic KG (2002) The Y chromosome of Drosophila melanogaster exhibits chromosome-wide imprinting. Genetics 162:1245–1258
19. Dorn R, Krauss V, Reuter G, Saumweber H (1993) The enhancer of position-effect variegation of Drosophila, E(var)3-93D, codes for a chromatin protein containing a conserved domain common to several transcriptional regulators. Proc Natl Acad Sci USA 90:11376–11380
20. Preis JI, Downes M, Oates NA, Rasko JE, Whitelaw E (2003) Sensitive flow cytometric analysis reveals a novel type of parent-of-origin effect in the mouse genome. Curr Biol 13:955–959
21. Sha K, Fire A (2005) Imprinting capacity of gamete lineages in Caenorhabditis elegans. Genetics 170:1633–1652

22. Goday C, Esteban MR (2001) Chromosome elimination in sciarid flies. Bioessays 23: 242–250
23. Breeuwer JA, Werren JH (1990) Microorganisms associated with chromosome destruction and reproductive isolation between two insect species. Nature 346:558–560
24. Nur U, Werren JH, Eickbush DG, Burke WD, Eickbush TH (1988) A "selfish" B chromosome that enhances its transmission by eliminating the paternal genome. Science 240:512–514
25. Dobson SL, Tanouye MA (1998) Evidence for a genomic imprinting sex determination mechanism in Nasonia vitripennis (Hymenoptera; Chalcidoidea). Genetics 149:233–242
26. Werren JH, Stouthamer R (2003) PSR (paternal sex ratio) chromosomes: the ultimate selfish genetic elements. Genetica 117:85–101
27. Prahlad V, Pilgrim D, Goodwin EB (2003) Roles for mating and environment in C. elegans sex determination. Science 302: 1046–1049
28. Komaru A, Kawagishi T, Konishi K (1998) Cytological evidence of spontaneous androgenesis in the freshwater clam Corbicula leana Prime. Dev Genes Evol 208:46–50
29. Ishibashi R, Ookubo K, Aoki M, Utaki M, Komaru A, Kawamura K (2003) Androgenetic reproduction in a freshwater diploid clam Corbicula fluminea (Bivalvia: Corbiculidae). Zoolog Sci 20:727–732
30. Baker BS (1975) Paternal loss (pal): a meiotic mutant in Drosophila melanogaster causing loss of paternal chromosomes. Genetics 80:267–296
31. Szabad J, Mathe E, Puro J (1995) Horka, a dominant mutation of Drosophila, induces nondisjunction and, through paternal effect, chromosome loss and genetic mosaics. Genetics 139:1585–1599
32. Szalontai T, Gaspar I, Belecz I, Kerekes I, Erdelyi M, Boros I, Szabad J (2009) HorkaD, a chromosome instability-causing mutation in Drosophila, is a dominant-negative allele of Lodestar. Genetics 181:367–377
33. Lewis EB, Gencarella W (1952) Claret and non-disjunction in Drosophila melanogaster. Genetics 37:600–601
34. Gao G, Cheng Y, Wesolowska N, Rong YS (2011) Paternal imprint essential for the inheritance of telomere identity in Drosophila. Proc Natl Acad Sci USA 108:4932–4937
35. Baulch JE, Lowe XR, Bishop JB, Wyrobek AJ (1996) Evidence for a parent-of-origin effect on sperm aneuploidy in mice carrying Robertsonian translocations as analyzed by fluorescence in situ hybridization. Mutat Res 372:269–278
36. Naumova AK, Leppert M, Barker DF, Morgan K, Sapienza C (1998) Parental origin-dependent, male offspring-specific transmission-ratio distortion at loci on the human X chromosome. Am J Hum Genet 62: 1493–1499
37. Croteau S, Andrade MF, Huang F, Greenwood CM, Morgan K, Naumova AK (2002) Inheritance patterns of maternal alleles in imprinted regions of the mouse genome at different stages of development. Mamm Genome 13:24–29
38. Pardo-Manuel de Villena F, de la Casa-Esperon E, Briscoe TL, Sapienza C (2000) A genetic test to determine the origin of maternal transmission ratio distortion. Meiotic drive at the mouse Om locus. Genetics 154:333–342
39. Puschendorf M, Terranova R, Boutsma E, Mao X, Isono K, Brykczynska U, Kolb C, Otte AP, Koseki H, Orkin SH, van Lohuizen M, Peters AH (2008) PRC1 and Suv39h specify parental asymmetry at constitutive heterochromatin in early mouse embryos. Nat Genet 40:411–420
40. Kitsberg D, Selig S, Brandeis M, Simon I, Keshet I, Driscoll DJ, Nicholls RD, Cedar H (1993) Allele-specific replication timing of imprinted gene regions. Nature 364:459–463
41. Bean CJ, Schaner CE, Kelly WG (2004) Meiotic pairing and imprinted X chromatin assembly in Caenorhabditis elegans. Nat Genet 36:100–105
42. Shiu PK, Raju NB, Zickler D, Metzenberg RL (2001) Meiotic silencing by unpaired DNA. Cell 107:905–916
43. Paigen K, Szatkiewicz JP, Sawyer K, Leahy N, Parvanov ED, Ng SH, Graber JH, Broman KW, Petkov PM (2008) The recombinational anatomy of a mouse chromosome. PLoS Genet 4:e1000119
44. Ng SH, Madeira R, Parvanov ED, Petros LM, Petkov PM, Paigen K (2009) Parental origin of chromosomes influences crossover activity within the Kcnq1 transcriptionally imprinted domain of Mus musculus. BMC Mol Biol 10:43
45. Billings T, Sargent EE, Szatkiewicz JP, Leahy N, Kwak IY, Bektassova N, Walker M, Hassold T, Graber JH, Broman KW, Petkov PM (2010) Patterns of recombination activity on mouse chromosome 11 revealed by high resolution mapping. PLoS One 5:e15340
46. LaSalle JM, Lalande M (1996) Homologous association of oppositely imprinted chromosomal domains. Science 272:725–728

47. Ling JQ, Li T, Hu JF, Vu TH, Chen HL, Qiu XW, Cherry AM, Hoffman AR (2006) CTCF mediates interchromosomal colocalization between Igf2/H19 and Wsb1/Nf1. Science 312:269–272
48. Yang J, Corces VG (2011) Chromatin insulators: a role in nuclear organization and gene expression. Adv Cancer Res 110:43–76
49. Sandhu KS, Shi C, Sjolinder M, Zhao Z, Gondor A, Liu L, Tiwari VK, Guibert S, Emilsson L, Imreh MP, Ohlsson R (2009) Nonallelic transvection of multiple imprinted loci is organized by the H19 imprinting control region during germline development. Genes Dev 23:2598–2603
50. Terranova R, Yokobayashi S, Stadler MB, Otte AP, van Lohuizen M, Orkin SH, Peters AH (2008) Polycomb group proteins Ezh2 and Rnf2 direct genomic contraction and imprinted repression in early mouse embryos. Dev Cell 15:668–679
51. Gribnau J, Hochedlinger K, Hata K, Li E, Jaenisch R (2003) Asynchronous replication timing of imprinted loci is independent of DNA methylation, but consistent with differential subnuclear localization. Genes Dev 17: 759–773
52. Kubai DF (1982) Meiosis in Sciara coprophila: structure of the spindle and chromosome behavior during the first meiotic division. J Cell Biol 93:655–669
53. Mayer, W., Smith, A., Fundele, R., and Haaf, T (2000) Spatial separation of parental genomes in preimplantation mouse embryos. J Cell Biol 148:629–634
54. Reed KM, Werren JH (1995) Induction of paternal genome loss by the paternal-sex-ratio chromosome and cytoplasmic incompatibility bacteria (Wolbachia): a comparative study of early embryonic events. Mol Reprod Dev 40:408–418
55. Verhulst EC, Beukeboom LW, van de Zande L (2010) Maternal control of haplodiploid sex determination in the wasp Nasonia. Science 328:620–623
56. Beukeboom LW, van de Zande L (2010) Genetics of sex determination in the haplodiploid wasp Nasonia vitripennis (Hymenoptera: Chalcidoidea). J Genet 89:333–339
57. Luedi PP, Dietrich FS, Weidman JR, Bosko JM, Jirtle RL, Hartemink AJ (2007) Computational and experimental identification of novel human imprinted genes. Genome Res 17:1723–1730
58. Babak T, Deveale B, Armour C, Raymond C, Cleary MA, van der Kooy D, Johnson JM, Lim LP (2008) Global survey of genomic imprinting by transcriptome sequencing. Curr Biol 18:1735–1741
59. Gregg C, Zhang J, Weissbourd B, Luo S, Schroth GP, Haig D, Dulac C (2010) High-resolution analysis of parent-of-origin allelic expression in the mouse brain. Science 329:643–648
60. Wolff P, Weinhofer I, Seguin J, Roszak P, Beisel C, Donoghue MT, Spillane C, Nordborg M, Rehmsmeier M, Kohler C (2011) High-Resolution Analysis of Parent-of-Origin Allelic Expression in the Arabidopsis Endosperm. PLoS Genet 7:e1002126
61. Gehring M, Bubb KL, Henikoff S (2009) Extensive demethylation of repetitive elements during seed development underlies gene imprinting. Science 324:1447–1451
62. Choufani S, Shapiro JS, Susiarjo M, Butcher DT, Grafodatskaya D, Lou Y, Ferreira JC, Pinto D, Scherer SW, Shaffer LG, Coullin P, Caniggia I, Beyene J, Slim R, Bartolomei MS, Weksberg R (2011) A novel approach identifies new differentially methylated regions (DMRs) associated with imprinted genes. Genome Res 21:465–476
63. Sapienza C, Peterson AC, Rossant J, Balling R (1987) Degree of methylation of transgenes is dependent on gamete of origin. Nature 328:251–254
64. Wolf JB, Cheverud JM, Roseman C, Hager R (2008) Genome-wide analysis reveals a complex pattern of genomic imprinting in mice. PLoS Genet 4:e1000091
65. Cheverud JM, Lawson HA, Fawcett GL, Wang B, Pletscher LS, R Fox A, Maxwell TJ, Ehrich TH, Kenney-Hunt JP, Wolf JB, Semenkovich CF (2011) Diet-dependent genetic and genomic imprinting effects on obesity in mice. Obesity (Silver Spring) 19: 160–170
66. Nolan CM, Killian JK, Petitte JN, Jirtle RL (2001) Imprint status of M6P/IGF2R and IGF2 in chickens. Dev Genes Evol 211: 179–183
67. Tuiskula-Haavisto M, Vilkki J (2007) Parent-of-origin specific QTL–a possibility towards understanding reciprocal effects in chicken and the origin of imprinting. Cytogenet Genome Res 117:305–312
68. Dunzinger U, Haaf T, Zechner U (2007) Conserved synteny of mammalian imprinted genes in chicken, frog, and fish genomes. Cytogenet Genome Res 117:78–85
69. Bongiorni S, Cintio O, Prantera G (1999) The relationship between DNA methylation and chromosome imprinting in the coccid Planococcus citri. Genetics 151:1471–1478

70. Bongiorni S, Prantera G (2003) Imprinted facultative heterochromatization in mealybugs. Genetica 117:271–279
71. Goday C, Ruiz MF (2002) Differential acetylation of histones H3 and H4 in paternal and maternal germline chromosomes during development of sciarid flies. J Cell Sci 115: 4765–4775
72. Greciano PG, Goday C (2006) Methylation of histone H3 at Lys4 differs between paternal and maternal chromosomes in Sciara ocellaris germline development. J Cell Sci 119:4667–4677
73. Joanis V, Lloyd VK (2002) Genomic imprinting in Drosophila is maintained by the products of Suppressor of variegation and trithorax group, but not Polycomb group, genes. Mol Genet Genomics 268:103–112
74. MacDonald WA, Menon D, Bartlett NJ, Sperry GE, Rasheva V, Meller V, Lloyd VK (2010) The Drosophila homolog of the mammalian imprint regulator. CTCF, maintains the maternal genomic imprint in Drosophila melanogaster. BMC Biol 8:105
75. Feng S, Jacobsen SE, Reik W (2010) Epigenetic reprogramming in plant and animal development. Science 330:622–627
76. Kota SK, Feil R (2010) Epigenetic transitions in germ cell development and meiosis. Dev Cell 19:675–686
77. Bongiorni S, Pugnali M, Volpi S, Bizzaro D, Singh PB, Prantera G (2009) Epigenetic marks for chromosome imprinting during spermatogenesis in coccids. Chromosoma 118:501–512
78. Hammoud SS, Nix DA, Zhang H, Purwar J, Carrell DT, Cairns BR (2009) Distinctive chromatin in human sperm packages genes for embryo development. Nature 460:473–478
79. Burton A, Torres-Padilla ME (2010) Epigenetic reprogramming and development: a unique heterochromatin organization in the preimplantation mouse embryo. Brief Funct Genomics 9:444–454
80. de la Casa-Esperon E, Roy A (2009) Mammalian gametogenesis to implantation. In: Reproduction and Development Biology, Encyclopedia of Biological, Physiological and Health Sciences, Encyclopedia of Life Support Systems(EOLSS). Eolss Publishers, Oxford ,UK
81. Han Z, Mtango NR, Patel BG, Sapienza C, Latham KE (2008) Hybrid vigor and transgenerational epigenetic effects on early mouse embryo phenotype. Biol Reprod 79:638–648
82. Arico JK, Katz DJ, van der Vlag J, Kelly WG (2011) Epigenetic Patterns Maintained in Early Caenorhabditis elegans Embryos Can Be Established by Gene Activity in the Parental Germ Cells. PLoS Genet 7:e1001391
83. Ferreira J, Carmo-Fonseca M (1997) Genome replication in early mouse embryos follows a defined temporal and spatial order. J Cell Sci 110(Pt 7):889–897
84. May A, Reifenberg K, Zechner U, Haaf T (2008) Asynchronous replication dynamics of imprinted and non-imprinted chromosome regions in early mouse embryos. Exp Cell Res 314:2788–2795
85. de la Casa-Esperon E, Sapienza C (2003) Natural selection and the evolution of genome imprinting. Annu Rev Genet 37:349–370
86. Engel N, Thorvaldsen JL, Bartolomei MS (2006) CTCF binding sites promote transcription initiation and prevent DNA methylation on the maternal allele at the imprinted H19/ Igf2 locus. Hum Mol Genet 15:2945–2954
87. Renfree MB, Hore TA, Shaw G, Graves JA, Pask AJ (2009) Evolution of genomic imprinting: insights from marsupials and monotremes. Annu Rev Genomics Hum Genet 10:241–262
88. Scott RJ, Spielman M (2006) Genomic imprinting in plants and mammals: how life history constrains convergence. Cytogenet Genome Res 113:53–67
89. Tourte Y, Kuligowski-Andres J, Barbier-Ramond C (1980) Different behaviour of paternal and maternal genomes during embryogenesis in the fern, Marsilea (author's transl). Eur J Cell Biol 21:28–36
90. Kermicle JL, Alleman M (1990) Gametic imprinting in maize in relation to the angiosperm life cycle. Dev Suppl, 9–14.
91. Vielle-Calzada JP, Baskar R, Grossniklaus U (2000) Delayed activation of the paternal genome during seed development. Nature 404:91–94
92. Bergstrom R, Whitehead J, Kurukuti S, Ohlsson R (2007) CTCF regulates asynchronous replication of the imprinted H19/Igf2 domain. Cell Cycle 6:450–454
93. Donohoe ME, Silva SS, Pinter SF, Xu N, Lee JT (2009) The pluripotency factor Oct4 interacts with Ctcf and also controls X-chromosome pairing and counting. Nature 460:128–132
94. Pardo-Manuel de Villena F, de la Casa-Esperon E, Sapienza C (2000) Natural selection and the function of genome imprinting: beyond the silenced minority. Trends Genet 16:573–579
95. Paldi A (2003) Genomic imprinting: could the chromatin structure be the driving force? Curr Top Dev Biol 53:115–138

96. Morison IM, Ramsay JP, Spencer HG (2005) A census of mammalian imprinting. Trends Genet 21:457–465
97. Barton SC, Surani MA, Norris ML (1984) Role of paternal and maternal genomes in mouse development. Nature 311:374–376
98. McGrath J, Solter D (1984) Completion of mouse embryogenesis requires both the maternal and paternal genomes. Cell 37:179–183
99. Surani MA, Barton SC, Norris ML (1984) Development of reconstituted mouse eggs suggests imprinting of the genome during gametogenesis. Nature 308:548–550
100. Johnston PG, Watson CM, Adams M, Paull DJ (2002) Sex chromosome elimination, X chromosome inactivation and reactivation in the southern brown bandicoot Isoodon obesulus (Marsupialia: Peramelidae). Cytogenet Genome Res 99:119–124

Index

Nora Engel (ed.), *Genomic Imprinting: Methods and Protocols*, Methods in Molecular Biology, vol. 925, DOI 10.1007/978-1-62703-011-3, © Springer Science+Business Media, LLC 2012

I

K

L

M

N

P

Q

R

S

T

U

Y

Z

MIX
Papier aus verantwortungsvollen Quellen
Paper from responsible sources
FSC® C105338

If you have any concerns about our products,
you can contact us on
ProductSafety@springernature.com

In case Publisher is established outside the EU,
the EU authorized representative is:
Springer Nature Customer Service Center GmbH
Europaplatz 3, 69115 Heidelberg, Germany

Printed by Libri Plureos GmbH
in Hamburg, Germany